Eco-Innovation and the Development of Business Models

Greening of Industry Networks Studies

VOLUME 2

Series Editors:

D.A. Vazquez-Brust
Joseph Sarkis

For further volumes:
http://www.springer.com/series/10444

Susana Garrido Azevedo
Marcus Brandenburg • Helena Carvalho
Virgílio Cruz-Machado
Editors

Eco-Innovation and the Development of Business Models

Lessons from Experience and New Frontiers in Theory and Practice

Springer

Editors
Susana Garrido Azevedo
Department of Management and Economics
University of Beira Interior
Covilhã, Portugal

Helena Carvalho
Virgílio Cruz-Machado
UNIDEMI, Department of Mechanical and Industrial Engineering
Faculty of Science and Technology (FCT)
Universidade Nova de Lisboa
Lisbon, Portugal

Marcus Brandenburg
Supply Chain Management
Faculty of Business and Economics
University of Kassel
Kassel, Germany

Department of Production Management
School of Economics and Management
Technische Universität Berlin
Berlin, Germany

ISBN 978-3-319-37772-8 ISBN 978-3-319-05077-5 (eBook)
DOI 10.1007/978-3-319-05077-5
Springer Cham Heidelberg New York Dordrecht London

Printed on acid-free paper

Springer is part of Springer Science+Business Media (www.springer.com)

Foreword

It is a pleasure that we write this short foreword to what we believe is an important book from a technological, policy, organizational, and sociological perspective. Environmental challenges such as pollution, climate change, water and natural resources depletion, and dwindling biodiversity are real threats to the survival of our civilization, and there is a need to learn how to act now. Fortunately this is exactly what this book does: presenting results from real-life cases and simultaneously providing theory, methodologies, and tools that can help us move forward toward increased awareness of how eco-innovation can help us support sustainable economic growth and save our planet for future generations.

The concept of eco-innovation is not one that will go away easily due to the many pressures we have put on ourselves and our environment. Human beings create most of the world's problems, but they are also the ones that have the ability to solve them. We need to realize that we all live on one world and that the resources and our time on this world are finite. There is no one to help us to live and continue to thrive on our planet, besides ourselves, as humans. Thus, human ingenuity, creativity, and innovation are needed, not only for our generation, but for future generations. This is why we wholeheartedly support this effort to address eco-innovation concerns and the many dimensions in which it appears.

The concept of eco-innovation is dynamic. We know that what is accepted practice today was very likely innovative at some time. This philosophy implies that what is innovative today is something we will live with for many generations. Eco-innovation with the focus on environmental sustainability is not just for convenience and making our lives easier, but is necessary for our long-term survival. Its role in society has only increased and gained in importance as anthropogenic causes to environmental degradation have increased. We have posited that green growth is necessary in many of our writings. We also have hope that ecological modernization through eco-innovation is a way to address the concerns of man's impact on the environment. But we should also critically examine these beliefs, our very own beliefs, to not become lax and hope that our crises will be solved on faith alone.

We should also be aware that man's ingenuity, although arguably limitless, will also require adoption, adaptation, and implementation. But our thoughts, actions, and ingenuity cannot be completed in a vacuum. Putting together a compendium of thoughts through an edited volume can help us to think more creatively. Each chapter in this book can convey a wonderful new insight or set of insights. We believe the reader should also look at the compendium of ideas that are fostered throughout and integrate them themselves. The book and its ideas together should be an eco-innovation to the creative minds of humans. To be able to link similar and disparate ideas presented is the process of creation and creativity. When reading through this book, do skip around, jump from chapter to chapter, and go back. Let your mind rewire and reorganize the works, and you will find that new ideas will flourish.

Whether you are a veteran scholar with decades of experience or knowledge or an ingénue in a very localized ecological drama just becoming aware of these topics for the first time, there is something for you. We cannot anticipate the perspectives of the readers of this book. We cannot control who will and will not read this book. But for those who do take the interest and time, we ask that you share your thoughts. Build on what you read in these pages, do your own research, and provide your own thoughts through writings and words. We know we live in a socially and virtually connected world, and this allows for greater and more rapid dissemination of the ideas of eco-innovation. Share your thoughts about these pages. Write a blog, post on one of your favorite social media sites, and leave a hard copy of this book on your living room table. The ideas presented here need to be diffused, and we encourage you to take advantage of our world's wonderful ideas from these pages and share.

The editors of this work, Susana Garrido Azevedo, Marcus Brandenburg, Helena Carvalho, and Virgílio Cruz-Machado, have established and continue to develop an astonishing record in eco-innovation and other organizational and natural environmental topics. As editors of the Springer series on Greening of Industry Networks, we are lucky and proud to have them agree to such a project. We have also learned much from their writings, and this compilation of high-quality chapters has made us anticipate the wonderful perspectives that we can integrate into our own thoughts. Critical awareness is an important step toward support for changes to policy and practice. This is why this book is important because it offers the basis for an integral, critical approach to eco-innovation, sustainability, and policy. Our hope and strong recommendation for the future is, therefore, that the readers of the book be inspired to seize this opportunity that eco-innovation offers.

Worcester, MA, USA — Joseph Sarkis
London, Egham Surrey, UK — Diego Vazquez-Brust

Contents

1 **Developments and Directions of Eco-innovation** 1
Susana Garrido Azevedo, Marcus Brandenburg, Helena Carvalho, and Virgílio Cruz-Machado

Part I Models and Frameworks Supporting Eco-innovation

2 **Managing Cross-Industry Innovations: A Search Strategy for Radical Eco-innovations** 19
Michaela Kloiber and Reinhold Priewasser

3 **How to Make Eco-innovation a Competitive Strategy: A Perspective on the Knowledge-Based Development** 39
Maria do Rosário Cabrita, Virgílio Cruz-Machado, and Florinda Matos

4 **A Framework for Developing and Assessing Eco-innovations** 55
Ida Gremyr, Jutta Hildenbrand, Steven Sarasini, and Hendry Raharjo

5 **Radical and Systematic Eco-innovation with TRIZ Methodology** 81
Helena V.G. Navas

Part II Application: Surveys and Case Studies on Eco-innovation Deployment

6 **Eco-innovation on Manufacturing Industry: The Role of Sustainability on Innovation Processes** 99
Patrícia A.A. da Silva, João C.O. Matias, Susana Garrido Azevedo, and Paulo N.B. Reis

7 **Portraying the Eco-innovative Landscape in Brazil: Determinants, Processes, and Results** ... 117
Flavia Pereira de Carvalho

8 **Contextual Factors as Drivers of Eco-innovation Strategies** ... 137
Marlete Beatriz Maçaneiro and Sieglinde Kindl da Cunha

9 **Conceptualizing Industry Efforts to Eco-innovate Among Large Swedish Companies** ... 163
Steven Sarasini, Jutta Hildenbrand, and Birgit Brunklaus

10 **Integrated Environmental Management Tools for Product and Organizations in Clusters** ... 179
Tiberio Daddi, Marco Frey, Fabio Iraldo, Francesco Rizzi, and Francesco Testa

11 **Toward Joint Product–Service Business Models: The Case of Your Energy Solution** ... 201
Andrea Bikfalvi, Rodolfo de Castro Vila, and Xavier Muñoz

12 **Business Model Innovation for Eco-innovation: Developing a Boundary-Spanning Business Model of an Ecosystem Integrator** ... 221
Anastasia Tsvetkova, Magnus Gustafsson, and Kim Wikström

Part III Future Directions: Eco-innovation Initiatives

13 **A New Methodology for Eco-friendly Construction: Utilizing Quality Function Deployment to Meet LEED Requirements** ... 245
William L. Gillis and Elizabeth A. Cudney

14 **Light Island Ferries in Scandinavia: A Case of Radical Eco-innovation** ... 275
Mette Mosgaard, Henrik Riisgaard, and Søren Kerndrup

15 **BioTRIZ: A Win-Win Methodology for Eco-innovation** ... 297
Nikolay Bogatyrev and Olga Bogatyreva

Erratum ... E1

Contributors

Andrea Bikfalvi Department of Business Administration and Product Design, University of Girona, Escola Politècnica Superior, Girona, Spain

Nikolay Bogatyrev BioTRIZ Ltd., University of Bath, Bath, UK

Olga Bogatyreva BioTRIZ Ltd., University of Bath, Bath, UK

Marcus Brandenburg Supply Chain Management, Faculty of Business and Economics, University of Kassel, Kassel, Germany

Department of Production Management, School of Economics and Management, Technische Universität Berlin, Berlin, Germany

Birgit Brunklaus Department of Energy and Environment, Division of Environmental Systems Analysis, Chalmers University of Technology, Gothenburg, Sweden

Helena Carvalho UNIDEMI, Department of Mechanical and Industrial Engineering, Faculty of Science and Technology (FCT), Universidade Nova de Lisboa, Lisbon, Portugal

Virgílio Cruz-Machado UNIDEMI, Department of Mechanical and Industrial Engineering, Faculty of Science and Technology (FCT), Universidade Nova de Lisboa, Lisbon, Portugal

Elizabeth A. Cudney Missouri University of Science and Technology, Rolla, MO, USA

Sieglinde Kindl da Cunha Environmental Management, Positivo University, Curitiba, PR, Brazil

Patrícia A.A. da Silva Faculty of Engineering, Department of Electromechanical Engineering, University of Beira Interior, Covilhã, Portugal

Tiberio Daddi Institute of Management – Sant'Anna School of Advanced Studies, Pisa, Italy

Flavia Pereira de Carvalho Fundação Dom Cabral, Nova Lima, Brazil

UNU-MERIT, Maastricht, The Netherlands

Rodolfo de Castro Vila Department of Business Administration and Product Design, University of Girona, Escola Politècnica Superior, Girona, Spain

Maria do Rosário Cabrita UNIDEMI, Department of Mechanical and Industrial Engineering, Faculty of Science and Technology (FCT), Universidade Nova de Lisboa, Lisbon, Portugal

Marco Frey Institute of Management – Sant'Anna School of Advanced Studies, Pisa, Italy

IEFE – Institute for Environmental and Energy Policy and Economics, Bocconi University, Milan, Italy

Susana Garrido Azevedo Department of Management and Economics, University of Beira Interior, Covilhã, Portugal

William L. Gillis Missouri University of Science and Technology, Rolla, MO, USA

Ida Gremyr Quality Sciences, Chalmers University of Technology, Gothenburg, Sweden

Magnus Gustafsson Industrial Management, Åbo Akademi University, Turku, Finland

PBI Research Institute, Turku, Finland

Jutta Hildenbrand Department of Energy and Environment, Division of Environmental Systems Analysis, Chalmers University of Technology, Gothenburg, Sweden

Fabio Iraldo Institute of Management – Sant'Anna School of Advanced Studies, Pisa, Italy

IEFE – Institute for Environmental and Energy Policy and Economics, Bocconi University, Milan, Italy

Søren Kerndrup Department of Development and Planning, Aalborg University, Aalborg, Denmark

Michaela Kloiber Institute for Environmental Management in Companies and Regions, Johannes Kepler University, Linz, Austria

Marlete Beatriz Maçaneiro State University Midwest – UNICENTRO, Guarapuava, PR, Brazil

João C.O. Matias Faculty of Engineering, Department of Electromechanical Engineering, University of Beira Interior, Covilhã, Portugal

Florinda Matos IC Lab Research Centre – ICAA – Intellectual Capital Accreditation Association, Santarém, Portugal

Mette Mosgaard Department of Development and Planning, Aalborg University, Aalborg, Denmark

Xavier Muñoz General Manager of Business DSET, Girona, Spain

Helena V.G. Navas UNIDEMI, Departamento de Engenharia Mecânica e Industrial, Faculdade de Ciências e Tecnologia, Universidade Nova de Lisboa, Caparica, Portugal

Reinhold Priewasser Institute for Environmental Management in Companies and Regions, Johannes Kepler University, Linz, Austria

Hendry Raharjo Quality Sciences, Chalmers University of Technology, Gothenburg, Sweden

Paulo N.B. Reis Faculty of Engineering, Department of Electromechanical Engineering, University of Beira Interior, Covilhã, Portugal

Henrik Riisgaard Department of Development and Planning, Aalborg University, Aalborg, Denmark

Francesco Rizzi Institute of Management – Sant'Anna School of Advanced Studies, Pisa, Italy

Steven Sarasini Department of Energy and Environment, Division of Environmental Systems Analysis, Chalmers University of Technology, Gothenburg, Sweden

Francesco Testa Institute of Management – Sant'Anna School of Advanced Studies, Pisa, Italy

Anastasia Tsvetkova Industrial Management, Åbo Akademi University, Turku, Finland

PBI Research Institute, Turku, Finland

Kim Wikström Industrial Management, Åbo Akademi University, Turku, Finland

PBI Research Institute, Turku, Finland

List of Figures

Fig. 1.1 Annual number of published papers 3
Fig. 1.2 Dimensions and impacts of eco-innovation 5

Fig. 2.1 Process model of using analogies in product development 26
Fig. 2.2 Opening up the solution space by abstraction from the underlying problem 27
Fig. 2.3 Cross-industry innovation process model strategy analysis 28
Fig. 2.4 The model of TRIZ problem-solving concept 31

Fig. 3.1 Pillars of KBD 43
Fig. 3.2 Related theories of knowledge economy 44
Fig. 3.3 Triple helix model of innovation 46
Fig. 3.4 Knowledge management key activities 49
Fig. 3.5 Building an integrative methodology for eco-innovation 50

Fig. 4.1 Key principles of the ECORE framework 62
Fig. 4.2 Key practices for our proposed framework 63
Fig. 4.3 Process map for beverage with central filling 67
Fig. 4.4 Stakeholder map for a product system based on decentralized beverage dispensing 68

Fig. 5.1 Steps of the TRIZ's algorithm for problem solving 84

Fig. 7.1 Main environmental impacts from innovations: results from PINTEC 124
Fig. 7.2 Partner in cooperative arrangements for eco-innovation 128
Fig. 7.3 Types of eco-innovation 130
Fig. 7.4 Main determinants of the eco-innovation 130
Fig. 7.5 Main results of the eco-innovations 132

Fig. 8.1 Cluster analysis and definition of taxonomy through the *eco-innovation strategies* construct 151
Fig. 8.2 Differences between the means of reactive and proactive strategies in the groups of firms. (**a**) Reactive strategies. (**b**) Proactive strategies 153
Fig. 8.3 Differences between the mean impacts of contextual factors in the groups of firms. (**a**) Environmental regulation. (**b**) Use of environmental and innovative incentives. (**c**) Reputational effects. (**d**) Top management support. (**e**) Technological competence. (**f**) Environmental formalization 156
Fig. 8.4 Cluster performance of contextual factors 157

Fig. 10.1 Methodological approaches employed in the five stages of the case study research 188

Fig. 11.1 Case outline – phases and critical junctures 205
Fig. 11.2 From knowledge to PSS • a phased approach 211
Fig. 11.3 Services offered 214
Fig. 11.4 Osterwalder's 9-point decomposition of a business model 215
Fig. 11.5 3S model of ICT for eco-innovation 217

Fig. 12.1 Target "biogas-for-traffic" ecosystem and the key stakeholders (the ecosystem integrator is marked with a *bold line*) 229
Fig. 12.2 Cooperation models in the target ecosystem 230

Fig. 13.1 Cx process flow chart 248
Fig. 13.2 QFD four-phase model 249
Fig. 13.3 HOQ model 250
Fig. 13.4 Adapted 4-phase model with LEED HOQ 257
Fig. 13.5 Pre-design HOQ 259
Fig. 13.6 LEED HOQ 262
Fig. 13.7 Design phase 265
Fig. 13.8 Construction phase 268
Fig. 13.9 O&M phase 270

Fig. 14.1 Sketch of the Eco Island Ferry 282
Fig. 14.2 Three interlinked optimization factors for ferry construction. In this case (Eco Island Ferry) energy efficiency and lightweight have been optimized, whereas the speed has been chosen so that it matches the reference ferry 287
Fig. 14.3 The interactive process of designing a ferry, where inputs from other sectors are illustrated in three external boxes. The development cycle is common for the development of other types of ferries as well 289

Fig. 15.1 How technology (*upper diagram*) and biology (*lower diagram*) address the challenges: "heat, beat, treat, and waste" vs. "ambient conditions, soft materials, problem prevention, and recycling" 302
Fig. 15.2 Step-by-step process of the design of a park/recreation area that restores surrounding damaged ecosystems. *Black zones* – areas with blooming plants, *central circle* – nesting zone is isolated from public. Wide ring foraging zone: *A* – concentric, *B* – radial. Large fields are subdivided into *C* (more radial beams) or *D* (concentric rings), *E* (single-spiral layout as a result of merging radial and concentric fragmentation), *F* (plain four-spiral layout), *G* (two-spiral layout with the interruption of spirals along "parallel and meridian") 312

List of Tables

Table 4.1 Stakeholders' interests and influence at different stages of the product life cycle ... 69
Table 4.2 Assessing stakeholders' needs for two critical components ... 71
Table 4.3 Assessing stakeholders' needs for the refined product ... 73
Table 4.4 Mapping needs into product's characteristics ... 74

Table 5.1 Ideality matrix for the camping stove ... 90
Table 5.2 Engineering parameters according to TRIZ ... 92
Table 5.3 Inventive principles of TRIZ ... 93
Table 5.4 Fragment of Altshuller's matrix of contradictions ... 94

Table 6.1 Classification of manufacturing sectors according to NACE Rev. 3 ... 105
Table 6.2 Hypothesis, independent variables definition, measures, and values ... 106
Table 6.3 Pearson's chi-squared test for H_1 ... 108
Table 6.4 Distribution of innovations with environmental benefits in the company according to the degree of innovation ... 108
Table 6.5 Spearman's rho (ρ) correlations for H_1 ... 109
Table 6.6 Pearson's chi-squared test for H_2 ... 110
Table 6.7 Distribution of innovations with environmental benefits resulting from the use of a product after-sales according to the degree of innovation ... 110
Table 6.8 Spearman's rho (ρ) correlations for H_2 ... 110
Table 6.9 Pearson's chi-squared test for H_3 ... 111
Table 6.10 Distribution of eco-innovations in response to external factors according to the degree of innovation ... 111
Table 6.11 Spearman's rho (ρ) correlations for H_3 ... 111
Table 6.12 Chi-squared test for H_4 ... 112

Table 6.13 Distribution of eco-innovations in response to internal factors according to the degree of innovation 112
Table 6.14 Spearman's rho (ρ) correlations for H_4 113

Table 7.1 Taxonomy of eco-innovations 120
Table 7.2 Determination of eco-innovations 121
Table 7.3 Variables in the survey 125
Table 7.4 Distribution of the sample by sectors 127

Table 8.1 Representativeness of sample companies 144
Table 8.2 Variables employed to measure the *eco-innovation strategies* construct 146
Table 8.3 Variables employed to measure the *environmental regulation* construct 147
Table 8.4 Variables employed to measure the *use of environmental and innovative incentives* construct 147
Table 8.5 Variables employed to measure the *reputational effects* construct 148
Table 8.6 Variables employed to measure the *top management support* construct 149
Table 8.7 Variables employed to measure the *technological competence* construct 149
Table 8.8 Variables employed to measuring the e*nvironmental formalization* construct 150
Table 8.9 Means of the cluster centers by variable of the *eco-innovation* construct in each cluster and ANOVA test 152
Table 8.10 Summary of the characteristics of each cluster as a result of their eco-innovation strategies 154
Table 8.11 Analysis of the differences between the mean values of the impact of contextual factors on each group of firms 155

Table 9.1 Dimensions and categories of eco-innovative activities 169
Table 9.2 Eco-innovative measures among large Swedish companies 169
Table 9.3 Eco-innovative measures: average per category and per company 172

Table 10.1 KPI (impact category) of leather and wool production from LCA 191
Table 10.2 KPI (impact category) of final products from LCA 192
Table 10.3 Environmental performance ranking 192

Table 12.1 The boundary-spanning business model of the biogas-for-traffic ecosystem integrator 235

Table 13.1 LEED certification levels and required points 246
Table 13.2 LEED-NC certification categories 247
Table 13.3 LEED credits and associated points 253
Table 13.4 Credits and points affected by the commissioning process 254

Table 14.1 Examples of eco-innovations in the ferry sector and their fuel saving potential 278
Table 14.2 Key differences between the carbon-fiber ferry and the reference ferry 282

Table 15.1 Living systems: desirable and harmful features for engineering 304
Table 15.2 Technology: advantages and shortcomings 304
Table 15.3 BioTRIZ axioms for eco-innovation 307
Table 15.4 Biological interpretation of engineering functions 310
Table 15.5 Ideality in ecological and engineering contexts 311

Chapter 1
Developments and Directions of Eco-innovation

Susana Garrido Azevedo, Marcus Brandenburg, Helena Carvalho, and Virgílio Cruz-Machado

Abstract To achieve business sustainability, companies must perform a set of changes not only in their internal processes but also in the large-scale level of business ecosystems. This change of paradigm from individual to macro level requires an understanding of new drivers of innovation as a way to improve and look for new business dynamics, to minimize the negative environmental externalities, and to keep the economic performance of companies. This chapter aims to summarize recent developments and future trends of eco-innovation in theory and application from its usual forms of incremental to more radical and systematic innovation. A brief introduction of eco-innovation and a terminological foundation consisting of adequate definitions of specific or ambiguous terms are given. It illustrates how models and frameworks for eco-innovation and its applications can support the ecological business modernization and how it can promote win-win situations among the different stakeholders. Beyond this, the chapter introduces and discusses the various remaining chapters in this book and presents summaries, insights, and linkages among them.

S. Garrido Azevedo (✉)
Department of Management and Economics, University of Beira Interior, Covilhã, Portugal
e-mail: sazevedo@ubi.pt

M. Brandenburg
Supply Chain Management, Faculty of Business and Economics, University of Kassel, Kassel, Germany

Department of Production Management, School of Economics and Management, Technische Universität Berlin, Berlin, Germany
e-mail: brandenb@uni-kassel.de; marcus.brandenburg@tu-berlin.de

H. Carvalho • V. Cruz-Machado
UNIDEMI, Department of Mechanical and Industrial Engineering, Faculty of Science and Technology (FCT), Universidade Nova de Lisboa, 2829-516 Lisbon, Portugal
e-mail: hmlc@fct.unl.pt; vcm@fct.unl.pt

S. Azevedo et al. (eds.), *Eco-Innovation and the Development of Business Models*, Greening of Industry Networks Studies 2, DOI 10.1007/978-3-319-05077-5_1,

Keywords Innovation • Sustainability • Ecological • Business development

1.1 Introduction

The rise of sustainability and green growth agendas in corporate management context is increasingly leading companies to revisit the concepts of value and profitability that drive their business models. Actually there are mainly two critical reasons for an increased need for green growth and sustainable development: (1) the emerging economies with their increasing demand for resources and (2) the necessity to decouple economic growth from the natural resources consumption by fostered resource productivity. However, the present pace of resource efficiency gains and incremental businesses improvements is too slow to respond to the need for decoupling since they are not sufficient to tackle the challenges. New or modified processes, techniques, practices, systems, and products are required to avoid or to reduce environmental harms and to promote the business sustainability. These factors have fostered the relevance of environmental aspects in innovation management, which can be comprehended as eco-innovation.

Developments and future directions of eco-innovation research and its application issues are outlined in this chapter. Selected publications on theory and applications of eco-innovations will be discussed and brought into context to the other chapters of the book at hand.

1.2 Developments of Eco-innovation Research

Only little research on the area of eco-innovation is found before 1990 (Schiederig et al. 2012). To assess the relevance of this scientific field, a keyword-based search[1] for related research publications has been performed in the Google Scholar database. The obtained results show that eco-innovation – including the issues of environmental or green innovation – has become a hot topic for researchers since 2000. The database search resulted in a total of 15,350 publications, out of which only 1,004 were published before 2000. Since then, the annual number of published papers shows a continuous growth and began to boost in 2007 (see Fig. 1.1).

Scientific literature on eco-innovation ranges from descriptions of innovation developments and diffusions over multilevel interactions between technological niches and socio-technological landscapes to the transition of management theories and coevolution of socioeconomic systems (Ekins 2010).

Schiederig et al. (2012) give a thorough review of related literature. The authors observe that nearly two thirds of all papers on eco-innovation are published in

[1] The search was based on the keywords "eco-innovation," "environmental innovation," and "green innovation" and performed on May 9, 2013.

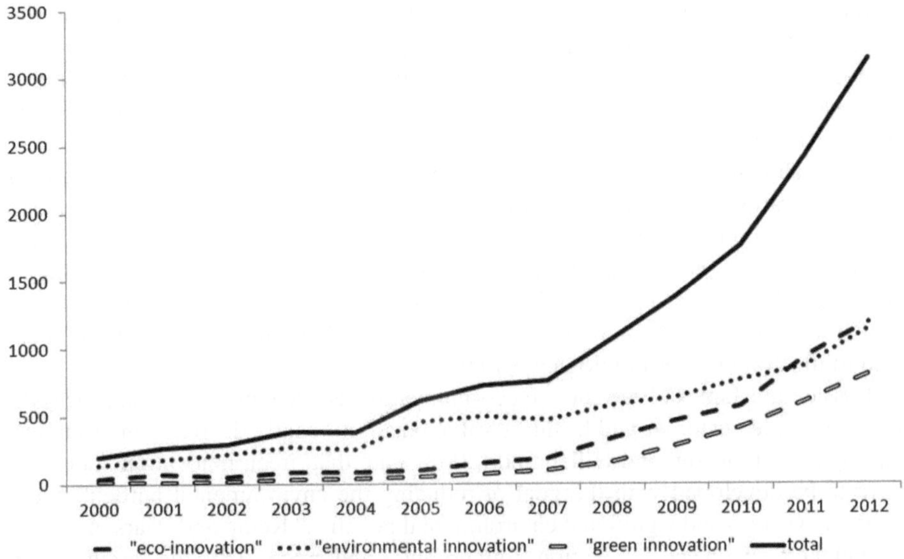

Fig. 1.1 Annual number of published papers

journals that are related to business, administration, finance, or economics, while only one fourth of the manuscripts were found in periodicals on social sciences, arts, or humanities. The remaining 10 % of all publications are related to engineering, computer sciences, and mathematics. Furthermore, Schiederig et al. (2012) identified that only 15 journals published more than 10 papers on eco-innovation and that *Journal of Cleaner Production* and *Business Strategy and the Environment* represent the periodicals with the highest relevance for this research area. Among the innovation management journals, the authors identify *Technological Forecasting and Social Change* and *Research Policy* as being highly relevant for eco-innovation research. Furthermore, the authors identify ten leading eco-innovation research institutions[2] and detect that eight of the ten are located in Germany, Italy, or the Netherlands.

1.3 Terminology and Characteristics of Eco-innovation

Although inventions and innovations are both characterized by novelty, the introduction into the marketplace, i.e., the economic utilization, is a distinctive feature of innovation (Rennings 2000). The Organisation of Economic Co-operation and Development (OECD) (1997) categorizes innovations into product, process, and organizational innovations. Referring to Schumpeter (1934), Hellström (2007) added novelties on demand and on supply market side as further categories. With regard to the degree of novelty, radical innovations are distinguished from

[2] Based on the number of publications written by the leading researcher.

incremental or marginal ones (Freeman and Soete 1997). Taking into account the extent of change that a system undergoes, component or modular innovations which do not modify the system but only some of its modules can be separated from architectural or systemic ones that result in more comprehensive system changes (Henderson and Clark 1990).

The inclusion of environmental aspects into the discussion of innovation leads to the comparably new area of environmental, green, or eco-innovation. Schiederig et al. (2012) found that the first expression was favored in the 1990s while the latter two notions were increasingly used within the last 5 years. Furthermore, the authors compared different scientific definitions that were suggested as a terminological basis for this research area. In the following, several definitions are briefly summarized to illustrate the terminological variety.

Fussler and James (1996) define eco-innovation as "new products and processes which provide customer and business value but significantly decrease environmental impacts." Driessen and Hillebrand (2002) state that green innovation "does not have to be developed with the goal of reducing the environmental burden" but it "does however, yield significant environmental benefits." Kemp and Pearson (2007) extend the view on eco-innovation from products and processes to the "assimilation or exploitation of a (…) service or management or business method" and include the consideration of the life cycle aspect. Oltra and Saint Jean (2009) define environmental innovation as "innovations that consist of new or modified processes, practices, systems and products which benefit the environment and so contribute to environmental sustainability." Arundel and Kemp (2009) emphasize that eco-innovations "can be motivated by economic or environmental considerations." These examples illustrate the broad variety of notations. The terms "eco-innovation," "environmental innovation," and "green innovation" are used interchangeably, and the perception of the related field has broadened from a focus on "greened" products or processes to the coherence of environmental aspects of physical items, virtual structures, and organizational procedures.

Taking into account the terminological heterogeneity arising from these various definitions, defining characteristics of eco-innovations are now distilled. Figure 1.2 illustrates dimensions and impacts of eco-innovations.

Rennings (2000) perceives eco-innovations as new approaches that help reduce environmental burdens or achieve ecological targets and differentiate between technological, organizational, social, and institutional ones. Including the economic perspective, Ekins (2010) considers an eco-innovation as being both economically and environmentally beneficial. Striving for more comprehensiveness, Schiederig et al. (2012) identify six defining characteristics of an eco-, environmental, or green innovation. The authors regard it as an object (product, process, service, or method) which is characterized by its market orientation (satisfaction of need or competitive market position) as well as its environmental benefit over its whole life cycle and which sets a new innovation or green standard to the company although its primary intention may be environmental or economic.

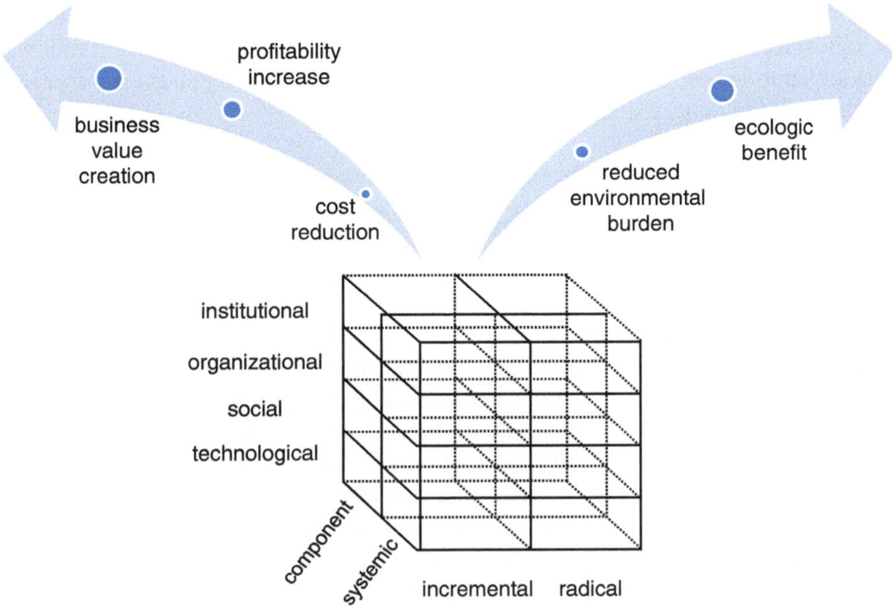

Fig. 1.2 Dimensions and impacts of eco-innovation

1.4 Models and Frameworks Supporting Eco-innovation

Emphasizing the need for methodological pluralism and multidisciplinarity, Rennings (2000) divides eco-innovation research into neoclassical economics that are deterministic and adequate to analyze incremental changes and into (co-)evolutionary concepts which cover uncertainties and enable the assessment of radical innovations. From neoclassical economy perspective, eco-innovation can be analyzed by two main pillars: the environmental economy and the innovation economy. According to the environmental economy, environmental policies should coerce the companies to realize the environmental innovations that create value on the market. Such policies emphasize the superiority of market-based instruments, such as taxes and marketable licenses, compared to cost-inefficient and quickly vanishing regulatory tools. Rennings (2000) points out that the resulting contributions to eco-innovation do not take into account the complexity of innovation management within a company. Innovation economics, the other pillar of neoclassical economics, is characterized by double externalities resulting from positive spillovers during the innovation and diffusion phases, i.e., reducing external environmental costs of production or products. Therefore, policies and regulations considering the environmental and innovation economies should be coordinated to promote eco-innovation among companies. Since companies are part of extended business ecosystem, with feedback cycles and evolution paths, (co-)evolutionary concepts are also used to explain the eco-innovation trajectories. This perspective links eco-innovation to the biological terms of selection and variation and to technological change and follows broader approaches. Rennings (2000) suggests that research on eco-innovation should combine neoclassical and evolutionary models.

1.4.1 Drivers of Eco-innovation

Rennings (2000) identifies technology push, regulatory push, and market pull as the three determinants of eco-innovation.

Technology push: Green technology mechanisms are related to the end-of-pipe technologies, process-integrated cleaner technologies, environmental research and development (R&D), or green product innovations (Demirel and Kesidou 2011; Ekins 2010). Machiba (2010) proposes a framework of eco-innovation that distinguishes primarily technology-related changes from primarily non-technological ones. In this framework, the incremental-radical categorization of eco-innovation mechanisms is narrowed down to the eco-innovation mechanisms of modification (resulting from small adjustments of products or processes), redesign (referring to significant changes of existing products), alternatives (substituting existing solutions), and creation (describing entirely new approaches). These mechanisms are assigned to eco-innovation targets, which are primarily related to either technological changes of products and processes or non-technological changes of marketing methods, organizations, and institutions. Machiba (2010) points out that non-technological change stemming from creation will lead to higher environmental benefits but also to more coordination difficulties.

Regulatory push: Ekins (2010) distinguishes between "hard" policy instruments comprising incentive-/market-based instruments and regulatory instruments and "soft" instruments consisting of voluntary/self-regulation agreements and information-/education-based instruments. The author compares the employment of these policies in Japan, the United States, and Europe based on several case studies from various industries. It is observed that a large majority of observations was related to product innovation technologies in context to classic regulation which was often complemented by incentive-/market-based instruments. In the three assessed US cases, classic regulations were always employed and in one case complemented by incentive-/market-based instruments resulting in mixed successes regarding the induced innovation. Better results regarding the induced innovation were observed in three cases from Japan, in which the two "hard" policy instruments were combined and in one case complemented by information-based policies. Employing one of the two "hard" policy instruments in different European countries induced innovation very well, while less promising results were achieved in Europe by only using voluntary policies.

In a survey of 289 UK companies, Demirel and Kesidou (2011) elaborate on the determinants of eco-innovations with regard to the companies' investments into green technologies. Regarding external factors, the authors detect that environmental regulations significantly influence the investments into end-of-pipe technologies or product innovations while process-integrated cleaner technologies remain unaffected. Contrastingly, no significant impact on any of the three green technologies resulted from environmental taxes. Regarding internal factors, efficiency significantly influences investments into end-of-pipe or process-integrated cleaner technologies (efficiency of machinery) as well as investments into product innovation and ecological R&D (cost efficiency), while corporate social responsibility failed to show significant influences on any of the eco-innovation investments. The observed effects of environmental management systems were comparable to those observed for environmental regulations.

Market pull: In a survey of more than 1,200 UK residents, Darnall et al. (2012) focus on the reasons why consumers choose to buy environmentally beneficial products. The authors detect that trust of government and environmental NGOs foster the consumers' willingness of green purchases while self-promoted green claims of private business are not able to increase a consumers' green consumption. Furthermore, the authors' findings support the conjecture that the consumers' awareness, knowledge, and affection, for environmental issues, can amplify their readiness for green consumption.

Jansson (2011) focuses on alternative fuel vehicles (AFVs) to assess the green buying behavior of Swedish consumers. In a survey of 642 car owners, the author conclude that, compared to users of "conventional" gasoline- or diesel-powered cars, owners of AFVs show higher levels of environmentally related personal and social norms. Unlike owners of "conventional" cars, AFV adopters have more positive attitudes towards "eco-cars," are more novelty seeking, and assess the risk of an AFV as being low. Furthermore, adopters perceive that AFVs offer higher relative advantages compared to "conventional" vehicles and that AFVs are more compatible to their personal values.

In another related survey of 1,832 car owners in Sweden, Jansson et al. (2010) detect that personal values, beliefs, and norms do not only influence the adoption to eco-innovation but also affect the green curtailment behaviors such as resource conservation or recycling.

1.4.2 New Approaches and Extensions

Many organizations implement eco-innovation in cooperation with other companies belonging to the same production value chain, such as suppliers and manufacturers, but also with actors belonging to the entire business ecosystem, e.g., universities or shareholders. Such alliances are seen as win-win opportunities for all groups of actors. This behavior highlights the importance of the corporate governance type to the eco-innovation implementation. In this context implementing cross-sector innovations or open innovations is particularly challenging since it requires the adoption of new technical, organizational, and also institutional arrangements by different stakeholders of the involved sectors. Part I "Models and Frameworks Supporting Eco-Innovation" of the book at hand comprises various approaches to face these challenges.

In Chap. 2, Kloiber and Priewasser provide insights on cross-industry innovations. A conceptual framework is proposed to support the systematical implementation of eco-innovation across industries. To obtain a detailed search strategy for environmental solutions that overcome organizational boundaries, theoretical considerations are added to cross-industry innovation models of Gassmann and Zeschky (2008) and Kalogerakis et al. (2005). The resulting framework consists of five major steps: (1) strategy analysis, (2) problem abstraction, (3) search for analogies, (4) assessment, and (5) adaptation. Furthermore, tools and methods are suggested to attain the objective of each phase.

Companies employ the open-innovation paradigm and interact with a range of different actors, from customers, suppliers, competitors, research institutes, and universities to support the creation and transference of knowledge among them. The open-innovation model is particularly critical as the innovation system influences the decision on the geographic sourcing concept of global value chains. Hence, there is a need to build knowledge on environmental aspects of innovation and to share it among the key actors of business ecosystems. In Chap. 3, Cabrita et al. discuss eco-innovations in a knowledge-based development perspective. The authors point out that acquiring, creating, developing, storing, and applying knowledge is crucial for a sustainable economic, social, and environmental development. A conceptual framework is proposed to integrate the eco-innovation concept and knowledge-based development perspective. According to these authors, innovation is extremely dependent on the availability of knowledge. Therefore, the complexity created by the explosion of richness and reach of knowledge has to be recognized and managed to ensure successful innovation. Knowledge management provides the tools, processes, and platforms that ensure the availability and accessibility of competences.

Chapters 2 and 3 highlight the necessity to adopt a systematic approach to eco-innovation development in order to continuously improve a company's ability to cope with regulations of its markets and environmental issues of its consumers. A proper work environment that fosters innovation within companies is an important driver of the sustainability of the business ecosystems. Moreover, the existence of frameworks for the development, evaluation, and assessment of eco-innovations is vital to understand and develop radical and systematic eco-innovation initiatives within a particular company and across different industrial sectors. In Chap. 4, Gremyr et al. propose a conceptual framework that picks up this managerial necessity and supports the development and assessment of radical eco-innovations. The framework named "Encore" synthesizes ideas and concepts from eco-innovation, quality management and life cycle assessment into a set of key principles and practices. It is illustrated by a hypothetical example related to the carbonated beverage industry.

In Chap. 5, Navas introduces the Theory of Inventive Problem Solving (TRIZ) method in context to eco-innovations. The author illustrates tools and techniques for the implementation and application of different TRIZ levels to support radical eco-innovations. Furthermore, engineering parameters and invention principles of TRIZ are discussed.

1.5 Application of Eco-innovation

Eco-innovation does not only represent a state-of-the-art in the theoretical fields of research, but also has developed to a considerable area of economic practice. In this section, the macroeconomic relevance of eco-innovation is illustrated and its applications and impacts on company level are outlined.

1.5.1 Eco-innovation in Macroeconomic Contexts

Motivated by regulatory pull, amplified by market push and enabled by technology pull, the environmental goods and service industry, has developed to a sector of considerable (macro)economic relevance. Most OECD countries consider eco-innovation as both environmentally and economically relevant (Ekins 2010). In 2005, the global market of the abatement goods and service industry, for instance, accounted for US $653 billion and approximately 4 million jobs with a potential of further growth (David and Sinclair-Desgagné 2010; Sinclair-Desgagné 2008).

Ekins (2010) gives a comprehensive overview over the European environmental industries. The author used the EU (2006) figures that categorize the environmental goods and services industry into two main groups: (1) "pollution management group" comprising eco-industry sectors that manage material streams from processes managed by humans (the technosphere) to nature, typically using "end of pipe" technology including cleaner technologies and products (it includes solid waste management and recycling, wastewater treatment, and air pollution control, among other industries), and (2) "resource management group" comprising a more preventive approach to managing material streams from nature to the technosphere. It includes water supply, renewable energy production, and eco-construction, among others industries. In 2004, the size of the environmental industry in the European Union (EU-25) was € 227 billion with Germany (€ 66 bn), France (€ 45 bn), UK (€ 21 bn) and Italy (€ 19 bn) as the four economies with the largest eco-industry sector (Ekins 2010). Two thirds of the sector revenues were related to pollution management activities, mostly to management of solid waste or of wastewater (both € 52 bn). Resource management activities, especially water supply (€ 45 bn), represented another one third (Ekins 2010). These revenue proportions match with the 3.4 million employees that are attributed to the eco-industries: about 2.2 m jobs are related to pollution management, mainly to management of solid waste (1.0 m jobs) or wastewater (0.8 m jobs), while the water supply sector (0.5 m jobs) accounts for the largest share of jobs in the resource management activities.

In nine in-depth case studies conducted at American, French, and German companies, Wagner and Llerena (2011) observe only few culture differences with regard to determinants of eco-innovations and conclude that sectoral contexts matter comparatively more. Furthermore, the authors emphasize the relevance of non-government organizations (NGOs) for eco-innovation and the need for more related research.

1.5.2 Eco-innovation on the Microeconomic Level

Wagner and Llerena (2011) identify that eco-innovation within companies is often a bottom-up activity which, however, needs subsequent support from top management. Informal mechanisms of eco-innovation can substitute a systematic integration of environmental aspects in the innovation processes. Hence, the authors identify a gatekeeper function of top management which is complemented by individual employees that actually work on green innovations.

In a survey of environmental new product development (ENPD) projects in North American companies of different industries, Pujari (2006) investigates on influencing factors of market performance. The author identifies statistically significant relationships between market performance and several influencing factors. In particular, positive impacts on ENPD performance stemming from cross-functional coordination, from supplier involvement, from activities of design-for-environment or life cycle analysis, and from market focus were observed.

Cortez and Cudia (2012) assess Japanese automotive and electronics companies to elaborate on the coherence of environmental innovations and financial performance. Based on data published between 2001 and 2010 in financial and sustainability reports of 18 companies, the authors detect a positive association between eco-innovations and net revenues. For the automotive industry, the authors find that eco-innovations positively affect net profit, assets, and equity and vice versa. For the electronics industry, positive influences between eco-innovations and long-term debts were observed.

1.5.3 Surveys and Case Studies on Eco-innovation Deployment

The understating of eco-innovation in macro- and microeconomic contexts is illustrated by a set of empirical studies in Part II "Application: Surveys and Case Studies on Eco-innovation Deployment" of the book at hand. Chapters 6, 7, 8, 9, 10, 11 and 12 exemplify the deployment of eco-innovation practices among companies from different sectors and countries by surveys, secondary data analyses, case studies, and clinical research.

In Chap. 6, da Silva et al. analyze secondary data from the Portuguese Community Innovation Survey to assess eco-innovations at Portuguese companies of various sectors. The authors elaborate on (1) the influence of the introduction of eco-innovation practices by manufacturing companies on their propensity to innovation and (2) the contribution of internal drivers, e.g., procedures to identify and reduce environmental impacts regularly, and external factors, e.g., existing environmental regulations or fiscal duties on pollution, for the eco-innovation behavior of the manufacturing industry. The results highlight four factors which stimulate or limit eco-innovative capability in manufacturing industry, namely, environmental benefits in the company, environmental benefits in after-sales, and external and internal factors.

In Chap. 7, Carvalho presents a cross-sector survey on eco-innovative activities in Brazilian companies carried out in 2012. The quantitative, descriptive, and explanatory methodology employs precise measurement to analyze eco-innovations of Brazilian companies and to characterize eco-innovative companies, the determinants, types, and results of eco-innovations and the existence of cooperative arrangements for eco-innovation. The results highlight the existence of two groups within the sample: (1) a largest group reactively responds to pressures from their main stakeholders; and (2) a smallest group performs radical innovations. The research reveals that organizational eco-innovations are the most often implemented type of eco-innovation initiatives, followed by the introduction of new technologies. The new products were the less representative type of eco-innovation. Moreover, the results show that most eco-innovative companies also conduct systematic in-house R&D activities and participate in cooperative arrangements for innovation. The author performs a detailed discussion of the results unfolding existing literature revealing the key variables related to eco-innovations.

Chapter 8 assesses the eco-innovation in Brazilian companies of the cellulose, paper, and paper product industry. Maçaneiro and Kindl conduct a survey at 117 companies and a cluster analysis to classify companies according to their eco-innovative strategy. As a result, a taxonomy for eco-innovation strategies comprising four categories is proposed: reactive organizations, indifferent organizations, proactive organizations, and eco-innovative organizations. The authors also provide a definition of external and internal factors that influence the adoption of these strategies.

Chapter 9 consolidates the existing conceptualizations by adapting a typology of eco-innovation under consideration of products, production processes, organizational processes, users, the value chain, and governance. Sarasini et al. analyze secondary data from annual and sustainability reports of 92 large Swedish companies. The authors highlight that large companies focus their eco-innovative effort on internal measures related to product and process changes. The authors detect that companies are less adept at collaborating with suppliers, users, and other external partners that can boost eco-innovation. As a conclusion, recommendations for policymakers are derived.

The rise of sustainability and green growth agendas in corporate management context is increasingly leading companies to revisit the concepts of value and profitability that drive their business models. There is a need for new or modified processes, techniques, practices, systems, and products to avoid or to reduce environmental harms and to promote the business sustainability. Coordinate actions between corporate environmental management systems are likely to increase effectiveness and efficiency of interorganizational exchanges. In this context, Daddi et al. describe a case study in Chap. 10 which is related to four industrial clusters of a fashion supply chain in the Tuscany region (Italy). A cluster approach based on environmental management systems and life cycle assessment tools is developed. The approach relies on a cooperative and modular life cycle management of common environmental problems among four clusters that are linked with each other as part of the same supply chain. The results highlight that companies located in the clusters under study are able to use this integrated process to improve their environmental competitiveness. Furthermore, companies can benefit from support tools that are derived from this cooperative approach.

The business model perspective offers a comprehensive way to understand how eco-innovation can be induced and diffused to enable systemic changes and business transformation. The focus on business models allows a better understanding of how environmental value is captured and turned into profitable products and services that deliver convenience and satisfaction to users. In this context, measuring and comparing environmental and economic benefits of eco-innovation is critical to support and promote new initiatives. As a consequence, the integration of eco-innovation with the existing managerial tools and methodologies becomes a less challenging task for managers and policy makers.

In Chap. 11, Bikfalvi et al. present a case study of a new technology business venture focusing on energy management systems and its evolution over a decade. The case study illustrates the evolution of innovation types and the company's business model trajectory through different stages, from pure product or service orientation, to joint product-service modes of operations, and ultimately to continuously evolving product-related services.

Companies need to redesign the value flows to make eco-innovation in the business ecosystem beneficial for all involved parties. Chapter 12 focuses on how to develop business models which promote eco-innovation. Tsvetkova et al. present the results of two research projects in a Finnish municipality. In these projects, a sustainable local biogas-for-traffic solution and the respective boundary-spanning business model were developed in cooperation with major ecosystem stakeholders. The authors combined clinical research and design science approaches to assure an iterative research process with tight cooperation among business actors. Challenges and opportunities in implementing a biogas-for-traffic ecosystem are revealed, and a sustainable business solution is designed under consideration of stakeholders' requirements.

1.6 Future Directions: Eco-innovation Initiatives

The final Part III "Future Directions: Eco-innovation Initiatives" of the book at hand provides novel perspectives about eco-innovation initiatives. In Chap. 13, Gillis and Cudney propose a new methodology for eco-friendly construction of buildings as an example of an organizational innovation to implement new management systems and tools. In the building commissioning process, the quality function deployment, a traditional quality management tool, is utilized to meet requirements of leadership in energy and environmental design (that promotes green, efficient, and sustainable design and construction). This new eco-innovative model provides a greater opportunity for the commissioning authority to ensure that the owner achieves their sustainability and efficiency goals with the final building construction.

In Chap. 14, Mosgaard et al. focus on modular product design in context to radical and systemic eco-innovation. In a case study based on action research in the Scandinavian shipbuilding industry, the authors show how to innovate traditional steel ferries by a transformative technology. In this context, organizational, economical, and institutional changes are required. A modified organization results from the integration of new actors, while open-source and non-commercial business approaches lead to economic adaptations. To address institutional changes, a purely competitive behavior between different actors is adapted to a more collaborative approach.

The creation of win-win situations that resolve conflictive economical and environmental objectives is in focus of Chap. 15. Bogatyrev and Bogatyreva outline BioTRIZ as an eco-innovation methodology which combines the concept of biomimetics, i.e., copying living nature by technology, and TRIZ. To this end they combine a set of BioTRIZ axioms considering biomimetics (e.g., axiom of simplification) and eco-engineering (e.g., axiom of maximization of useful function) with a set of BioTRIZ rules for eco-innovation. The authors propose this methodology as a way to deal with contradictions between biology and technology, because its main mechanism is based on revealing conflicting requirements and a win-win resolution. The implementation of the BioTRIZ axioms is illustrated at a case example of architectural design and layout of an eco-park which is pollinated by bumblebees.

1.7 Conclusion

Despite its novelty, eco-innovation has developed to a decisive area of scientific research and industrial application. Driven by technology and regulatory push as well as by market pull, eco-innovation enables economic success without compromising on environmental issues.

Highlights of the book comprise state-of-the-art models for a comprehensive management of eco-innovations and its applications as well as novel findings from empirical research. The suggested models allow for implementing eco-innovation processes on a cross-industry level and extending the topical focus from purely ecological factors to the multidimensional concept of sustainability which combines economic and environmental criteria with social aspects. From the academic perspective, focused approaches to eco-innovation are broadened to more general ones. From the practitioners' point of view, these approaches support the implementation of eco-innovation practices on an interorganizational level while taking into account industry-specific particularities of each company.

Empirically, the book highlights novel insights and trends of eco-innovation management from various European and Latin American countries and several industry sectors. In this context, determinants, drivers, and influencing factors of eco-innovations are elaborated, and a typology of eco-innovation strategies and processes is derived.

The book at hand offers new approaches, illustrates various application areas, and points towards future directions of this highly relevant topic thereby providing new insights into eco-innovation for researchers and practitioners. Limitations of the book point towards future research perspectives on eco-innovation. From the conceptual perspective, the suggested conceptual frameworks can be disaggregated to greater detail. This would foster the implementation of these models in managerial practice. Furthermore, empirical research is recommended to test the suggested frameworks in industrial application or to elaborate on eco-innovations in other cultural or geographical contexts. Beyond this, future eco-innovation research could take into account grassroots innovation or eco-innovation in and for developing countries or prevalent technical bifurcations. Besides, formal modeling and operations research represent methodologies that are worth being employed in context to eco-innovation.

References

Arundel A, Kemp R (2009) Measuring eco-innovation. United Nations University – Maastricht Economic and Social Research and Training Centre on Innovation and Technology. Maastricht, UNU-MERIT #2009-017

Cortez MAA, Cudia CP (2012) Environmental innovations and financial performance of Japanese automotive and electronics companies. In: Vazquez-Brust DA, Sarkis J (eds) Green growth: managing the transition to a sustainable economy. Springer, Heidelberg, pp 173–190

Darnall N, Ponting C, Vazquez-Brust DA (2012) Why consumers buy green. In: Vazquez-Brust DA, Sarkis J (eds) Green growth: managing the transition to a sustainable economy. Springer, Heidelberg, pp 287–308

David M, Sinclair-Desgagné B (2010) Pollution abatement subsidies and the eco-industry. Environ Resour Econ 45:271–282

Demirel P, Kesidou E (2011) Stimulating different types of eco-innovation in the UK: government policies and firm motivations. Ecol Econ 70:1546–1557

Driessen P, Hillebrand B (2002) Adoption and diffusion of green innovations. In: Nelissen W, Bartels G (eds) Marketing for sustainability: towards transactional policy-making. IOS Press Inc., Amsterdam, pp 343–356

EC (European Commission) (2006) Eco-industry, its size, employment, perspectives and barriers to growth in an enlarged EU. Final report to DG environment from Ernst & Young, European Commission, Brussels. Available at: http://ec.europa.eu/environment/enveco/eco_industry/pdf/ecoindustry2006.pdf

Ekins P (2010) Eco-innovation for environmental sustainability: concepts, progress and policies. Int Econ Econ Policy 7:267–290

Freeman C, Soete L (1997) The economics of industrial innovation, 3rd edn. Pinter, London

Fussler C, James P (1996) Driving eco-innovation: a breakthrough discipline for innovation and sustainability. Pitman, London

Gassmann O, Zeschky M (2008) Opening up the solution space: the role of analogical thinking for breakthrough product innovation. Creat Innov Manage 17:97–106

Hellström T (2007) Dimensions of environmentally sustainable innovation: the structure of eco-innovation concepts. Sustain Dev 15:148–159

Henderson RM, Clark KB (1990) Architectural innovation: the reconfiguration of product technologies and the failure of established firms. Admin Sci Q 35:9–30

Jansson J (2011) Consumer eco-innovation adoption: assessing attitudinal factors and perceived product characteristics. Bus Strategy Environ 20:192–210

Jansson J, Marell A, Nordlund A (2010) Green consumer behavior: determinants of curtailment and eco-innovation adoption. J Consum Mark 27(4):358–370

Kalogerakis K, Herstatt C, Lüthje C (2005) Generating innovations through analogies. An empirical investigation of knowledge brokers. Working paper 33, Hamburg-Harburg

Kemp R, Pearson P (2007) Final report of the MEI project measuring eco innovation. UM Merit, Maastricht

Machiba T (2010) Eco-innovation for enabling resource efficiency and green growth: development of an analytical framework and preliminary analysis of industry and policy practices. Int Econ Econ Policy 7:357–370

OECD (1997) OECD proposed guidelines for collecting and interpreting technological innovation data – Oslo-manual. OECD/Eurostat, Paris

Oltra V, Saint Jean M (2009) Sectoral systems of environmental innovation: an application to the French automotive industry. Technol Forecast Soc 76:567–583

Pujari D (2006) Eco-innovation and new product development: understanding the influences on market performance. Technovation 26:76–85

Rennings K (2000) Redefining innovation – eco-innovation research and the contribution from ecological economics. Ecol Econ 32:319–332

Schiederig T, Tietze F, Herstatt C (2012) Green innovation in technology and innovation management – an exploratory literature review. R&D Manage 42(2):180–192

Schumpeter JA (1934) The theory of economic development. Harvard University Press, Cambridge, MA (First published in German, 1912)

Sinclair-Desgagné B (2008) The environmental goods and services industry. Int Rev Environ Resour Econ 2:69–99

Wagner M, Llerena P (2011) Eco-innovation through integration, regulation and cooperation: comparative insights from case studies in three manufacturing sectors. Ind Innov 18(8):747–764

Part I
Models and Frameworks Supporting Eco-innovation

Part I
Models and Frameworks Supporting Eco-innovation

Chapter 2
Managing Cross-Industry Innovations: A Search Strategy for Radical Eco-innovations

Michaela Kloiber and Reinhold Priewasser

Abstract Companies can enhance their internal knowledge and develop innovations by working together with external partners from their own value chain. Besides this common open innovation strategy, the cooperation with partners from distant industries is becoming ever more important for companies as this concept can be a significant source for radical eco-innovations. These cross-industry innovations (CII) can lead to lower development time, lower project risks, and higher growth rates and margins owing to the discovery of radical innovations. We have found that due to a lack of existing process models and methods, companies implement cross-industry innovation strategies – in the most part – unsystematically and that the TRIZ database approach, the knowledge broker approach, and a creativity workshop with external experts and lead users are the most appropriate methods to creating cross-industry eco-innovations. The contribution of this chapter is, firstly, to analyze why the cross-industry innovation framework can have significant influence on the creation of radical eco-innovations and, secondly, to present a step-by-step process which should help companies – and especially innovation managers – to systematically implement eco-innovations across industries in the fuzzy front end of the innovation process.

Keywords Cross-industry innovation • Radical eco-innovation • Analogical problem solving

M. Kloiber (✉) • R. Priewasser
Institute for Environmental Management in Companies and Regions, Johannes Kepler University, Altenbergerstrasse 69, A-4040 Linz, Austria
e-mail: michaela.kloiber@jku.at; reinhold.priewasser@jku.at

S. Azevedo et al. (eds.), *Eco-Innovation and the Development of Business Models*, Greening of Industry Networks Studies 2, DOI 10.1007/978-3-319-05077-5_2,

2.1 Introduction

Global trade, rapid technology change, and short product life cycles have all led to increasing competition over the past decades. These developments are forcing companies to rethink their traditional internal innovation strategy in order to differentiate themselves from their competitors (Lichtenthaler 2011; Stötzel et al. 2011). Consequently, open innovation has been proposed as a new concept for the management of innovation (Chesbrough 2003; Gassmann 2006). Empirical research demonstrates that collaborations across organizational boundaries – by the use of a wide range of external actors and sources – have a positive impact on a company's innovation performance (Laursen and Salter 2006).

The eco-industry consists of companies who "produce goods and services to measure, prevent, limit, minimize or correct environmental damage" and which are less environmentally harmful than the use of relevant alternative products and services (OECD/Eurostat 1999). Due to a high degree of maturity[1] in many fields of this industry, breakthrough innovations are required. These so-called radical innovations are defined as innovations that go beyond incremental changes in products, processes, and business models. New methodologies and frameworks are needed to move toward green growth because incremental improvements are not enough to counterbalance the population growth and the global increase in resource consumption (Jones et al. 2001; von Weizsäcker et al. 1999).

Creating innovations between companies of distant industries is a new phenomenon in both theory and practice in respect to a step-change open innovation approach. These so-called cross-industry innovations (CII) can be technologies, patents, specific knowledge, business processes or whole business models (Enkel and Gassmann 2010). Imitation and reworking of already existing solutions from other industries can also contribute significantly to the development of breakthrough innovations in the eco-industry. Highly novel innovations lead to lower development time, lower project risks, and higher growth rates and margins. Despite these advantages, companies tend to implement the cross-industry innovation approach very rarely and/or unsystematically (Enkel and Dürmüller 2011). Consequently, this phenomenon has led to the research question of this paper as being: "How can companies implement cross-industry innovations systematically in the fuzzy front end of the eco-innovation process?"

This chapter is structured as follows: Firstly, we define the term radical eco-innovation. In a subsequent step, we review current literature on open innovation and cross-industry innovations, and we illustrate an example of a company who developed a radical eco-innovation successfully by implementing the "cross-industry innovation framework." Then, we present a step-by-step process which helps

---

[1] For example, eco-products in the area of air pollution prevention, water quality control, material recycling and energy efficiency improvements are well established and have reached advanced technological levels.

companies, and especially innovation managers, to systematically implement eco-innovations across industries in the fuzzy front end of the innovation process.

2.2 Theoretical Background on Cross-Industry Innovation

2.2.1 Eco-innovation: A Conceptual Approach

One way of classifying innovation in general, and one that particularly suits eco-innovation classification, is to refer to the target area of which is aimed at being changed. Referring to the Oslo Manual (OECD/Eurostat 2005), we can distinguish between:

- Product innovation
- Process innovation
- Organizational or business model innovation
- Marketing innovation

Furthermore, innovations can also be distinguished as to their degree of originality:

- *Incremental innovations* aim at modifying and improving existing technologies or processes to a higher level of resource efficiency or emissions reduction, without fundamentally changing the relevant technology base. Presently, this is the dominant type of innovation in enterprises (OECD 2011).
- *Radical innovations*, however, lead to a profound change, either in specific fields of technology processes and products or to larger-scale shifts in the overlying systemic context (Gjoksi 2011; OECD 2011). Both radical technology and systems changes can also be referred to as *transformative innovations* (Stirling et al. 2009). Radical innovations, in terms of environmental benign technologies, represent technical solutions which are significantly different from their predecessors and are able to raise the technological system to a new economic-ecological equilibrium (Gjoksi 2011). Examples of radical innovations include the following: the development of automatically functioning biomass pellet heating devices as an alternative for oil-fired heating installations, the creation of biopolymers from starch or sugar plants instead of the use of petrochemical-based polymers or the generation of biodiesel and other liquid biogenic energy carriers from cellulose (as opposed to biofuels from oil plants). A number of other examples are given by OECD et al. (2012).
- *Systemic innovations* often involve technological breakthrough innovations and lead to an increased organizational complexity (Nuij 2001). They typically include a reconfiguration, for example, of complementary infrastructures or product-service systems. Taking place beyond the boundaries of companies or organizations, they represent a multi-actor process (Stirling et al. 2009). Thus, innovations of this type often require a social and cultural change, adopting new values and behavioral patterns, on behalf of both the producer and the consumer

(OECD et al. 2012). Examples of systemic innovations are as follows: the "cradle-to-cradle concept," comprising of closed resource input to waste output loops, or fundamentally new urban mobility systems that are primarily based on public transportation or bike traffic. Even the recently promoted implementation of e-mobility requires essential preconditions that can create changes in the traffic system.

Originally, the concept of eco-innovation focused on progress in the field of technology and process, with an intention to reduce environmental impacts of economic activities. These environmental technologies applied to managing pollution (e.g., air pollution control, waste management), lessening of polluting, and lessening of resource-intensive products and services (e.g., fuel cells) and ways of managing resources more efficiently (e.g., water supply, energy-saving technologies) (European Commission 2004).

In time, eco-innovation awareness has essentially broadened, particularly where the following aspects are concerned:

- Eco-innovation does not only apply to clean and resource-efficient technologies that are specifically aimed at reducing environmental harm. Every product or service generating an environmental benefit (reduced use of natural resources and lower use of noxious emissions and waste) in relation to relevant alternatives should be recognized as an eco-innovation (Kemp and Pearson 2007).
- Eco-innovation encompasses all environmental improvements across the whole product life cycle, concerning the way they are designed, produced, used, reused, and recycled (EIO 2011).
- Eco-innovation, in a broader perspective, also embraces environmentally orientated organizational and marketing approaches, including eco-innovative business models which can have effects on the consumer behavior (EIO 2013).
- Finally, eco-innovation can be established on a cross-entrepreneurial systems level. Here, eco-innovations improve or create entirely new product or service systems with a reduced overall environmental impact. Examples of this are as follows: new housing concepts, new mobility systems, and even up to modeling "green cites" (EIO 2013).

Concentrating on most of these additional aspects, and with reference to the OECD general definition of innovation described in the Oslo Manual (OECD 2005), the expert group of the Eco-Innovation Observatory defines eco-innovation as the following:

> Eco-innovation is the introduction of any new or significantly improved product (good or service), process, organizational change or marketing solution that reduces the use of natural resources (including materials, energy, water, and land) and decreases the release of harmful substances across the whole life-cycle. (EIO 2011)

From a political perspective, eco-innovation closely corresponds to the environmental sustainability approach, which is based on increasing resource efficiency, dematerialization, and reduction of waste and emissions and finally leading to a well-maintained balance between the economy and the ecological system (Pujari 2004).

A specifically eco-innovation-related and better functioning taxonomy, reflecting the different roles of eco-innovative approaches on a greening market, is delivered by *Andersen*, who suggests five categories of eco-innovations (European Environment Agency 2006; Andersen 2008):

- *Add-on eco-innovations.* These innovations represent improvements in pollution and resource controlling or managing technologies and services. Progress in those fields normally has limited influence on existing production and consumption practices.
- *Integrated eco-innovations.* This kind of innovations deals with cleaner technological processes and more environmentally sound products. These innovations bear an increase of energy and resource efficiency, lead to a higher level in product or material recycling or promote the substitution of environmentally harmful substances.
- *Alternative product eco-innovations.* They offer new technological paths in the production or product design. Examples of alternative product eco-innovations are e-cars and respectively e-bikes as an alternative to conventionally driven vehicles or heat pumps opposed to conventional heating devices.
- *Macro-organizational eco-innovations.* These innovations entail new solutions for more environmentally friendly ways of organizing production and consumption on systemic level of economic or social structures. New functional structures of that kind are, for instance, industrial symbiosis (meaning energy, resource, and recycling networks of companies) or new ways of mobility such as car sharing.
- *General-purpose eco-efficient innovations.* Certain technologies affect the economy fundamentally as they lay behind and feed into other technological innovations, leading to an entirely new techno-economic paradigm. In their fundamental character, changes in those general-purpose technologies will also have major effects on eco-innovations. This holds true, for example, in biotechnology, nanotechnology, or information and communication technology.

Stricter environmental specifications to processes and products, combined with an increasingly tight global competition, are the main reasons why the key strategy of many companies is to create breakthrough eco-innovations. Therefore, the next chapter concentrates on the creation of radical cross-industry innovations in technologically and environmentally friendly products.

2.2.2 Cross-Industry Innovation Through Analogical Problem Solving

In the past, most companies were focused on internal research and development (R&D) activities and managing their internally developed products which were distributed by the organization. This type of innovation strategy, which limited interaction with external partners, is defined as closed innovation strategy. In contrast to this closed strategy, a new innovation framework has evolved. The open innovation concept can be seen as an open system, as innovative companies use a wide range of external actors and sources to help them to create new innovations (Laursen and Salter 2006; Chesbrough 2006a, b). Chesbrough (2006a) defines open innovation as "the use of purposive inflows and outflows of knowledge to accelerate internal innovation, and expand the markets for external use of innovation, respectively" (Chesbrough 2006a, b).

There are diverse potentials of open innovation. Firstly – through accommodating for the flow of external knowledge transfer into the company – it could lead to shorter innovation cycles, lower development costs and the opening up of new markets and to the reduction of market and technological uncertainty in the innovation process. Secondly, profits can be generated by licensing intellectual property and/or multiplying technology through the transfer of internal ideas to other companies (Gassmann and Enkel 2004; Leimeister et al. 2010). These numerous potential benefits signify different open innovation strategies.

Gassmann and Enkel (2004) identified three core open innovation processes:

1. *Outside-in process*: the aim is to enhance the company's knowledge base by integrating the innovative knowledge of external actors.
2. *Inside-out process*: earning profits by transferring ideas to the outside environment.
3. *Coupled process*: linking outside-in and inside-out processes by working in alliances with complementary companies.

Recent empirical studies demonstrate that an open innovation strategy provides the ideas and resources that help organizations gain and exploit innovative opportunities to a much greater extent (Laursen and Salter 2006).

Companies can interact with a range of different external actors, and they can transfer external knowledge from customers, suppliers, competitors, research institutes, and universities. Analogical problem solutions and novel ideas can also be successfully found among distant industries (Enkel and Gassmann 2010; Reichwald and Piller 2006). This open innovation strategy leads to cross-industry innovations.

Innovations can be described as the outcome of the linkage of different elements of knowledge that have not been connected before (Hargadon 2002; Geschka et al. 1994). Up until now companies have often concentrated on internal knowledge when searching for new innovations. As it is difficult for the employees to overcome established thinking patterns, they search mainly for solutions in the close proximity of the problem and not outside of their limited section of environment. Additionally, experience from former projects can hinder the creative innovation process. A fundamental cognitive mechanism which can overcome these difficulties is analogical thinking (Schild et al. 2004; Birch and Rabinowitz 1951).

Searching for analogies is a promising methodology to create new combinations of knowledge. Analogies are characterized through similar aspects of two objects of different domains. Drawing analogies beyond the borders of one's own industry can evolve cross-industry innovations. Here, a solution is found in one industry and applied to solve a problem in another. On the one hand, the creation of cross-industry innovations may reduce uncertainty as potential solutions have already proved to function in a similar context. On the other hand, analogies applied across industries may entail breakthrough innovations owing to the combination of different pieces of knowledge (Enkel and Dürmüller 2011; Kalogerakis et al. 2005; Gassmann and Zeschky 2008).

Where some analogies are obtained from knowledge bases that are similar to the solution, other analogies are drawn from bases that are inconsistent. Hence, analogies can be differentiated between near and far analogies. If an analogy is closely related to the base domain, it is termed a near analogy, and if the analogy comes from a distant domain, it is a far analogy. Near analogies normally lead to incremental innovations as they represent smaller conceptual distances between present solutions and the new idea and may be seen as less original. In comparison, far analogies are considered the main drivers of truly innovative thought. As they have a greater potential to foster creativity, radical innovations are more likely to result from far analogies between distant domains (Dahl and Moreau 2002; Schild et al. 2004; Gassmann and Zeschky 2008).

As in many other industries, the products, services, and business models of the eco-industry are largely shaped by the mindset of their own business sector. As this chapter argued, analogical problem solving and search for technological solutions in distant industries can open up interesting new perspectives and can be a significant source for radical eco-innovations.

The following list presents advantages which can apply to the eco-industry:

- The time to market is shorter and the project risk is lower due to the integration of an already tested and utilized technology in other industries.
- Analogies from another industry can generally be utilized without competitive conflicts.
- Stronger differentiation of the product in comparison to the competitors can lead to higher growth rates and margins.
- The collaboration with new partners and the combination of complementary knowledge can help a company to enhance its innovative capacity.

In practice, it is possible to identify successful cross-industry innovation projects where a systematic approach is applied, as well as projects where the solution evolved less systematic. Generally, in the search for analogies, a meaningful combination and balance of methodical-systematic and creative-chaotic approaches is aspired to. Thereby, teams with representatives from different fields, sectors, cultural areas, and personalities play an essential role (Enkel and Dürmüller 2011).

There are a multitude of examples of technology spillovers from other industries. The path-breaking interface iDrive from the BMW Group is based on the tried-and-tested technology of the joystick of the video game industry. Another famous example is the Aeroccino from Nespresso. The company adapted the established stir principle used in labs, which uses a contact-free driven beater with magnetic torque transmission. A milk creamer that is easy to clean was the innovative result (Enkel and Gassmann 2010; Enkel and Dürmüller 2011).

The waterless urinal from Geberit is a good example of an eco-innovation, which was developed through the implementation of a systematic process. The task was to create a new urinal that would be able to operate without the supply of external energy and chemicals and yet would feature a reliable odor seal. In the search for the solution, it was important to enlarge the system boundaries in comparison to the previous in-house operations. In the search for new solution principles, the energy flow for a repetitive and lossy system was demonstrated in depth. In the second stage, the project team assigned the corresponding status changes with possible mechanical energy forms. After an initial assessment, four solution principles were developed. These were constructively enhanced, and, following this, a prototype was produced and tested in a rapid prototyping procedure. Ultimately, one principle, which was based on an earlier invention, namely, the Erlenmeyer flask found in chemical laps, stood out among other considerations. This solution concept allows for significantly lower maintenance requirements than competitive products (Enkel and Gassmann 2010; Enkel and Dürmüller 2011).

2.3 Models of Cross-Industry Innovations

In the following section, two famous process models are presented which are often employed as a means to finding analogies and in the transfer of company knowledge. The first process was created by Kalogerakis et al. (2005) (see Fig. 2.1).

The procedure proposed by Kalogerakis et al. (2005) is based on four phases: (1) definition of the search field, (2) search for analogies, (3) verification and evaluation of analogies, and (4) development of the solution via the transfer of analogies. The first phase comprises of making a detailed description of the abstraction of the problem along with a review of the constraints and general conditions. In the next phase, it is recommended that companies search via people and/or via databases for innovative solutions. The search for analogies is completed by an evaluation of the

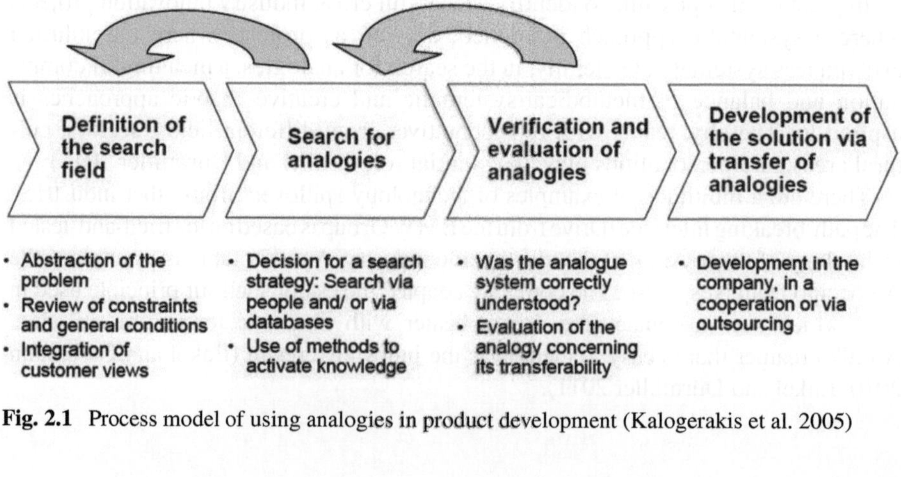

Fig. 2.1 Process model of using analogies in product development (Kalogerakis et al. 2005)

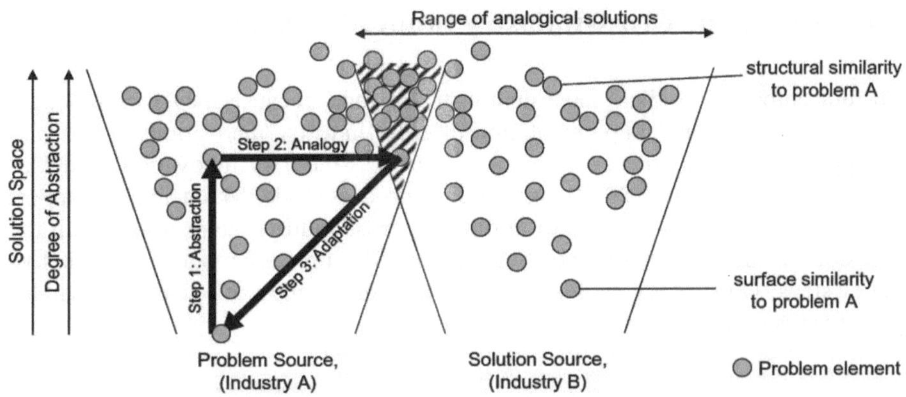

Fig. 2.2 Opening up the solution space by abstraction from the underlying problem (Gassmann and Zeschky 2008)

founded analogy. This evaluation and verification should be conducted in team discussions. The last phase – transfer – does not present an in-depth description. The authors only recommend that companies should consider problems of intellectual property rights or convents with other clients during the transfer. The process is well structured, and the authors provide detailed practical ideas and useful methods for each phase.

Based on a multiple case study, even Gassmann and Zeschky (2008) proposed a model for the development of product innovations by means of analogical thinking (see Fig. 2.2). It is aimed at targeting early innovation challenges by eliciting highly novel solutions.

This cross-industry innovation model includes three major steps: (1) abstraction, (2) analogy, and (3) adaptation. Firstly, a problem must be analyzed in detail, and key terms have to be abstracted. Secondly, analogies from different industries need to be found. In the last phase, the relevant knowledge technology has to be transferred and adapted. Gassmann and Zeschky describe the three core process phases of a systematic cross-industry innovation model, but they do not describe the search for analogies in detail. A description with recommended tools and methods would be helpful for management to integrate the model in a company's innovation process. The main strength of this model is that it is based on empirical findings and has been applied successfully in four engineering companies.

These two cross-industry innovation models are rare examples of existing frameworks. The developed process phases of both models are crucial for each CII process. However, they do not include a strategy phase, and the analogy search is not described in enough detail. Therefore, the purpose of this paper is to present a cross-industry innovation process which combines the major phases of the two presented processes and which leads to a recommended detailed search strategy. This progressive model will help companies in the eco-industry to find systematically promising eco-analogies for their technological problems and to implement them in the developing phase.

2.4 Systematic Cross-Industry Innovation Model

Current empirical studies[2] have demonstrated that, because they lack practical advice, managements' implementation of cross-industry innovations is likely to be applied unsystematically. Extending the cross-industry innovation model of Gassmann and Zeschky (2008), we present a process model which aims to support companies in the eco-industry to implement cross-industry innovations with the "outside-in" dimension of open innovation. The process model includes five major steps: (1) strategy analysis, (2) problem abstraction, (3) search for analogies, (4) assessment and (5) adaptation (see Fig. 2.3).

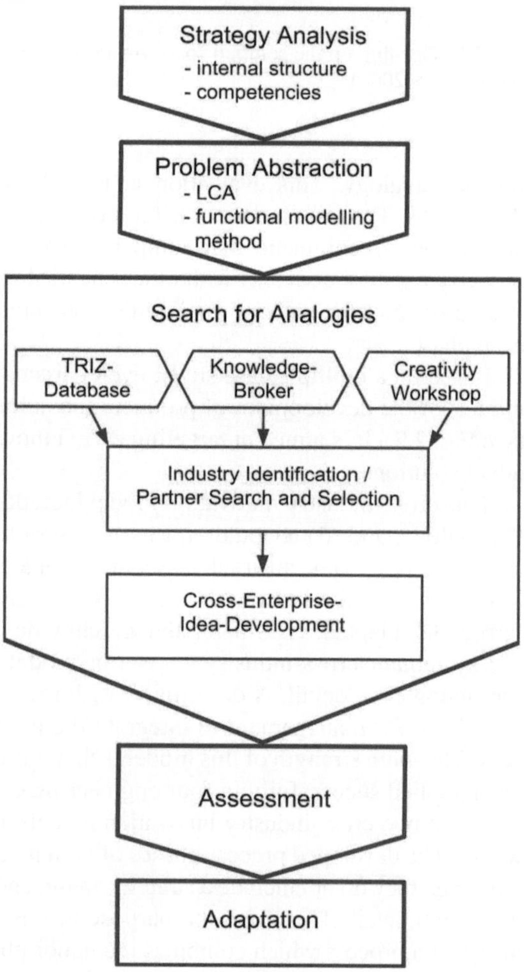

Fig. 2.3 Cross-industry innovation process model strategy analysis

[2] For example, Brunswicker and Hutschek (2010).

The cross-industry innovation strategy with its open and transparent character can clash with the existing closed corporate strategies. It is not sufficient to adapt or enhance the traditionally internal innovation strategy – companies must rethink their internal strategy in order to cooperate with external complex organization networks and its own rules. The objective of the strategy analysis phase is to analyze the internal corporate strategy and its potential strengths, as well as its weaknesses in detail. In order to determine a cross-industry innovation strategy, the management must foster an open mindset. Employees should be allowed to question their own products and technologies and also need to be aware of external developments and innovations in other industries (Gassmann and Zeschky 2008; Ertl 2010).

The competency analysis is a crucial part of the strategy analysis. The companies have to analyze their existing technological knowledge base and competencies critically and with an open mind in order to enhance and extend them with new solution inputs from outside of the company. Other helpful methods are the SWOT analysis (strengths, weaknesses, opportunities, and threats analysis) or Porter's five forces analysis.

2.4.1 Problem Abstraction

If the objective of the cross-industry innovation process is to filter and select solutions of an eco-innovative design problem, the project team has to assess the environmental impact loads at each stage of the product's life cycle. LCA[3] tools and other eco-design tools can be useful to determine the required improving elements of the existing product. Furthermore, companies can also apply the eco-efficiency guidelines from the WBCSD (World Business Council for Sustainable Development) to identify any problems with their eco-product (WBCSD 2000):

- Reduce material intensity
- Reduce energy intensity
- Reduce dispersion of toxic substances
- Enhance recyclability
- Maximize the sustainable use of renewable resources
- Extend product durability

The problem has to be defined, analyzed in detail and abstracted in order to find successful solutions that cross-organizational boundaries. Integrating the views and needs of the customers is also an important success factor. This phase links to the creation of solution ideas between different domains (Enkel and Horvàth 2010). The problem abstraction phase provides one of the greatest challenges as the project team of the company must break down the problem into its various functions and subproblems. This method only works if the team members are willing to rethink their

[3] The International Organization for Standardization (ISO) defines LCA (life cycle assessment) as a "compilation and evaluation of inputs, outputs, and the potential environmental impacts of a product throughout its life cycle" (ISO 14040 2006).

expertise with an open mindset. Those who possess a curios manner, partake in diverse personal pursuits and hold open view of the world can also have a positive effect on the creativity process. To ensure that the problem is defined in its different parameters, the project team should consist of creative and highly experienced employees, sourced from various departments of the company (e.g., engineers, designers, marketing and sustainability managers, etc.). Finally, good communication skills toward the employees are essential. Team members must be able to communicate their experiences and knowledge to the project team easily and directly (Kalogerakis et al. 2005).

A number of abstraction methods may be implemented. The functional modeling method, for instance, helps to analyze and abstract the functions of a product or a problem in order to identify technologies used in other industries with similar functions. The produced model demonstrates the systems functionality and the logical interconnections between that functionality. It can help the project team to overview the system as a whole and to gain more understanding as to what the customers expect (Burge 2009).

2.4.2 Search for Analogies

The search for analogies should start internally. During brainstorming sessions, the employees can find analogies to past projects or experiences from leisure activities, education, or any other area. In order to generate radical innovative ideas such as cross-industry innovations, the pure knowledge of the internal employees is generally not sufficient (Schild et al. 2004). Additionally, there are several methods available which help to create cross-industry innovations and where a focus on the eco-industry is concerned; of these, the following three concepts are deemed most appropriate: the TRIZ database approach, the knowledge broker approach, and a creativity workshop with external experts and lead users.

The *TRIZ method* (Theory of Inventive Problem Solving) was developed in the former Soviet Union by Genrich Altshuller. It was based on the idea that all technical problems have already been solved by someone, somehow (Altshuller 1984). The main findings of TRIZ are (Mann n.d.):

- That the same problems and solutions appear again and again across different industries, but most organizations tend to reinvent the wheel rather than look outside their own experiences or the experiences of their direct competitors.
- That the most powerful solutions are those ones which successfully eliminate the compromises and trade-offs conventionally viewed as inherent in systems.
- That there are only a small number of possible strategies for overcoming such contradictions.
- That the most powerful solutions also make maximum use of resources. Most organizations are highly inclined to solve problems by adding things rather than making the current things work more effectively or transforming the things viewed as harmful into something useful.
- That technology evolution trends follow highly predictable paths.

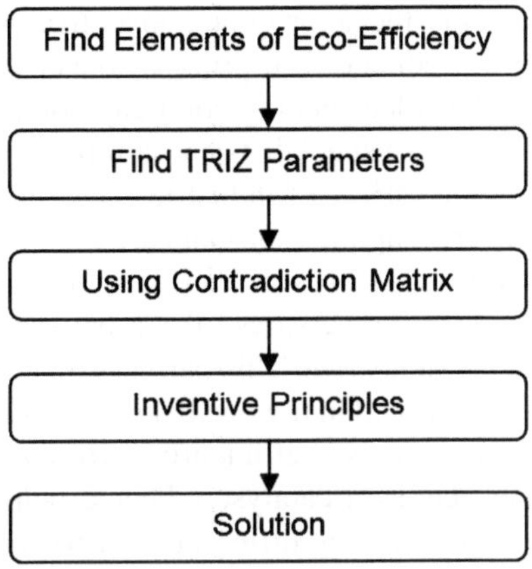

Fig. 2.4 The model of TRIZ problem-solving concept (Adapted from Chen and Chen 2007)

Altshuller created a database which evaluated nearly three million patents of technical and business innovations spanning many different fields. The information of the patents is condensed into 40 inventive principles. After analyzing the technical problem, the corresponding contradictions which need to be resolved must be defined. Finally, it has to be verified which of the principles match to the identified contradictions in order to find suitable solutions. TRIZ offers many other concepts and tools which aid the problem solver in finding analogies. A very important element of the TRIZ toolkit is the contradiction matrix which contains 39 engineering parameters. This evolved as an extension of the 40 inventive principles. Each identified contradiction is matched with the meaning of two appropriate parameters to find the most frequently used principles. The advantages of TRIZ are that one can search systematically in a limited search space and that the method is effectively supported by software (Mann n.d.; Mann and Dewulf n.d.; Terninko et al. 1998).

During the last decade, more and more academics have discussed the ability of TRIZ methodology to help create step changes to products and processes, which are currently required in the field of eco-innovation (Jones and Harrison 2000; Chen and Liu 2001; Chen and Chen 2007). The easiest way to create eco-innovations with the help of TRIZ is to determine the contradictions and to find appropriate inventive principles from the contradiction matrix.[4]

In the following section, we present an eco-innovative design methodology which uses two different TRIZ toolkits (contradiction matrix and inventive principles) to solve technical or physical contradiction and to find radical eco-innovations (see Fig. 2.4).

As previously mentioned in the problem abstraction phase, the required improvement elements of eco-efficiency based on environmental regulations or the LCA evaluation results of the product need to firstly be identified. Next, the relationship of each eco-efficiency element with the corresponding engineering parameters of TRIZ can be examined. With the help of the contradiction matrix, the TRIZ user quickly

[4] An example for contradictions is a zero waste emission process.

obtains the appropriate inventive principles. Finally, the user can innovate new eco-product or eco-process concepts (Chen and Liu 2001; Chen and Chen 2007).

The idea-seeking company could also form a partnership with a *knowledge broker*. These specialists are familiar with many knowledge domains as they work with a wide variety of industries. Therefore, they are able to offer valuable insights into different applications and are able to transfer a solution from one domain to another. Knowledge brokers can be found typically in consulting companies, design agencies and universities or can be diversified development partners who work across many industries. This partnership can bring out a leveraging effect in cross-industry innovation. On the one hand, this cooperation allows project knowledge and experience to be used efficiently across borders, and on the other hand, a knowledge broker acts as a catalyst – he/she helps companies to make contact with interesting partners among different industries (Enkel and Dürmüller 2011; Schild et al. 2004; Kalogerakis et al. 2005; Hargadon 2002).

Companies could also organize a *creativity workshop* with experts of different industries or even lead users to search together for analogies. Lead users are highly qualified and forward-looking users. They are very motivated to take part in the product development process because they profit directly from the new innovations which will solve their problems and satisfy their needs (von Hippel 1988). If the external experts and the lead users are highly diverse and if they come from different areas, the probability is high that completely new ideas and solutions can be generated. Such a creativity workshop is a great opportunity for interesting discussions and recombination of ideas and knowledge (Enkel and Horvàth 2010). A certain degree of cognitive distance can have positive effects on the cross-industry innovation process. "When people with different knowledge backgrounds and perspective interact, they stimulate and help each other to stretch their knowledge for the purpose of bridging and connecting this diverse knowledge" (Nooteboom et al. 2007). Another important process factor is that the participating experts are generally highly motivated. Experts from different domains are often very interested in cross-industry innovation workshops because they can enhance their network and receive insights into a completely different sector which can be helpful in developing their own products. The idea-seeking company should invite experts and lead users who work in high potential analogy areas to work together with them. For example, a company in the renewable energy industry could gain advantages from experts of the chemistry industry, the automotive industry and/or the aviation industry. Many different creativity methods are conducted during the workshop, and this needs to be moderated from a skilled facilitator. In order to answer specific questions about the technical problem and the product group, internal managers, and specialist staff work together with the external experts in the workshop (Enkel and Horvàth 2010). Enkel and Horvàth (2010) recommend that a maximum of 15 participants should take part in this creativity process. Companies should invite two scientists, two lead users, and three to four external experts from three to four analogy areas.

Companies do not have to exclusively choose only one of these methods. They could also combine the three methods to gain successful ideas. Besides these concepts, the authors recommend company employees to regularly visit exhibitions of other industries in order to open up their own mindset. Another important innovation factor is the effective management of human resources. The company has to recruit new employees with experiences from different industries who can give the cross-industry innovation process new impetus.

With the help of the cross-industry innovation methods, the company is able to *identify* an *industry* and in addition a *partner* who would be interested to cooperate with them. The potential candidates should have similar organizational structures and strategic goals to the seeking company, and the technological distance should lie within acceptable boundaries. Additionally, cultural and language issues should also be kept in mind.

At the next stage – *cross-enterprise-idea-development* – a workshop should be conducted where both parties (the idea seeker and the solution provider) could generate ideas for novel solution principles in an interactive manner.

2.4.3 Assessment

The objective of this step is to acquire a pool of assessed ideas from which the company can filter the concepts most relevant for them. It is very important to understand the structure and the functions of the analogies in order to evaluate them (Kalogerakis et al. 2005; Gassmann and Zeschky 2008). Majchrzak et al. (2004) recommend a two-step approach which helps to evaluate analogies. The first evaluation is a little coarse. The project team has to evaluate three categories for the analogy: (1) reliability, (2) relevance, and (3) adaptability of the analogous solution to the target problem. The analogies which do not fail in the first evaluation step are evaluated in more detail in the second. Here, the fulfillment of the target, the attractiveness for the market and the cost of the implementation are assessed in detail. Another method which helps to assess analogies is the "idea funnel" of Wheelwright and Clark (1992). This approach is very similar to the evaluation process of Majchrzak et al. (2004).

2.4.4 Adaptation

After the assessment phase, the best identified potential solution can be transferred to the knowledge base and applied to rectify the company's problem. An analogy can be transferred on a number of different levels (Hill 1999):

1. Direct transfer of an existing technology to a new context
2. Transfer of structural aspects
3. Partial transfer from functional principles
4. Use of an analogy for idea stimulation

For example, sometimes it is not possible to transfer the identified analogous solution in its entirety, but relevant knowledge about special aspects of an analogy can be transferred and applied to the technological problem (Gassmann and Zeschky 2008). Before the transfer, it has to be considered if problems will arise due to intellectual property rights. If this problem exists, a more abstract transfer might be a solution (Kalogerakis et al. 2005). In order to avoid resistance within the company, the competencies, and culture of the solution provider have to be kept in mind (Schild et al. 2004). In addition, literature has further argued that trust is a crucial factor that should be considered especially in the case of uncertainty, as it is the case at the transfer of analogies of distant industries (Gassmann et al. 2010).

2.5 Limitations

This chapter presents concrete insights and reflections when applying a managerial framework for cross-industry innovations. Furthermore, a systematic process model is proposed. However, there are some limitations that need to be kept in mind.

Since the proposed process model is conceptual, it has to be tested empirically. Subsequently, the authors of this chapter will be applying the CII model to various companies of the eco-industry to gain further insights into the process and further developing it if deemed necessary. Further insights can be drawn from additional case studies in other industries, and these findings could better improve the existing framework. In addition, this process model has been created for large companies, as the recommended analogy search methods – such as the TRIZ database approach – can be very complex and expensive. Further research is needed to propose a systematic cross-industry model for small- and medium-sized companies.

2.6 Conclusion

This chapter presents a new framework which helps to develop eco-innovations. We state that the search for analogies plays an important role in the fuzzy front end of the eco-innovation process. Furthermore, far analogies and particularly cross-industry innovations increase the chance to generate radical eco-innovation.

We aim to extend the cross-industry innovation models of Gassmann and Zeschky (2008) and Kalogerakis et al. (2005) by adding a detailed search strategy. This new process model can help managers to search and implement cross-industry eco-innovations in a systematic way with limited risk and cost. The CII model includes five major phases:

1. First, companies have to analyze their internal corporate strategy and their potential strengths. It is necessary that management foster the search for external solutions and allow the employees to rethink established thinking patterns in order to find suitable analogies.

2. The problem has to be defined and analyzed in detail. The abstraction of the problem might be difficult for companies who have already established products and processes. These problems could be overcome when the employees are willing to rethink their expertise with an open mindset.
3. We found that after abstracting the technological problem, analogies can be searched best by TRIZ database, knowledge broker and/or workshops with external experts and lead users in order to create eco-innovations.
4. Companies must understand the structure and the function of the analogies in order to evaluate what knowledge is valuable and thus is subject for transfer. The methods of Majchrzak et al. (2004) and Wheelwright and Clark (1992) can help to find the concept most relevant for the companies.
5. After the assessment phase, the best identified potential solution can be transferred to the knowledge base and applied to rectify the company's problem.

References

Altshuller G (1984) Creativity as an exact science. The theory of the solution of inventive problems. Gordon and Breach Science Publishers, New York

Andersen MM (2008) Eco-innovation – towards a taxonomy and theory. Paper to be presented at the 25th DRUID celebration conference 2008 entrepreneurship and innovation – organizations, institutions, systems and regions, Copenhagen, 17–20 June 2008

Birch HG, Rabinowitz HJ (1951) The negative effect of previous experience on productive thinking. J Exp Psychol 47(2):121–125

Brunswicker S, Hutschek U (2010) Crossing horizons: leveraging cross-industry innovations search in the front-end of the innovation process. Int J Manage 14(4):683–702

Burge S (2009) The systems engineering tool box. http://www.burgehugheswalsh.co.uk/uploaded/documents/FM-Tool-Box-V1.0.pdf. Accessed 24 May 2013

Chen JL, Chen W-C (2007) TRIZ based eco-innovation in design for active disassembly. In: Takata S, Umeda S (ed) Advances in life cycle engineering for sustainable manufacturing businesses. Proceedings of the 14th CIRP conference on life cycle engineering. Waseda University, Tokyo, Springer, pp 83–87

Chen JL, Liu C–C (2001) An eco-innovative design approach incorporating the TRIZ method without contradiction analysis. J Sustain Prod Des 1:263–272

Chesbrough HW (2003) Open innovation: the new imperative for creating and profiting from technology. Harvard Business School Press, Boston

Chesbrough HW (2006a) Open innovation. The new imperative for creating and profiting from technology. Harvard Business School Press, Boston

Chesbrough HW (2006b) Open innovation: a new paradigm for understanding industrial innovation. In: Chesbrough HW, Vanhaverbeke W, West J (eds) Open innovation: researching a new paradigm. Oxford University Press, Oxford, pp 1–12

Dahl DW, Moreau P (2002) The influence and value of analogical thinking during new product ideation. J Mark Res 34(1):47–60

EIO – Eco-Innovation Observatory (2011) The eco-innovation challenge: pathways to a resource-efficient Europe. The way to a green economy through eco-innovation. In: O'Brien M et al (ed) Annual report 2010, DG Environment, Brussels. http://www.eurosfaire.prd.fr/7pc/doc/1308928736_eco_report_2011.pdf. Accessed 29 May 2013

EIO – Eco-Innovation Observatory (2013) Europe in transition. Paving the way to a green economy through eco-innovation. Annual report 2012, Meghan O'Brien/Michal Miedzinski, Eds.

Funded by the European Commission, DG Environment, Brussels. http://www.europarl.europa.eu/sides/getDoc.do?type=REPORT&reference=A6-2005-0141&language=EN. Accessed 29 May 2013
Enkel E, Dürmüller C (2011) Cross-industry-innovation: Der Blick über den Gartenzaun. In: Gassmann O, Sutter P (eds) Praxiswissen innovations management. Von der Idee zum Markterfolg. Carl Hanser Verlag, Munich, pp 215–236
Enkel E, Gassmann O (2010) Creative imitation. Exploring the case of cross-industry innovation. R&D Manage 40:256–270
Enkel E, Horvàth A (2010) Mit Cross-Industry-Innovation zu radikalen Neuerungen. In: Ili S (ed) Open Innovation umsetzen. Prozesse, Methoden, Systeme, Kultur. Symposion Publishing, Düsseldorf, pp 293–314
Ertl M (2010) Strategiebildung für die Umsetzung von Open Innovation. In: Ili S (ed) Open Innovation umsetzen. Prozesse, Methoden, Systeme, Kultur. Symposion Publishing, Düsseldorf, pp 61–84
European Commission (2004) Stimulating technologies for sustainable development: an environmental technologies action plan for the European Union. (Communication from the Commission to the European Parliament), COM (2004) 38 final, Brussels
European Environment Agency (2006) Eco-innovation indicators. European Environment Agency, Copenhagen
Gassmann O (2006) Opening up the innovation process: towards an agenda. R&D Manage 36(3):223–228
Gassmann O, Enkel E (2004) Towards a theory of open innovation: three core process archetypes. R&D management conference (RADMA). Lisbon
Gassmann O, Zeschky M (2008) Opening up the solution space: the role of analogical thinking for breakthrough product innovation. Creat Innov Manage 17:97–106
Gassmann O, Zeschky M, Wolff T, Stahl M (2010) Crossing the industry line: breakthrough innovation through cross-industry alliances with, non-suppliers'. Long Range Plann 43:639–654
Geschka H, Moger S, Rickards T (1994) Creativity and innovation: the power of synergy. Geschka & Partner, Darmstadt
Gjoksi N (2011) Resource policies and the innovation dimension. European Sustainable Development Network, ESDN Case Study No. 5
Hargadon A (2002) Brokering knowledge: linking learning and innovation. Res Organ Behav 24:41–85
Hill B (1999) Naturorientierte Lösungsfindung: Entwickeln und Konstruieren von biologischen Vorbildern. Expert-Verlag, Renningen-Malmsheim
ISO 14040 (2006) Environmental management – life cycle assessment – principles and framework. International Organisation for Standardisation (ISO), Geneva
Jones E, Harrison D (2000) Investigating the use of TRIZ in eco-innovation. TRIZ Journal, September. http://www.triz-journal.com/archives/2000/09/b/default.asp
Jones E, Darell M, Harrison D, Stanton N (2001) An eco-innovation case study of domestic dishwashing through the application of TRIZ tools. Creat Innov Manage 10(1):3–14
Kalogerakis K, Herstatt C, Lüthje C (2005) Generating innovations through analogies. An empirical investigation of knowledge brokers. Working paper 33. Hamburg-Harburg
Kemp R, Pearson P (2007) Measuring eco-innovation. Final report MEI project about measuring eco-innovation. Project No. 044513. http://www.oecd.org/env/consumption-innovation/43960830.pdf. Accessed 28 May 2013
Laursen K, Salter A (2006) Open for innovation: the role of openness in explaining innovation performance among U.K. manufacturing firms. Strateg Manage J 27(2):131–150
Leimeister JM et al (2010) Theses about managing open innovation processes. In: Jacobsen H, Schallock B (eds) Innovationsstrategien jenseits traditionellen Managements. Fraunhofer Verlag, Stuttgart, pp 200–211
Lichtenthaler U (2011) Open innovation: past research, current debates, and future directions. Acad Manage Perspect 25(1):75–93

Majchrzak A, Cooper LP, Neece OE (2004) Knowledge reuse for innovation. Manage Sci 50(2):174–188
Mann D (n.d.) TRIZ: an introduction. Systematic Innovation Ltd. http://www.systematic-innovation.com/Articles/02,%2003,%2004/July04-TRIZ-%20AN%20INTRODUCTION.pdf. Accessed 29 May 2013
Mann D, Dewulf S (n.d.) Updating the contradiction matrix. http://www.systematic-innovation.com/Articles/02,%2003,%2004/Jan03-Updating%20the%20Contradiction%20Matrix.pdf. Accessed 28 May 2013
Nooteboom R et al (2007) Optimal cognitive distance and absorptive capacity. Res Policy 36:1016–1034
Nuij R (2001) Eco-innovation helped or hindered by integrated product policy. J Sustain Prod Des 1:49–51
OECD (2011) Fostering innovation for green growth, Green growth studies. OECD Publishing, Paris
OECD, European Commission & Nordic Innovation Joint Network (2012) The future of eco-innovation: the role of business models in green transformation. OECD background paper
OECD/Eurostat (1999) The environmental goods and services industry. Manual for data collection and analysis. OECD Publications Service, Paris
OECD/Eurostat (2005) Oslo manual. Guidelines for collecting and interpreting innovation data, 3rd edn. OECD/Eurostat, Paris
Pujari D (2004) Eco-innovation and new product development: understanding the influences on market performance. Technovation xx:1–10. http://www.sciencedirect.com. Accessed 28 May 2013
Reichwald R, Piller F (2006) Interaktive Wertschöpfung: Open Innovation, Individualisierung und neue Formen der Arbeitsteilung. Gabler Verlag, Wiesbaden
Schild K, Herstatt C, Lüthje C (2004) How to use analogies for breakthrough innovations. Working paper 24
Stirling A et al (2009) Transformative innovation. A report to the Department for Environment, Food and Rural Affairs. University of Sussex, Brighton
Stötzel M, Wiener M, Amberg M (2011) Key differentiators of open innovation platforms – a market-oriented perspective. Wirtschaftsinformatik proceeding. Paper 60
Terninko J, Zusman A, Zlotin B (1998) Systematic innovation: an introduction to TRIZ (theory of inventive problem solving). St Lucie Press, Boca Raton
von Hippel E (1988) The source of innovation. Oxford University Press, New York
von Weizsäcker E, Lovins AB, Lovins LH (1999) Factor four: doubling wealth – halving resource use. Earthscan Publications Limited, London
WBCSD – World Business Council for Sustainable Development (2000) Eco-efficiency. Creating more value with less impact. http://www.wbcsd.org/web/publications/eco_efficiency_creating_more_value.pdf. Accessed 29 May 2013
Wheelwright S, Clark K (1992) Revolutionizing product development: quantum leaps in speed, efficiency, and quality. The Free Press, New York

Chapter 3
How to Make Eco-innovation a Competitive Strategy: A Perspective on the Knowledge-Based Development

Maria do Rosário Cabrita, Virgílio Cruz-Machado, and Florinda Matos

Abstract Eco-innovation is currently a fuzzy concept in need of theoretical clarification. It is difficult to define because of the complexity of the subject. Nevertheless, eco-innovation can be described as innovation that consists of new or modified products, processes, practices, systems and business methods which aims to prevent or reduce environmental risks and can contribute to environmental sustainability. In this sense, the term eco-innovation has called attention to the positive contribution that industry can make to sustainable development and a competitive economy. Hence, eco-innovation is understood as the combined improvement of economic and environmental performance of society. If eco-innovation is considered a dynamic of our society, economic growth, social development and environmental integrity are essential ingredients of sustainability. Performance is the result of complex socio-economic process and has different dimensions (environmental and economic) and levels (micro and macro). A systematic analysis of eco-innovation dynamic should consider – at micro level – social processes between knowledge, institutions and firms and look for causal links among them. At the macro level, it is necessary to understand environmental, economic and social dimensions in which eco-innovation strategies may develop. The new economy is not only a knowledge economy but also an economy based on responsible behaviour. In the knowledge-based development (KBD), the key to growth and prosperity relies on the issues of acquiring, creating, developing, storing and applying knowledge for a sustainable economic, social and environmental development. A knowledge-based

M.R. Cabrita (✉) • V. Cruz-Machado
UNIDEMI, Department of Mechanical and Industrial Engineering, Faculty of Science and Technology (FCT), Universidade Nova de Lisboa, 2829-516 Lisbon, Portugal
e-mail: m.cabrita@fct.unl.pt; vcm@fct.unl.pt

F. Matos
IC Lab Research Centre – ICAA – Intellectual Capital Accreditation Association, Rua dos Bombeiros Voluntários, 4 R/C-Esq, 2000-205 Santarém, Portugal
e-mail: florinda.matos@icca.pt

S. Azevedo et al. (eds.), *Eco-Innovation and the Development of Business Models*, Greening of Industry Networks Studies 2, DOI 10.1007/978-3-319-05077-5_3,

perspective may provide a holistic approach on wealth creation where eco-innovation is seen as a concept which provides direction and vision for pursuing the overall societal changes needed to achieve sustainable development. This chapter aims to explore the foundations of KBD and propose a framework to integrate the two key issues, eco-innovation and KBD.

Keywords Eco-innovation • Knowledge-based development • Sustainable development

3.1 Introduction

In recent decades, the expansion of economic activity has been accompanied by the growing global environmental concerns, such as climate change, energy security and increasing scarcity of resources. New economic laws and social trends have arisen, and many challenged drivers have started to transform markets. In fact, we are witnessing a burgeoning public consciousness of the role of business in helping to cultivate and maintain highly ethical practices in society and particularly in the natural environment. A differentiated feature in the context of knowledge economy (KE) is that not only economic laws drive the concept of excellence (Peters and Waterman 1982; Ironica et al. 2010), but corporate social responsibility and ethics will also shape it in the future (Viedma and Cabrita 2012). In the KE, sustainable competitive advantages are usually based on intangibles or intellectual capital and have to be achieved within an ethically and socially responsible business models.

In the global knowledge economy, innovation is essential for the creation of wealth, new jobs and achieving societal goals. Leveraging innovation is particularly important today, in what is the most severe global economic crisis since the Great Depression of the 1930s. By all indicators, this crisis will last longer than most past crisis because it is global and the risk is systemic. Nevertheless, this is also a time of hope. History has shown that times of crisis are also times of innovation, when institutional, mental and other obstacles are more easily removed. In this time, governments, organisations and individuals have a key role to play in the transformation of socio-economic development strategies embracing a new vision, ethical behaviours and adequate policies that promote business and job opportunities oriented to an improvement in both an economic and environmental performance (win-win solution). The flagship initiative "resource-efficient Europe" of the Europe 2020 strategy (European Commission 2010) has become an "umbrella" issue included in various policy agendas (Eurostat 2009) and contexts (EurObserv'ER 2009). As a response, a "green recovery" policy is being followed in several countries.

Rising demand for a better environment has led to an expanding supply of environmentally friendly techniques, products and services in both the industrialised and developing countries (European Commission 2009). In response, industrial companies have increasingly shown more interest in sustainable production and have adopted certain corporate social and environmental responsibility initiatives (European Commission 2011). If we consider eco-innovation as a dynamic of our society, both economic growth and decoupling are essential ingredients of sustainability. Eco-innovation can contribute to new business opportunities which could make firms, sectors and countries more competitive. These days, eco-innovation presents interesting growth perspectives for an ever-greater number of businesses, thanks to a wide variety of niche market opportunities. Roland Berger Strategy Consultants (2009) expects 3.1 trillion in global sales generated by eco-industries by 2020, i.e. more than a doubling, and calls eco-technologies the *twenty-first-century lead industry*. The Eco-innovation Action Plan (EcoAP), launched by the European Commission in December 2011, is a significant step forward for eco-innovation moving the EU beyond green technologies and fostering a comprehensive range of eco-innovative processes, products and services. The ambitious plan will also focus on developing stronger and broader eco-innovation actions across and beyond Europe.

A broad view of innovation refers to something that is new relative to a given context. Innovation may be new to the country in which it appears, to the region or the sector in which it takes place or to the firm that develops or adopts it. What matters is the diffusion of this relative novelty as a source of wealth, jobs and welfare. Then, innovation policy should, in priority, aim to capture global knowledge and technology and to adapt and disseminate them in local contexts (World Bank 2010). Another important issue is the fundamental role of pro-poor innovations or inclusive innovation. Four billion people, a majority of the world's population, form the bottom of the economic pyramid. Innovation should be encouraged to benefit poor communities. The range and success of such innovations have flourished, adapting existing technologies and knowledge to local contexts (e.g. India's Aravind Eye Hospital[1] or the Malaria Research and Training Centre in Bamako).[2]

A knowledge-based development (KBD) perspective, integrating the concepts of economic performance, social well-being and environmental principles, can provide a holistic approach to understand the nature and scope of eco-innovation and its rising role in the globalising economy. In its approach this chapter points to the need of linking up the concept of eco-innovation to knowledge-based competitiveness.

[1] India's Aravind Eye Hospital deals with blindness in general and the elimination of needless blindness in particular in rural India. They use Internet kiosks in remote locations in Madurai to screen people's eyes under the supervision of a paramedic. The information is sent by the Internet to a clinic of diagnosis. The Aravind eye-care system treats 1.4 million patients a year, and since its inception, it has performed over 2 million operations.

[2] The Malaria Research and Training Centre in Bamako, Mali, created in 1992, is internationally recognised for its contributions to research on malaria and the improvement of public health standards. It works with traditional doctors to create a source of immediate care in the Bandiagara region and has helped to reduce the mortality rates of young children significantly.

3.2 A Knowledge-Based Development Perspective

Knowledge-based development (KBD) as a field of study and practice is receiving a growing amount of attention from governments and public entities. The origins of KBD can be traced back to influent disciplines such as economics, urban studies and planning, anthropology, psychology, social sciences, architecture, political economy, innovation management, information and technology management and knowledge management. Recent literature provides various approaches on the attempts to integrate those disciplines contributing to KBD understanding. The concept is anchored in knowledge management (KM), which primarily emerged in business as a response to the need of identifying, valuing and capitalising all factors of value creation, namely, knowledge-based factors. Later on, this stream of research expands to individual-, organisational- and social-based development, covering broader contexts such as cities, regions, clusters and nations.

The World Bank has introduced a knowledge-based framework at the national level called the knowledge-based economy which provides the foundations to develop strategies for the countries to follow in developing knowledge economies. Such framework consists of four pillars – an educated and entrepreneurial population; a dynamic information systems structure; an economic and legislative environment that favours knowledge transfer, entrepreneurship and information and communication technology (ICT) infrastructure; and an efficient innovation system. The Asian Development Bank expanded this knowledge-based economy framework to include the application of knowledge management to sociocultural and natural-environmental domains. The resulting framework is called knowledge-based development (KBD) which intends to reflect the combination of two powerful development paradigms: sustainable development and knowledge-based management.

Knowledge has always been important for development. However, knowledge needed in KE is wider than technological knowledge; it includes cultural, social and managerial knowledge. This knowledge also refers to culture, values, emotions, relationships and other intangibles that coming together form a coherent framework for the company's future value. The organisation's knowledge is about its capability in integrating information with expertise to take action. Knowledge is then assumed as "capacity to act" (Sveiby 2001) or a capacity (actual or potential) to take effective action in varied and uncertain situations.

Since the early 1990s, researchers and policymakers in advanced economies have focused on KBD as a means to stimulate economic growth (OECD 1996) and sustain industrial competitiveness. A major concern has been to gain in consistency along a number of lines that seem to be shaping this emerging field: interdisciplinary, conceptual frameworks, systems approach and, in particular, strategic perspective. Several KBD models have been proposed and criticised (Carrillo 2006; Yigitcanlar and Velibeyoglu 2008), emerging a key and common feature that is the focus on the creation, exchange and application of knowledge to solve problems and drive economic growth, social development and ecological responsibility.

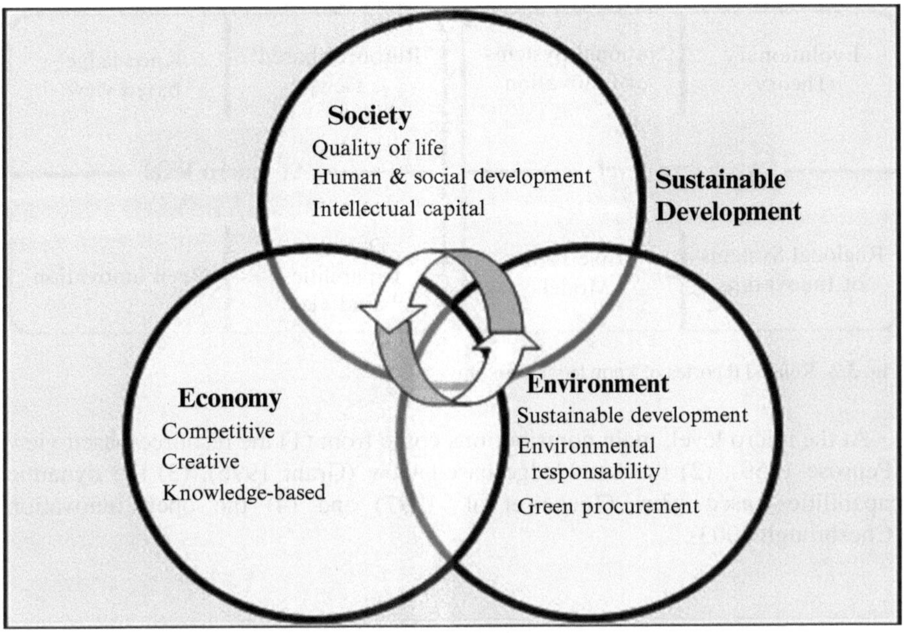

Fig. 3.1 Pillars of KBD (Source: Adapted from Yigitcanlar and Velibeyoglu 2008)

The challenge is to think and promote a sustainable society based on three target areas: social solidarity, economic efficiency and ecological responsibility, as depicted in Fig. 3.1.

In response, governments in both advanced industries and developing economies have embarked on a number of initiatives focused on national capacity building, education, entrepreneurial skills and innovation to create a favourable context for value-creating activities and improved standards of living. The literature provides some evidences in the field of KBD at different levels: urban (Singapore, Barcelona), regional (Veneto, Basque Country), national (Denmark, Australia, New Zealand) and supranational (European Union).

3.3 Related Theories of Knowledge Economy

Since the 1960s, when knowledge was recognised as being increasingly important in economic activities, many economic theories at both macro and micro levels (Fig. 3.2) have emerged to examine the phenomenon.

At the macro level (cities, regions and nations), main contributions come from (1) the evolutionary theory (Nelson and Winter 1982), (2) the national systems of innovation (Lundvall 1985), (3) the regional systems of innovation (Cooke 1992) and (4) the triple helix model (Etzkowitz and Leydesdorff 2000).

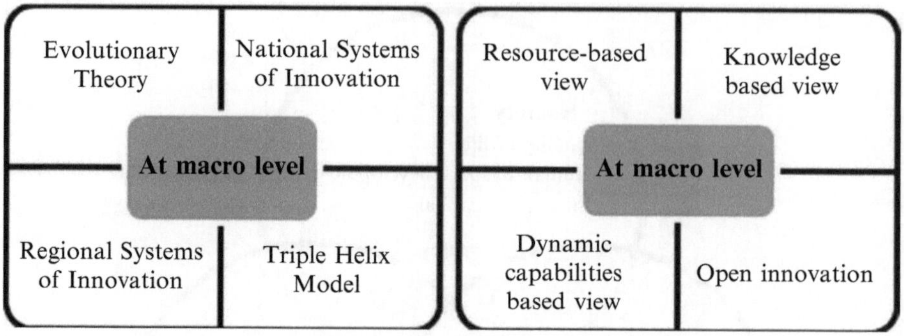

Fig. 3.2 Related theories of knowledge economy

At the micro level, main contributions come from (1) the resource-based view (Penrose 1959), (2) the knowledge-based view (Grant 1996), (3) the dynamic capabilities-based view (Teece et al. 1997) and (4) the open innovation (Chesbrough 2003).

3.3.1 At the Micro Level

Value creation process in the context of knowledge economy is directly linked to the intelligence, the speed and the agility that comes from a host of latent intangibles which represent a reservoir of potential talent and innovation that provides a source of competitive advantage. This suggests that the value generated is a function of the way in which resources are managed. The key to value creation lies with the effectiveness of knowledge transfers and conversions (Sveiby 2001). In response, new models of business are emerging where the value chains have their hard nucleus in the creation, dissemination, application and leverage of intellectual resources.

3.3.1.1 The Resource-Based View (RBV)

The focus of resource-based view is on the relationship between firm resources and firm performance. Following the seminal work of Penrose (1959), the RBV proposes that firms consist of bundles of productive resources and it is the heterogeneity of skills and capabilities available from its resources that gives each firm its uniqueness.

3.3.1.2 The Knowledge-Based View (KBV)

The KBV of the firm considers knowledge as the most strategically significant resource of a firm, and in that sense this perspective is an extension of the RBV. Value is then created through complex dynamic exchanges between tangibles (goods and money) and intangibles (cognition processes, intelligence and emotions) where individuals, groups or organisations engage in a value network by converting what they know, both individually and collectively, into tangible and intangible value.

3.3.1.3 The Dynamic Capabilities-Based View

The concept of dynamic capabilities has evolved from the RBV of the firm. Dynamic capabilities have lent value to the RBV arguments as they transform what is essentially a static view into one that can encompass competitive advantage in a dynamic context. Teece et al. (1997) defined dynamic capability as a firm's ability to integrate, build and reconfigure competence.

3.3.1.4 The Open Innovation

The key idea behind open innovation is that, in a world of widely distributed knowledge, companies cannot afford to rely entirely on their own research, but should instead buy or license processes or inventions (i.e. patents) from other companies.

3.3.2 At the Macro Level

Innovation processes germinate and develop within what are called "innovation systems". These are made up of private and public organisations and actors that connect in various ways and bring together the technical, commercial and financial competencies and inputs required for innovation. A number of models have been proposed for modelling the production process of university-industry-government relations.

3.3.2.1 Evolutionary Theory

The evolutionary theory of economic changes observes the economy in an evolutionary process. In their pioneering work, Nelson and Winter (1982) equated firms with living organisms. Firms have capabilities as "routines", just like "genes" in living organisms, and are heterogeneous in capabilities. When doing businesses, firms repeat their routines and imitate other firms' routines deemed suitable to the market. In this process of performing these routines, innovation occurs naturally as the unpredictable "mutation" of routines, giving some advantages to the innovative firms. As such, firms' capabilities evolve and so does the economy. The evolutionary models share some common features of the dynamic and non-deterministic economic processes which never end in a stable state of equilibrium.

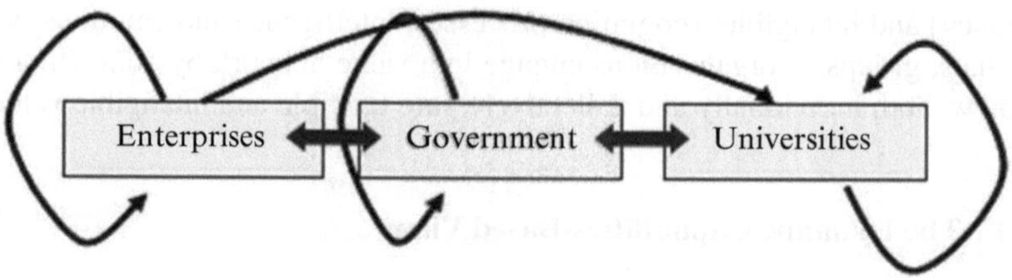

Fig. 3.3 Triple helix model of innovation

3.3.2.2 National System of Innovation (NIS)

The National System of Innovation (NIS) may be conceptualised as a means by which a country seeks to create, acquire, diffuse and put into practice new knowledge that will help that country and its people achieve their individual and collective goals. The approach finds broad applications in policy contexts – by regional authorities and national governments as well as by international organisations such as the OECD, the European Union, the UNCTAD and the UNIDO.

3.3.2.3 The Regional Innovation Systems (RIS)

Similar to NIS, the concept encourages the application and dissemination of knowledge, skills and best practices in regions. One of the assumptions of the RIS approach is that many innovative firms operate within regional networks, cooperating and interacting not only with other firms such as suppliers, clients and competitors but also with research and technology resource organisations, innovation support agencies, venture capital funds and local and regional government bodies. Innovation is a learning process that benefits from the proximity of organisations that can trigger this process.

3.3.2.4 Triple Helix Model of Innovation

The triple helix model (Etzkowitz and Leydesdorff 2000; Leydesdorff 2006) set out the generative principles of knowledge-based economies based on three basic elements: (1) industry as the locus of production, (2) government as the source of relations that guarantee stable interactions and exchange and (3) universities as source of new knowledge and technological developments. As demonstrated in Fig. 3.3, the helices interact among one another leading to the dynamics of the whole system.

Based on the review of theories that support the KBD, it is clear that innovation is strictly related to the application of knowledge. The pervasive use of new technologies in all industries and activities requires new skills and new types of knowledge. Higher levels of education and greater flexibility in policies and institutions are necessary to take advantage of the innovation potential of such advances and to build the foundations of the so-called knowledge economy (World Bank 2007).

3.4 The Eco-innovation Challenge

According to *Oslo Manual*, innovation is described as "the implementation of a new or significantly improved product (good or service), or process, a new marketing method, or a new organizational method in business practices, workplace organization or external relations" (OECD and Eurostat 2005: 46). Although this definition generally applies to eco-innovation, the concept goes beyond the product/process/practice towards a broader societal context including two further significant characteristics: (1) eco-innovation is innovation that seeks to find new solutions to existing problems, as well as offering opportunities of new activities with emphasis on a reduction of environmental impact, whether such an effect is intended or not, and (2) it is not limited to innovation in products, processes, marketing methods or organisational practices, but also includes innovation in social and institutional structures.

The concept of eco-innovation emerged in recent literature with Fussler and James (1996). Eco-innovation is currently a fuzzy concept in need of theoretical consistency. James (1997) defines eco-innovation as new products and processes that provide customer and business value but significantly decrease environmental impacts. The OECD (2007) defines eco-innovation as "activities which produce goods and services to measure, prevent, limit, minimize or correct environmental damage to water, air, soil as well as problems related to waste, noise and ecosystems. This includes technologies, products, and services that reduce environmental risks and minimise pollution". Eco-innovation becomes an emerging priority in EU, relating to different aspects of almost all industries. In line with the Competitiveness and Innovation Framework Programme (CIP), eco-innovation is a fairly recent business and technology area which describes products, services and processes aiming at reducing environmental impacts.

According to international official statistics, eco-innovation covers a wide range of activities including areas such as the following: alternative energy, including energy storage and supply infrastructure; energy savings; consultancy and innovative project/business engineering and finance services; environmental damage remediation, including brownfield rehabilitation; transport; recycling; eco-innovative product engineering, i.e. factoring recycling from development; new ways of leveraging natural resources; construction, eco-construction and urban regeneration; new products, processes and business models and even possibly new uses and adaptations of existing products and materials (eco-design and eco-products), as well as new materials; environmentally friendly agriculture, including production and breeding of natural organisms (ladybirds, earthworms, etc.); spatial planning; zero-energy housing, intelligent water management housing and housing built with sustainable construction products; and the wellness industry, which in some regions can also include the development and processing of organic products, eco-tourism and therapeutic tourism as well as preventive medicine and medical care for the elderly.

To date, the promotion of eco-innovation has focused mainly on environmental technologies; however, tendencies to broaden the scope of concept are now emerging. In Japan, the government's Industrial Science Technology Policy Committee defines eco-innovation as a new field of techno-social innovations that focuses less in products and more in the environment and people (METI 2007). Social innovation is innovation that considers the human element integral to any discussion on resource consumption. It includes market-based dimensions of behavioural and lifestyle change and the ensuing demand for green goods and services. The *Forum on Social Innovation* defines it as innovation that "concerns conceptual, process or product change, organizational change and changes in financing, and can deal with new relationships with stakeholders and territories". Eco-innovation is then seen as a concept which provides direction and vision for pursuing the overall societal changes needed to achieve sustainable development.

3.5 Eco-innovation and Knowledge Management

The OECD Project on Sustainable Manufacturing and Eco-innovation was launched in 2008. Its aim is the acceleration of sustainable manufacturing production through the diffusion of existing knowledge and the facilitation of the benchmarking of products and production processes. It also aims to promote the concept of eco-innovation and to stimulate the development of systemic solutions to global environmental challenges.

At the micro level, organisations should leverage their knowledge-based ecosystems. Van der Borgh et al. (2012) state that value creation by knowledge-based ecosystems draws on the dynamics of single firms (interacting and partnering) as well as the ecosystem at large. The authors identified two key drivers of value creation in the ecosystem: (1) facilitation of the innovation processes for individual firms (transactions of resources between members of ecosystem) and (2) creation of an innovation community (macro-culture, shared beliefs/assumptions that all ecosystem members hold on innovation).

Knowledge-based view (KBV) identifies in knowledge, which is characterised by scarcity and difficult to transfer and replicate, a critical resource for achieving competitive advantage. Knowledge management (KM) implies the development, transfer and application of knowledge within the organisation so as to attain and maintain competitive advantage. KM relates to the processes and practices through which organisations create knowledge-based value. Key KM activities are as follows: knowledge acquisition (from customer, supplier, competitor and partner relations), knowledge development (directed towards creation of new skills and products, better ideas and improved processes), knowledge distribution (exchange and dissemination of knowledge from an individual to a group or the organisation),

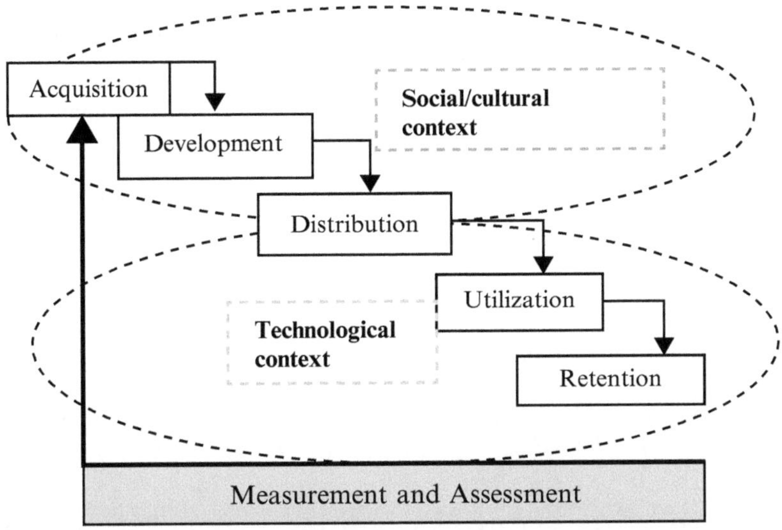

Fig. 3.4 Knowledge management key activities

knowledge utilisation (productive use for the benefit of the organisation), knowledge retention (selection, storage and updating of information, documents and experience) and measurement and assessment of knowledge. The flow of knowledge depends on people and the social environment they operate in. Figure 3.4 illustrates this conceptual linkage between sociocultural context (practical approach) and technological context (process approach).

KM thus represents a systematic approach towards searching and using the knowledge on behalf of creating value. These basic principles of KM can be applied in production firms, financial entities, business organisations and also government institutions, representing also a critical area to be developed in the eco-innovation industries (Cabrita et al. 2010). Building upon KM processes and linking them to the eco-innovation frameworks enable organisations to capture, reconcile, store and transfer knowledge in an efficient manner and at the same time enhance their business performance and competitive advantage.

3.6 Integrating Eco-innovation Frameworks and Knowledge-Based Development Perspective

Based on an extension of the definition of innovation in the OECD *Oslo Manual* and on the existing literature (e.g. OECD 2009), eco-innovation can be understood and analysed according to its *targets* (the main focus), its *mechanisms* (methods for introducing changes in the target) and its impacts (the effects on environmental conditions).

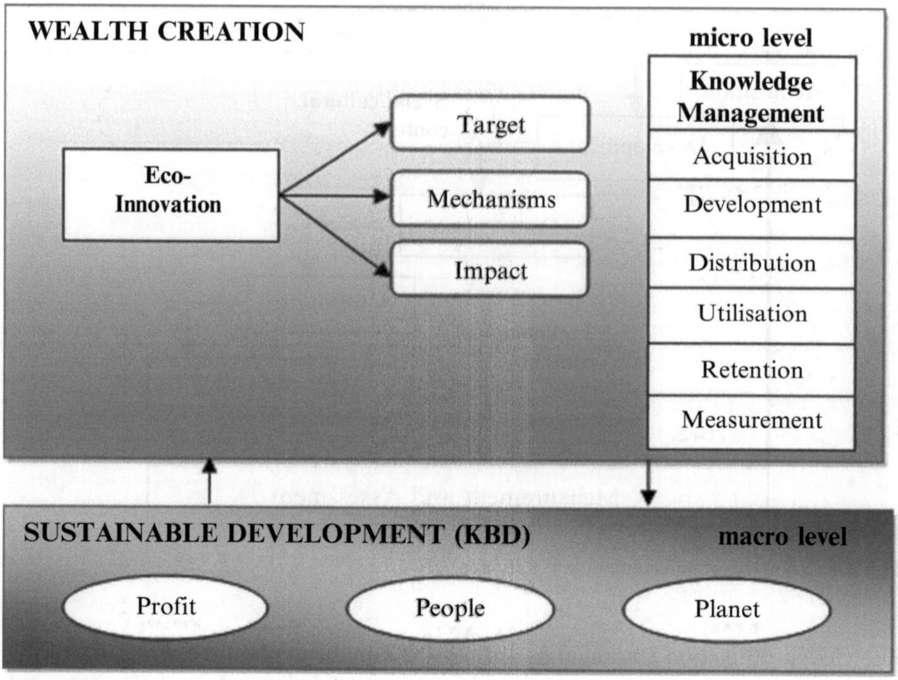

Fig. 3.5 Building an integrative methodology for eco-innovation

Following the *Oslo Manual* the target of an eco-innovation may be products (goods and services), processes (e.g. production method), marketing methods (pricing and promotion or other market-oriented strategy), organisations (e.g. structure of management) and institutions (broader societal area). The target of eco-innovation can be technological (products and services) or non-technological (marketing, organisations and institutions).

Mechanisms refer to the method by which the change is introduced. Four basic mechanisms are identified: modification (progressive product or process adjustments), redesign (significant changes in products, processes, structures, etc.), alternatives (introduction of goods and services that can fulfil the same functional need and can operate as substitute of other products) and creation (design and introduction of entirely new products, processes, procedures, organisations and institutions). Finally, impact refers to the eco-innovation's effect on the environment, across its life cycle or some other focus area.

A tentative template for an integrative methodology for eco-innovation on a knowledge-based development perspective might be illustrated in Fig. 3.5.

The general structure of the integrative methodology for eco-innovation is grounded in knowledge-based development perspective. Sustainable development platform represents economic (profit), social (people) and environmental (planet) drivers in business activities, which influence management decisions in what concerns innovation target and mechanisms. Impacts are measured at a micro (leveraging organisational knowledge) and at a macro level (assuring sustainable development). Potential environmental impacts stem from the eco-innovation's target and mechanism and their interplay with its socio-technical surroundings.

Innovation is extremely dependent on the availability of knowledge, and therefore, the complexity created by the explosion of richness and reach of knowledge has to be recognised and managed to ensure successful innovation. Knowledge management provides the tools, processes and platforms to ensure knowledge availability and accessibility, e.g. through structuring of the knowledge base. Knowledge management can facilitate collaboration as mechanism to foster innovation through provision of technological platforms and tools to enable knowledge sharing. Knowledge management also ensures the flow of knowledge used in the innovation process.

In essence, this approach represents the intricacies of sustainable development principles of economic growth, social well-being and environmental responsibility that can boost eco-innovators' wealth creation capacity.

3.7 Conclusion

Innovation is at the heart of economic development, social welfare and protection of the environment. In a dynamic and global environment, the need for innovation is greater than ever, and the challenge to make these three objectives compatible is vital. Innovation should be understood as something new to a given context that improves economic performance, social well-being or the environmental setting. It can be new to the firm, new to the economy or new to the world. Every society has to find the ways and means to innovate that correspond to its needs and capabilities (World Bank 2010).

KBD is a strategic approach that aims to make countries compatible with the KE helping to address sustainable development strategies. Eco-innovation presents interesting growth perspectives for an ever-greater number of businesses, thanks to a wide variety of niche market opportunities. Integrating eco-innovation strategies and a KBD approach can provide a holistic view of eco-innovation drivers and leverage the positive contribution that industry can make to sustainable development and a competitive economy.

Acknowledgement We gratefully acknowledge the support given by UNIDEMI, R&D Unit in Mechanical and Industrial Engineering of the Faculty of Science and Technology (FCT), New University of Lisbon, Portugal.

References

Cabrita MR, Cruz-Machado V, Grilo A (2010) Leveraging knowledge management with the balanced scorecard. In: Proceedings of international conference on industrial engineering and engineering management, Macau, 7–10 December

Carrillo F (2006) The century of knowledge cities. In: Carrillo F (ed) Knowledge cities: approaches, experiences and perspectives. Butterworth-Heinemann, New York

Chesbrough HW (2003) Open innovation: the new imperative for creating and profiting from technology. Harvard Business School Press, Boston
Cooke P (1992) Regional innovation systems: competitive regulation in the new Europe. Geoforum 23:365–382
Etzkowitz H, Leydesdorff L (2000) The dynamics of innovation: from national systems and mode 2 to a triple helix of university-industry-government relations. Res Policy 29(2):109–123
EurObserv'ER (2009) The state of renewable energies in Europe. 2009 Edition
European Commission (EC) (2009) European's attitudes towards the issues of sustainable consumption and production. Flash Eurobarometer 256 – the Gallup Organisation, Hungary. Available at: http://ec.europa.eu/public_opinion/flash/fl_256_en.pdf
European Commission (EC) (2010) Communication from the Commission: Europe 2020 – a strategy for smart, sustainable and inclusive growth. COM (2010) 2020, Brussels
European Commission (EC) (2011) Attitudes of European entrepreneurs towards eco-innovation. Flash Eurobarometer 315 – The Gallup Organization, Hungary
Eurostat (2009) Sustainable development in the European Union. 2009 monitoring report of the EU sustainable development strategy. Statistical Office of the European Communities, Luxembourg
Fussler C, James P (1996) Driving eco-innovation: a breakthrough discipline for innovation and sustainability. Pitman PUB, London/Washington, DC
Grant R (1996) Towards a knowledge-based view of the firm. Strateg Manage J 17:109–122
Ironica A, Baleanu V, Edelhauser E, Irimie S (2010) TQM and business excellence. Annals of the University of Petrosani. Economics Issue 10(4):125–134
James P (1997) The sustainability circle: a new tool for product development and design. J Sustain Prod Des 2:52–57, http://www.cfsd.org.uk/journal
Leydesdorff L (2006) The knowledge-based economy: modeled, measured, simulated. Universal Publishers, Boca Raton
Lundvall B-Å (1985) Product innovation and user-producer interaction, industrial development, Research series 31. Aalborg University Press, Aalborg
Ministry of Economy, Trade and Industry, Japan (METI) (2007) The key to innovation creation and the promotion of eco-innovation. Report by the Industrial Science Technology Policy Committee of the Industrial Structure Council, METI, Tokyo
Nelson R, Winter S (1982) An evolutionary theory of economic change. The Belknap Press of Harvard University Press, Cambridge, MA
Organization for Economic Co-operation and Development (OECD) (1996) The knowledge-based economy. Science, technology and industry outlook. OECD, Paris
Organization for Economic Co-operation and Development (OECD) (2007) Measuring material flows and resource productivity. The OECD guide ENV/EPOC/SE (2006)1/REV3, Environment Directorate, Paris
Organization for Economic Co-operation and Development (OECD) (2009) Sustainable manufacturing and eco-innovation: framework, practices and measurement. Structural Policy Division. OECD Directorate for Science, Technology and Industry
Organization for Economic Co-operation and Development (OECD) and Statistical Office of the European Communities (Eurostat) (2005) Oslo manual: guidelines for collecting and interpreting innovation data, 3rd edn. OECD, Paris
Penrose ET (1959) The theory of the growth of the firm. Oxford University Press, New York
Peters TJ, Waterman RH (1982) In the search of excellence: lessons from America's best-run companies. Harper and Row, New York
Roland Berger Strategy Consultants (2009) Green Business. Available at: http://www.think-act.com
Sveiby K (2001) A knowledge-based theory of the firm to guide in strategic formulation. J Intell Cap 2(4):344–358
Teece DJ, Pisano G, Shuen A (1997) Dynamic capabilities and strategic management. Strateg Manage J 18(7):509–533
Van der Borgh M, Cloodt M, Romme A (2012) Value creation by knowledge-based ecosystems: evidence from a field study. R&D Manage 42(2):150–169

Viedma JM, Cabrita MR (2012) Entrepreneurial excellence in the knowledge economy: intellectual capital benchmarking system. Palgrave Macmillan, Great Britain
World Bank (2007) Building knowledge economies: advanced strategies for development. World Bank, Washington, DC
World Bank (2010) Innovation policy. A guide for developing countries. World Bank, Washington, DC
Yigitcanlar T, Velibeyoglu K (2008) Knowledge-based strategic planning. In: 3rd international forum on knowledge asset dynamics, Matera, pp 296–306

Chapter 4
A Framework for Developing and Assessing Eco-innovations

Ida Gremyr, Jutta Hildenbrand, Steven Sarasini, and Hendry Raharjo

Abstract This chapter presents a framework entitled "ECORE," which aims to assist in developing and assessing radical eco-innovations. Our proposed framework seeks to address theoretical gaps and unresolved problems from three research fields – eco-innovation, quality management, and life cycle assessment. ECORE synthesizes ideas and concepts from these three fields into a set of key principles and practices that can further integrate sustainability into business practices. These key principles are based on the idea that stakeholder interactions should form the basis of eco-innovation, that a life cycle perspective should be adopted in the design stage of eco-innovation, and that stakeholder needs must be translated into eco-innovation characteristics throughout the design process. We illustrate our framework with a hypothetical example that focuses on reducing the environmental impacts of carbonated beverage consumption. The chapter concludes by presenting the views of practitioners that were invited to provide feedback on our proposals.

Keywords Quality management • Eco-innovation • Life cycle assessment • Stakeholders • Sustainability

---

I. Gremyr (✉) • H. Raharjo
Quality Sciences, Chalmers University of Technology, Gothenburg, Sweden
e-mail: ida.gremyr@chalmers.se; hendry.raharjo@chalmers.se

J. Hildenbrand • S. Sarasini
Department of Energy and Environment, Division of Environmental Systems Analysis, Chalmers University of Technology, Gothenburg, Sweden
e-mail: jutta.hildenbrand@chalmers.se; steven.sarasini@chalmers.se

S. Azevedo et al. (eds.), *Eco-Innovation and the Development of Business Models*, Greening of Industry Networks Studies 2, DOI 10.1007/978-3-319-05077-5_4,

4.1 Introduction

Companies increasingly understand the importance of sustainable development. The sustainability challenge has been translated partly into a need to develop eco-innovations, which are "new products and processes which provide customer and business value but significantly decrease environmental impacts" (James 1997, p. 53). In order to become eco-innovative, companies must operationalize new competences, concepts, and tools that can assist in both maximizing economic returns to environmental activities and integrating sustainability into business practices.

Many scholars (e.g., Luttropp and Lagerstedt 2006; Maxwell and Van der Vorst 2003) argue that sustainability should be integrated into existing managerial toolboxes. One such toolbox is found within quality management (QM). QM focuses mainly on customer expectations (Dean and Bowen 1994), and sustainability is becoming an explicit customer need. However, companies that focus only on customers may limit their potential to be eco-innovative. This is because innovations that result from interactions with customers tend to be of an incremental nature (Biemans 1991), whereas interactions between companies and other partners such as suppliers and universities can stimulate more radical innovations (Liyanage 1995; Baiman et al. 2002). Hence partnerships with a range of stakeholders could help companies develop eco-innovations with a greater potential to reduce environmental impacts.

A further problem is that it is difficult to quantify the actual environmental impacts of eco-innovation. Life cycle assessment (LCA), which seeks to quantify the environmental impacts of a product system from "cradle to grave" (Baumann and Tillman 2004), is one potential means to resolve this problem. The application of LCA in industry is notable, with life cycle assessment applied by the majority in some industry contexts (Brunklaus et al. 2012). However, LCA is usually applied on a post hoc basis. Adopting a life cycle perspective at the design stage could both provide useful guidance to product developers and enhance interactions with stakeholders that assist in developing eco-innovations.

This chapter develops a framework entitled "ECORE," whose main aim is to (1) explore ways to promote interactions with stakeholders that can assist in developing eco-innovations, (2) explore ways to reconfigure QM tools such that they support radical eco-innovation, and (3) examine ways to measure and quantify the environmental benefits of eco-innovation. Here we perform state-of-the-art reviews of three research fields (eco-innovation, QM, and life cycle assessment). We also draw on research from the wider field of innovation studies to suggest ways to assist practitioners in developing and evaluating eco-innovations. Our aim is to identify research gaps and to synthesize ideas and concepts from these three fields as a basis for the ECORE framework.

This chapter comprises six sections of which this was the first. The next section outlines our methodological approach. Our state-of-the-art reviews are presented in Sect. 4.3, and Sect. 4.4 describes a set of key principles and practices that underpin the ECORE framework. In Sect. 4.5, we exemplify the use of our framework by considering potential changes to carbonated beverage consumption. Section 4.6

concludes by considering the views of ten practitioners that provided valuable comments and feedback regarding our proposals.

4.2 Methodology

This chapter is based on a conceptual method (Meredith 1993) where our proposed framework integrates "a number of different works on the same topic, summarizes the common elements, contrasts the differences, and extends the work." Our contribution integrates eco-innovation, QM, and LCA using MacInnis' (2011) four conceptual goals (envisioning, explicating, relating, and debating). Relating is based on the specific goal of integrating, i.e., "to see previously distinct pieces as similar, often in terms of a unified whole" (MacInnis 2011: 138).

Our state-of-the-art reviews[1] provide the basis for a framework that assists in developing and assessing eco-innovations. The state-of-the-art reviews were derived through an iterative process capitalizing on the authors' expertise from the three areas (eco-innovation, QM, and LCA). As a first step, each researcher searched for literature in their respective fields through searches in Google Scholar and Web of Science and via discussions with colleagues in the respective fields. We searched for papers that are widely cited and highly regarded within each research field. These papers were presented and discussed at a research workshop that resulted in a first skeleton of a framework. The framework was later refined by including a hypothetical example and further discussed and documented in a second research workshop.

In the process of developing this framework, we also invited ten practitioners, mostly from the Swedish automotive industry, to comment and provide feedback on our proposals at a workshop. The practitioners each had different levels of experience, ranging from 5 to 30 years. The workshop resulted in further refinements of the framework, mainly to clarify key elements such as the continuous and iterative use of LCA and the integration of sustainability concerns into daily engineering practice.

4.3 State of the Art

This section reviews research on eco-innovation, QM, and LCA. We identify research gaps, common themes, and potential synergies between these three fields as a conceptual basis for our proposed framework. We also focus on studies that address interdisciplinary issues that are key to our aim: to develop a framework that assists managers seeking to make products and processes more environment friendly.

[1]We are not able to include references to all reviewed works due to space requirements. A full bibliography is available upon request.

4.3.1 Eco-innovation

Eco-innovation is a term that is often used interchangeably with others such as environmental technology and eco-efficiency (Hellström 2007). However, eco-innovation is not limited to technology. While definitions vary, eco-innovation typically relates to the commercialization of novel products and processes that can reduce environmental impacts (James 1997; Rennings 2000).

Researchers have adopted management-, industry- and society-level perspectives to study eco-innovation. Research that has adopted a management perspective has focused primarily on four themes. The first theme relates to management tools that can facilitate eco-design (e.g., Ferrer et al. 2012). Various tools have been utilized for this purpose, including design for environment, case-based reasoning, and TRIZ modeling. Despite these efforts, scholars have earmarked eco-design as an under-researched topic within the field of eco-innovation (Carrillo-Hermosilla et al. 2009). Another under-researched area relates to the assessment of the environmental impacts of eco-innovation. Some work has been done on this second theme (e.g., Bocken et al. 2012), utilizing tools such as LCA, simplified LCA, and life cycle planning. A third theme develops typologies for categorizing different types of eco-innovation in terms of new products and processes, value chain, and network and governance measures (e.g., Carrillo-Hermosilla et al. 2009).

The fourth and most researched theme focuses drivers and barriers of eco-innovation. This theme resonates with other fields of research such as corporate environmentalism, where scholars have utilized different theoretical approaches to examine why companies engage in environmental activities. These include new institutional theory (Hoffman and Ventresca 1999), stakeholder theory (Delmas and Toffel 2004), the resource-based view of the firm (Sharma and Vredenburg 1998; Hart 1995), supply chain management (Handfield et al. 2005), neo-Gramscian theory (Levy and Egan 2003), and strategic management (Nehrt 1996). Studies of drivers and barriers are thus not exclusive to management-level studies but extend to the industry and society levels. A commonality as regards environmental activities in general and eco-innovation in particular is that drivers and barriers typically emanate from different stakeholders. These include the government (via public policies) (e.g., Sierzchula et al. 2012; Veugelers 2012), customers (e.g., Tsai et al. 2012), external stakeholders (e.g., Stevens and Stevenson 2012; Oltra 2011), and internal stakeholders with key competences (e.g., Kesidou and Demirel 2012).

However, studies that adopt industry- and society-level perspectives identify an additional range of factors that support eco-innovation. These include participation in innovation networks, for example, which can spur both the development and dissemination of knowledge and innovative ideas (Yang et al. 2012). Similarly, some studies have identified the importance of systemic interactions between actors such as universities, research institutes, companies, suppliers, governments, and consumers that provide key resources for innovation. These studies utilize approaches such as socio-technical systems (Geels 2004), the multilevel perspective (Markard and Truffer 2008a; Geels and Schot 2007), and technological innovation systems. The latter approach has been applied to environmental technologies such as alternative transport fuels (Sandén and Hillman 2011), solar photovoltaics (Dewald and Truffer 2011), and stationary fuel cells (Markard and Truffer 2008b). By interacting with different types of stakeholders, companies can gain access to resources that can be critical for developing eco-innovations. These resources include knowledge and competence, access to markets, legitimacy, finance, and human capital. These resources are particularly useful for companies that pursue radical eco-innovations given their inherent risks.

4.3.2 *Quality Management*

The field of QM focuses on developing and improving product offerings to meet and exceed customer expectations. The term quality relates to "what we think, feel, or sense as a result of the objective reality…there is a subjective side of quality" (Shewhart 1931). Some scholars have elaborated on the more objective aspects of quality. Taguchi (1986), for example, defines quality loss as "the loss a product causes to society after being shipped, other than any losses caused by its intrinsic functions." Quality thus has societal dimensions beyond (subjective) customer preferences. The subjective aspects of quality imply that QM would benefit from a broadened view of the customer that extends the current focus on buyers/users to encompass diverse stakeholders.

Such a broadened view can have various benefits. By considering the needs and potential inputs of diverse stakeholders, QM can both develop more radical innovations (Liyanage 1995; Baiman et al. 2002) and account for noncustomers' needs (Garvare and Johansson 2010). Stakeholder interaction can thus assist in making QM serve society at large, as advocated by Taguchi (1986), and in making businesses more environmentally sustainable. However, despite facing rising pressures to enroll in sustainability initiatives, practitioners require support to integrate sustainability into engineering practices (Wilkinson et al. 2001). Some scholars have argued that sustainability should be integrated into existing toolboxes instead of developing new ones (Kitazawa and Sarkis 2000; Luttropp and Lagerstedt 2006; Maxwell and Van der Vorst 2003). QM appears useful in this respect because "[it] can serve as a bridge between the concept of environmental sustainability and concrete applications by managers" (Rusinko 2005: 59). The rationale that supports this statement is that QM is "well known, corroborated and integrated into most organizations' management processes, familiar to most managers and also very easy to adapt to an environmental program" (Silva et al. 2013).

4.3.3 *Life Cycle Assessment*

LCA encompasses a set of tools that are typically used to examine the environmental impacts of product systems. Its central attributes include a life cycle perspective (covering raw material acquisition, production, use, disposal, and potential recycling/reuse loops) and the consideration of general impact categories (resource availability, human health, and ecological consequences). While LCA is typically performed on a post hoc basis to evaluate the impacts of existing or newly developed products and product systems, results can be also used as a basis for decision-making in development projects. According to the eco-design paradox (Poudelet et al. 2012), choosing environmentally favorable options in the early stages of product development can help to reduce environmental impacts but is forgone mostly due to lack of data. There is, however, a lack of empirical support for this claim (Baumann et al. 2002).

There are two main approaches for incorporating LCA tools and methods into product design. The first incorporates LCA into green product development by focusing on material selection. This is part of a top-down approach that combines environmental properties with technological feasibility, recyclability, work conditions, and cost assessment (e.g., Fleischer and Schmidt 1997). A lack of information at the initial design stages could potentially be resolved by funneling processes that consider mass fractions and toxicity or other properties that are potentially harmful for the environment (Tischner et al. 2000). These require substantial expertise regarding different material properties. The second approach seeks to incorporate green options into product design while acknowledging time and other resource constraints. To this end, Luttropp and Lagerstedt (2006) suggest forgoing cumbersome quantitative assessments and instead argue for the use of qualitative rules which translate results from prior assessments into eco-design principles.

However, eco-design tools and life cycle thinking are not fully implemented in industry. This is mainly due to the time, expertise, and volume of data required for conducting LCA and interpreting results. Moreover, the design process is iterative and dynamic, whereas practitioners tend to see LCA as a static and retrospective means to assess environmental impacts (Poudelet et al. 2012). In other words, life cycle thinking is not fully integrated at the design stage of product development. One reason for this is that LCA-based eco-design tools do not meet designers' expectations as they tend to highlight key issues in terms of the main aspects of products and product systems that contribute to environmental degradation, but fail to help designers resolve them (Birch et al. 2012).

Other researchers address this by focusing on the organization and argue that environmental specialists should be included in product development networks in order to supply the competences required to support "green subprojects" (Johansson and Magnusson 2006). However, subprojects can lead to confusion about the responsibility for environmental performance and contradict the idea that sustainability should be integrated as a core principle of product development (Silva et al. 2013).

4.4 The ECORE Framework

In this section, we aim to create synergies between eco-innovation, QM, and LCA by outlining a set of principles and practices (Dean and Bowen 1994) that form the basis for our proposed framework. The proposed framework "ECORE" aims to make environmental sustainability a core principle of product development that supports eco-innovation.

QM includes a set of well-established toolboxes that focus on identifying and satisfying customer needs. One of the main challenges is to broaden this focus to include a wider range of stakeholders in order to make sustainability a core eco-design principle. Research on eco-innovation identifies a need for tools for eco-design and for evaluating environmental impacts. LCA represents a potential solution for environmental assessment, but its application is limited because of time and other resource constraints.

Our framework focuses on stakeholder interaction as a means to resolve these issues. From a QM perspective, this serves to broaden the concept of customer from end users or buyers to diverse stakeholders. The eco-innovation and LCA literatures identify a range of stakeholders that are useful for both the creation of eco-innovations and their environmental sustainability. Life cycle thinking, for instance, requires innovators to consider the environmental impacts of eco-innovation from cradle to grave. This means that eco-innovations must be assessed in terms of the environmental impacts of materials, energy sources, emissions, and waste, for instance, that are produced and consumed during the manufacturing, distribution, and disposal phases of products' life cycles. Considering product life cycles is not only important for the assessment of eco-innovations but can also provide useful inputs to the design process. Collaborations with stakeholders responsible for waste management, for instance, can help to identify secondary uses for waste at the end of the (first) use phase, effectively prolonging the life cycle.

By referring to these actors as stakeholders, we reiterate the traditional definition of the term as "any group or individual who can affect or be affected by the achievement of an organization's objectives" (Freeman 1984). Our framework seeks to regard stakeholders not only as "affected actors" – they can provide useful inputs for eco-innovation at the design stage. In the next section, we outline a set of key principles that can encourage fruitful stakeholder interaction.

4.4.1 Key Principles

Our proposed framework is based on three key principles (Fig. 4.1). While eco-innovation is commonly driven by policy and customer demands, the literature on innovation networks shows that collaborations with key suppliers, competitors, and science partners (universities, research institutes, etc.) within forums such as industry associations can provide useful inputs and resources to the innovative process such as key competences and skills. A first key principle is that *companies should aim to collaborate with diverse stakeholders in order to generate ideas for eco-innovation.* Stakeholder collaborations are however limited by various factors, including company size and location, time constraints, lack of networking competences, and individuals' social networks. This key principle can thus be modified such that it is restated as an aim, whereby practitioners seek to develop stakeholder collaborations over time.

Our second key principle is that *life cycle thinking should form the basis for developing and assessing eco-innovations.* LCA is thus restated as a useful component of the design process. We propose that practitioners perform simplified LCA to

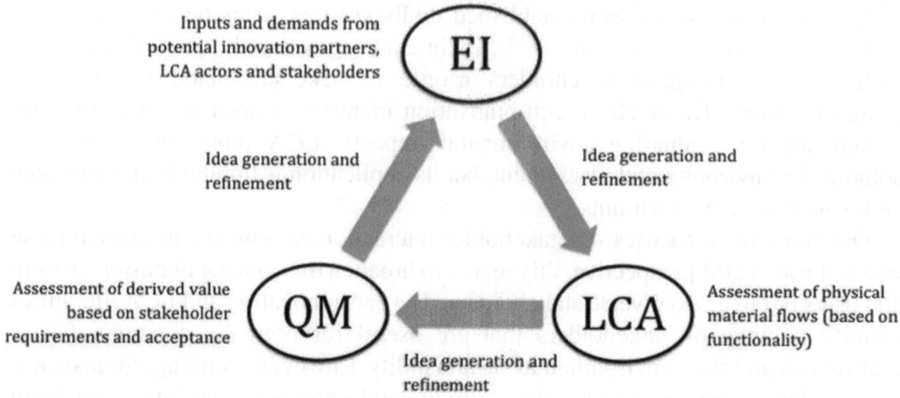

Fig. 4.1 Key principles of the ECORE framework

assess ideas and as a means to compare proposed eco-innovations with functional alternatives. By this, we mean that an innovation such as email should be evaluated in terms of any functional equivalent (a traditional postal service, text messaging, or shared documenting using a Web-based service) in terms of physical and material flows. A crucial point at that stage is the definition of functional requirements, which also supports decisions about which options are equivalent and which other products or services could provide them as a means to stimulate new ideas. Performing LCA in this way can also help to reduce environmental impacts at the design stage by providing environmental information regarding, for example, different materials or production techniques. Ideas can thus be continually refined and improved. Again the comprehensiveness of the assessment is limited here by time constraints, lack of detailed information, etc.

A third key principle is that *a focus on stakeholders' needs, and translation of those into eco-innovation characteristics, must be maintained throughout the design process.* In order to operationalize eco-innovations, it is important to assess and ascribe value to different stakeholders' needs and demands as a means to select an appropriate course of action. A final key principle is that this is a continuous process that is aligned with the QM notion of continuous improvements (Dean and Bowen 1994). This notion is also applied in environmental management (Burström von Malmborg 2002). Eco-innovations can be continually refined and improved as new ideas are generated, assessed, and evaluated based on the key principles outlined above.

4.4.2 Key Practices

In this section, we propose a set of key practices that assist in operationalizing the principles described below (see Fig. 4.2). The overall aim is to make continuous environmental improvements to products and processes.

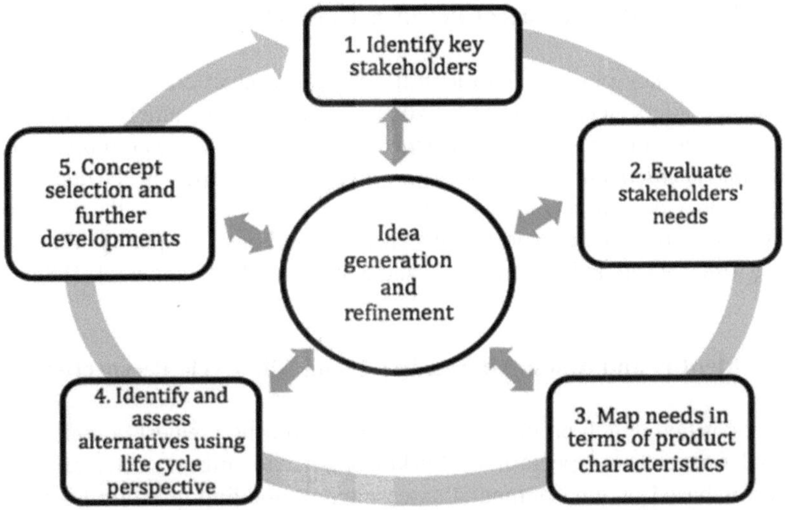

Fig. 4.2 Key practices for our proposed framework

4.4.2.1 Identify Key Stakeholders

The first key practice is to identify key stakeholders that can affect or be affected by existing or proposed eco-innovations. This practice is based on three guiding questions: *who is affected, who can contribute and what is their influence*? These questions are intended to assist in identifying key stakeholder groups and whether they support or oppose the eco-innovation. They also assist in evaluating the stakeholders' influence or salience in terms of the types of resources they control (such as social or political legitimacy, finance, knowledge and competences, etc.). Furthermore, stakeholders' influence can also be assessed throughout the product life cycle (raw material extraction and refinement, production and manufacturing, sales and distribution, use, and end of life). There are several tools and methods that may be of use for this key practice, including stakeholder mapping (Mitchell et al. 1997) and social life cycle assessment (Benoît-Norris et al. 2011).

4.4.2.2 Evaluate Stakeholders' Needs

This key practice evaluates stakeholders' needs as regards an existing or proposed product or process. A guiding question is: *what are stakeholders' needs in terms of the overall functionality of the product or process*? The aim is to examine the diverse range of stakeholder needs that must be met through eco-innovation. Note that some stakeholders may have similar needs, whereas others may have conflicting needs. Market research methods can be utilized for this practice (e.g., Griffin and Hauser 1993).

4.4.2.3 Map Needs in Terms of Products Characteristics

The aim of this key practice is to identify critical product characteristics that can satisfy stakeholders' needs. There are two guiding questions, namely, *how can stakeholder needs be translated into functional specifications* and *which product characteristics affect critical functional specifications*? This can be performed quantitatively in order to provide concrete assessments that can be fed back into the development process. Tools that are suitable for this include the house of quality[2] (Hauser and Clausing 1988; Akao and Mazur 2003) and Kano's model (Kano et al. 1984).

4.4.2.4 Identify and Assess Alternatives Using Life Cycle Perspective

Having identified which product characteristics offer the greatest potential to meet stakeholder needs and expectations, this key practice identifies a range of functional alternatives based on environmental assessment. A guiding question for this practice is: *which alternatives provide the greatest environmental improvement from a life cycle perspective*? If, for instance, the aim is to reduce the amount of toxic materials in a given product, it is important to consider the range of less toxic alternatives in terms of their entire life cycles in order to minimize environmental burdens. Tools that are suitable for the identification part of this practice include theory of inventive problem-solving (TRIZ) (Altshuller 1999) and simplified and streamlined LCA (Fleischer and Schmidt 1997; Graedel 1998).

4.4.2.5 Concept Selection and Further Developments

The environmental assessment in practice #4 helps to identify "hotspots," which are aspects of the product life cycle with a high environmental impact. It also makes trade-offs between different types of impact more visible. This key practice focuses on feeding this information back into the development process such that informed decisions can be made regarding overall design characteristics and requirements for a given product or process. Two guiding questions for this practice are: *how can environmental impacts be further reduced*, and *how can environmental criteria receive the right priority*?

Pugh concept selection (Pugh 1991) and analytic hierarchy process (AHP) (Saaty 1980) are suitable methods for assessing alternatives based on the information generated in previous steps. This approach allows for qualitative comparisons of the various alternatives based on numerous criteria, including environmental impacts.

At this stage, the design team may choose to develop a particular product or process. Alternatively, they may decide to restart the cycle in order to gather more

[2] The house of quality commonly maps needs on a scale of 1-3-9 in order to strongly discriminate weak, medium, and strong associations. This has been discussed by Ghiya et al. (1999) and further by Raharjo (2013).

information that can further contribute to the innovative process, possibly following the rejection of a previous idea and the development of a new design based on the information generated in the steps above. It may be suitable to identify new stakeholders that can act as innovation partners given their expertise in a particular technical area (e.g., materials science). Collaborations with stakeholders responsible for particular functions in a product life cycle (e.g., waste management) may also prove fruitful. For more radical eco-innovations, it may be beneficial to collaborate with partners that control key resources such as finance (e.g., venture capitalists), competences (e.g., universities), and legitimacy (e.g., governments).

4.5 An Illustrative Example

In this section, we illustrate our proposed framework using an illustrative example. Here we focus on reducing the environmental impacts of carbonated beverage containers. We selected carbonated beverage containers because they are a relatively simple and well-known product. It is however possible to apply our framework to products that are more complicated than beverage containers. We focus specifically on ways to improve plastic bottles that are at present widely available in shops and supermarkets. While plastic bottles of this type are recyclable, their environmental impacts are by no means negligible. We thus apply our framework to identify ways of reducing the environmental impacts of carbonated beverage consumption.

We adopt the perspective of a beverage company and aim to introduce life cycle thinking as a means to both develop and assess alternatives to plastic bottles that have lower environmental impacts. By adopting the perspective of a single firm, we hope to make our example understandable to product developers. However, to incorporate life cycle thinking, it is important to consider the company as a single actor in the product life cycle and examine how relevant emissions relate to its own processes (Baumann et al. 2011).

Despite the apparent simplicity of a beverage container, a diverse range of actors is involved in the production, distribution, retail, and after-sales phases of its life cycle, or product system. It is thus necessary to consider the needs of these actors in any new development. This is especially the case if a new alternative is derived from an existing product system. We outline every step of our procedure in detail, but in order to illustrate the main features of our framework, we focus on a limited set of product features. Costs and expected revenues, for instance, are excluded from our description despite the fact that it is necessary to include such information in a real development project.

We start by defining the functional specifications of the product system to highlight properties that are key to fulfilling customer needs. Existing beverage containers typically consist of resealable plastic bottles that are used as portable, disposable flasks. We treat resealability as a key functional property. Plastic bottles also preserve carbonated beverages for at least 6 months. This implies that a container must not alter the taste or add pollutants to beverages, and carbonation levels must be maintained.

By considering the life cycle impacts of plastic bottles, it is possible to fulfill these functional characteristics using alternatives. Plastic bottles are an oil-based product, and where fossil oil is used to produce bottles, this depletes natural resource stocks and results in greenhouse gas emissions. Carbonated beverages are typically bottled at centralized plants and then distributed to retailers. This further depletes oil stocks and contributes to global warming given the transport sector's heavy reliance on fossil fuels. Following their disposal, plastic bottles can be recycled to allow plastics to be reused. While this is considered to be more environment friendly than landfill disposal, recycling is both energy and water intensive.

This list of environmental implications over a life cycle is not exhaustive, but it does illustrate some of the potential areas for improvement. As an illustrative example, we treat disposable plastic bottles with screw caps as a baseline product system for comparison with a functional alternative. The alternative we propose consists of a beverage-dispensing machine located at sales points. The proposed alternative dispenses carbonated beverages into reusable containers that are either sold separately by retailers or purchased by customers via other channels. This means that the bottling process takes place at the sales point rather than at a centralized location, potentially reducing the environmental impacts in the distribution phase. Once the dispensing machine is installed, there is still a need to transport syrup and carbon dioxide cartridges to sales points where they are mixed with water as beverages are sold. For the sake of our argument, we assume that our functional alternative serves to reduce the total environmental impacts of beverage consumption. In practice, a simplified LCA can be performed to ascertain whether this is really the case. We now apply the ECORE framework in an iterative fashion to further develop and evaluate this idea.

4.5.1 Identify Key Stakeholders

In Sect. 4.4.2, we highlighted the importance of identifying key stakeholders in terms of *who is affected, who can contribute, and what is their influence*. Various stakeholders may influence or be affected by our functionally equal alternative means to sell and dispense carbonated beverages. Even for a relatively simple product, it is necessary to expand to a product system that includes several actors and their contribution vis-à-vis the required functionality. A basic process layout (Fig. 4.3) serves several purposes: it can help to identify which processes diverge between the existing option and a new alternative, and it can also help to identify which actors are involved.

Figure 4.3 shows the aggregated production processes for our proposed alternative product system. Some components in the system such as refillable bottles and cartridges can be reused. Some processes are unaffected by our proposed change, such as syrup production and water treatment, whereas the filling process, bottle production and distribution undergo changes with several actors involved.

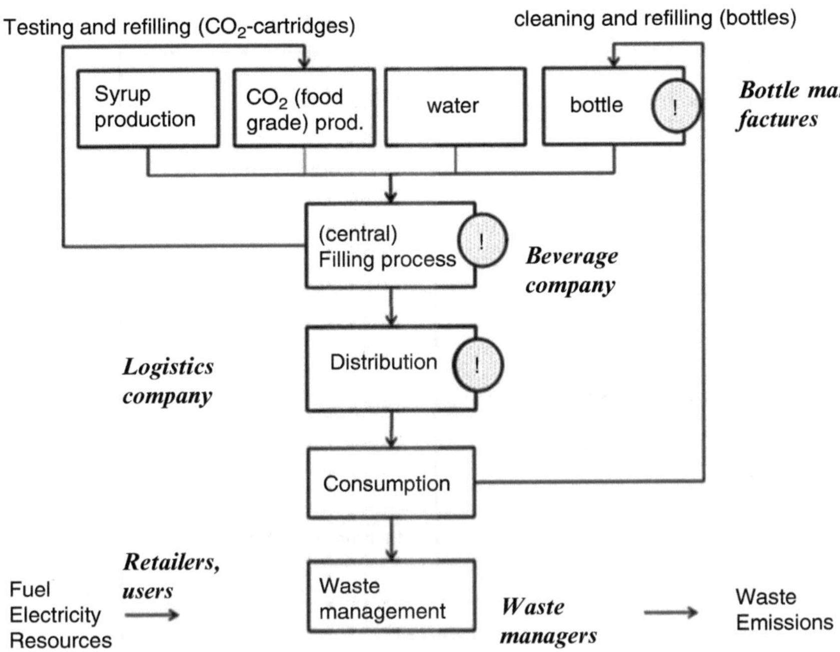

Fig. 4.3 Process map for beverage with central filling

Figure 4.3 shows that incorporating life cycle thinking means considering a rather complex process layout, even for relatively simple changes to existing product systems. However, a basic process layout serves at least two purposes. First, it can help to point out which processes diverge between the existing option and a new alternative. Second, Fig. 4.3 reveals some of the stakeholders that are key to implementing our proposed changes. A more exhaustive list of stakeholders that are either interested or affected by our changes includes consumers, beverage companies, packaging producers, bottle manufacturers, retailers, distributors, government authorities, waste managers, and environmental organizations. Some stakeholders could potentially contribute to the proposed innovation in terms of technological developments. Machine manufacturers, for instance, could assist in providing technical knowledge and inputs that assist in reducing the impacts of beverage dispensers.

Figure 4.4 maps stakeholders in terms of their power and interest. In what follows, we prioritize the four stakeholders that appear in to be more interested and powerful: customers, retailers, machine manufacturers, and beverage companies (Table 4.1).

We assume that consumers can strongly influence the success of our proposed alternative because they are key actors in the use phase. Despite the fact that the dispensing machine does not influence demand for carbonated beverages, it does imply changes to user behavior. Hence consumer willingness to adapt is key to their decision to purchase beverages in this manner. Consumers are also key to the

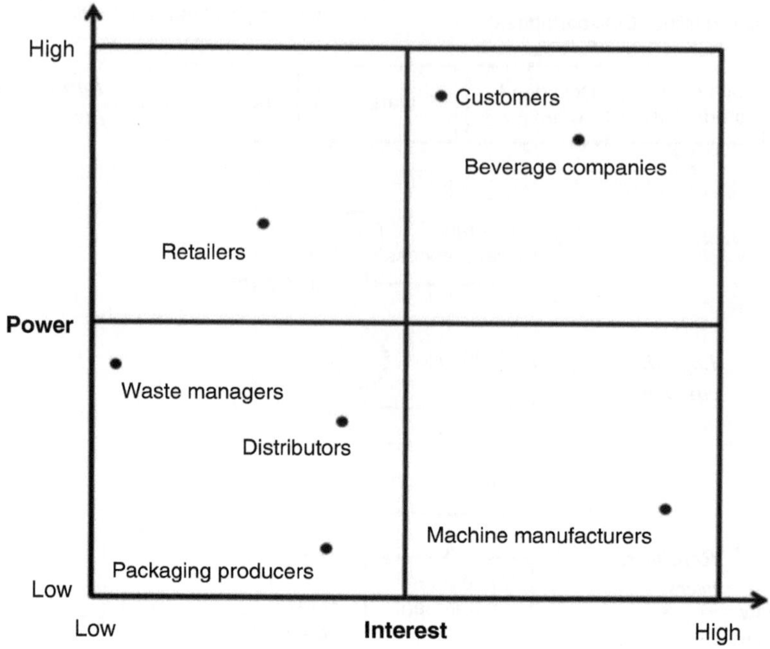

Fig. 4.4 Stakeholder map for a product system based on decentralized beverage dispensing (This stakeholder map is for illustrative purposes only, i.e., it is based on authors' estimations in order to exemplify the possible results of real-world stakeholder mapping. In practice, different methods can be used to map stakeholders' power and interest such as surveys and focus group discussions. We anticipate that the application of such methods will vary depending on the type and scale of the project in question)

end-of-life phase in that they must take responsibility to make reusable containers available for recycling when they eventually perish. Retailers are similarly important in the use phase but for different reasons. Retailers must make space for beverage dispensers and take responsibility for cleaning and daily maintenance in the use phase. However, we assume that their influence is only medium since they are largely unaffected by the proposed change. Machine manufacturers are similarly influential, with interests that relate to several life cycle phases. We assume that machine manufacturers will support the proposed alternative since it represents an innovation opportunity. However, interactions with machine manufacturers could potentially reduce the environmental impacts of the product service system since they are responsible for material selection and design for disassembly. By leasing machines to retailers, machine manufacturers can ensure that low-impact materials are used and take responsibility for recycling key components in the end-of-life phase.

In contrast, we assume that beverage companies will strongly influence our proposed alternative. Dispensers are currently used in bars and restaurants, and most carbonated beverage companies are thus able to produce syrups for mixing at

Table 4.1 Stakeholders' interests and influence at different stages of the product life cycle

Stakeholder	Raw material	Production	Distribution	Use	End of life	Priorities	Normalized priorities (%)
Customer				Decision to buy	Recycling	Strong (9)	37.5
Retailer				Space, daily maintenance		Medium (3)	12.5
Machine manufacturer	Material selection	Design for disassembly		Machine maintenance	Design for environment	Medium (3)	12.5
Beverage company			Syrup	Loss of branding	Less littering	Strong (9)	37.5

sales points. By promoting the use of reusable containers, our proposed alternative can also reduce littering, which represents an opportunity for eco-branding among beverage companies. However, the existing plastic bottle system represents a branding opportunity for beverage companies that may be lost if reusable containers without their company logos are used. For the sake of our argument, we assume that beverage companies are not in favor of our proposed alternative because of this branding loss.

In addition to mapping these stakeholder interests onto a power/influence grid as is commonplace of stakeholder analysis, we map them according to different stages of the product life cycle. By adding a life cycle perspective, we aim to display necessary interactions and causal relationships between different stakeholders. Feedback loops become visible and can be addressed, ideally together with involved stakeholders. An example is the use of secondary plastic material after a recycling process. Successfully closing a loop and saving virgin raw material is only possible if an upstream production process is capable of accommodating secondary material. This may depend on material properties like purity that can be influenced by the design of the recycling process. A timely consideration of stakeholders' needs thus allows for a balanced product life cycle design. We also quantify stakeholders' influence on a scale of "1-3-9" to denote "weak," "medium," and "strong," respectively. This method is adopted from the house of quality (Akao and Mazur 2003) and allows us to easily prioritize stakeholders whose needs must be further addressed in the next step. The normalized priorities, which are shown in the last column of Table 4.1, are derived using the formula:

$$w_i^{norm} = \frac{w_i}{\sum_{i=1}^{m} w_i}, \ \forall \quad i = 1,2,\ldots,m$$

where m denotes the number of stakeholders and w_i denotes the assigned priority or weight of the ith stakeholder.

Note that normalized priorities are useful for the sake of practical interpretation of the priorities. The Delphi method (Okolo and Pawlowski 2004) or the analytic hierarchy process (AHP) can also be employed here (see Raharjo et al. (2008) for a more extensive example of using AHP in the house of quality).

4.5.2 Evaluate Stakeholders' Needs

Having identified stakeholders' influence, we proceed with identification of "hotspots" or "critical components" of the product system. This is because changes to critical components can influence the success of the proposed change. From a life cycle perspective, the critical components of this particular product system include the beverage container, dispenser machine, and logistics. Focusing on critical

Table 4.2 Assessing stakeholders' needs for two critical components

		Beverage container			Dispenser machine		
		Stakeholder needs					
	Influence	Portable	Avoid littering	Branding opportunity	Low mainte-nance	Easy to operate	Machine recyclability
Consumers	37.5 %	Strong	Medium	Weak		Strong	
Retailers	12.5 %	Weak			Strong	Medium	Medium
Machine producers	12.5 %				Strong		Strong
Beverage company	37.5 %		Strong	Strong		Medium	
Absolute importance		3.50	4.88	3.75	2.25	4.88	1.50
Relative importance		16.9 %	23.5 %	18.1 %	10.8 %	23.5 %	7.2 %
Summed relative importance			*58.5 %*			*41.5 %*	

components and prioritizing them in decisions facilitates the design of a product system that is accepted by prioritized stakeholders.

To keep our case simple, we focus on two critical components – the beverage container and the dispenser machine. In Table 4.2, three stakeholders' needs are listed as regards each component. Each of those needs is then assessed in terms of the four most influential stakeholders. The relationship between a certain need and a certain stakeholder is quantified using the same scale as previously. Finally, the absolute importance (AI) is calculated based on the sum product of the relationship strength column and the need column using the following formula:

$$AI_j^k = \sum_{i=1}^{m} R_{ij}^k w_i^{norm}, \ \forall \quad j = 1,2,\ldots,n; \ k = 1,2,\ldots,p$$

where R_{ij}^k is the relationship strength between the ith stakeholder and the jth need of the kth critical component. In this case, there are four stakeholders ($m = 4$), two critical components ($p = 2$), and three needs for each component ($n = 3$).

The relative importance (RI) of the absolute importance is computed by the following formula:

$$RI_j^k = \frac{AI_j^k}{\sum_{k=1}^{p}\sum_{j=1}^{n} AI_j^k}, \forall \quad j = 1,2,\ldots,n; \ k = 1,2,\ldots,p$$

Table 4.2 shows that the relative importance of the beverage container is higher than that of the dispenser machine (16.9 + 23.5 + 18.1 = 58.4 %). Hence we now focus on appeasing stakeholders' needs vis-à-vis the beverage container.

Here we narrow our focus to the two most influential stakeholders as regards beverage containers – consumers and beverage companies. Table 4.2 reiterates that consumers and beverage companies are the two most influential stakeholders and that they each have the same level of importance. This is reflected in the influence column value in Table 4.3. Here we focus more specifically on the needs and interests of these two stakeholders as a means to further identify the key functional aspects of beverage containers that must be developed.

In Table 4.3, we consider as many functional requirements of beverage containers as possible in order to identify potential trade-offs. A similar process of assessing stakeholders' needs is repeated here for the refined selection of functional requirements. Since new requirements have been added and others from the original selection have been excluded, the results are not exactly similar. The aim is to quantify the absolute and relative importance of each functional requirement. We identify "avoid littering," "branding opportunity," and "portable" as the three most important needs.

4.5.3 Map Needs in Terms of Products Characteristics

In this step, the stakeholders' needs that were identified in the step above are broken down into product characteristics. These are the properties that can be changed by product developers and are thus helpful in directing possible actions. Stakeholders' needs are mapped into product components in terms of its required functional specifications. Five product characteristics are treated as key to the required functionality: material type, weight, shape, salvage value, and resealability. Similar steps as before were repeated, and the results are shown in Table 4.4.

Our analysis indicates that the two most important product characteristics of the containers are "material type" and "salvage value." Subsequent efforts to develop beverage containers should take these factors into account. In other words, it has now become clear that the main focus of product development should be put on developing the material of the container while considering its salvage value if the alternative product is to be implemented.

4.5.4 Identify and Assess Alternatives Using Life Cycle Perspective

This step focuses on quantifying the environmental impacts of the proposed alternative product system (reusable bottles plus a drink dispenser) for comparison with the existing baseline product system (drinks sold in disposable plastic bottles).

Table 4.3 Assessing stakeholders' needs for the refined product

	Influence	Light weight	Recycle	Portable	Transparent	Durable surface	Branding opportunity	Avoid littering
Consumers	50 %	Medium	Medium	Strong	Medium	Medium		Medium
Beverage company	50 %						Strong	Strong
Absolute importance		1.5	1.5	4.5	1.5	1.5	4.5	6
Relative importance		7.14 %	7.14 %	21.43 %	7.14 %	7.14 %	21.43 %	28.57 %

Table 4.4 Mapping needs into product's characteristics

	Need importance	Material type	Weight	Shape	Salvage value	Resealability
Light weight	7.14 %	Strong	Strong	Medium		
Recyclable	7.14 %	Strong			Strong	Weak
Portable	21.43 %	Weak	Strong	Weak		Strong
Transparent	7.14 %	Strong				
Durable	7.14 %	Strong			Medium	
Branding	21.43 %	Weak		Strong		Medium
Littering	28.57 %				Strong	
Absolute importance		3.00	2.57	2.36	3.43	2.64
Relative importance		21.43 %	18.37 %	16.84 %	24.49 %	18.88 %

To do this, we assume that a product system is defined which could serve as basis for calculating a life cycle inventory. From a stakeholder perspective, "material type" and "salvage value" were previously identified as the most important product characteristics. A lightweight recyclable material for bottles as part of a deposit system could thus be a suitable alternative. To illustrate the use of an LCA perspective, we further assume that decentralized dispenser systems are available and that they can be used to refill plastic bottles made from polyethylene terephthalate (PET). PET is used to manufacture the disposable bottles currently in circulation. However, for our proposed alternative, we assume that bottles contain more PET since they must be more durable and reusable. The bottle weight for the refillable bottles is thus assumed to be higher. We further assume that reusable bottles are rinsed at the dispenser before filling, and hence water, energy, and detergents must be considered in the calculation.

Refined process maps for the baseline and alternative system can be drawn using this information. Note that we focus on one alternative in order to illustrate our example. It is however possible to include other options. To create process maps, input and output flow data for the identified processes are collected to establish a life cycle inventory of resources and emissions. Since time constraints may be critical to product development, we suggest the use of generic data from databases like EcoInvent for widely used processes (e.g., transport and logistics or electricity). We also recommend focusing mainly on processes that diverge between the baseline product system and our proposed alternative. Syrup production, for instance, is unaffected by the change and can probably be excluded. In contrast, bottle production, reuse, and recycling must be considered. Since at this stage not all processes are fully defined, best- and worst-case scenarios can be created to identify most relevant characteristics from an environmental perspective. To simplify the environmental assessment further, fuel and electricity consumption can be used as proxies instead of calculating a full LCA with several impact criteria. At this stage, it is important that the structure of the product system is maintained to identify where contributions to the overall environmental load are generated and which possible trade-offs appear.

4.5.5 Concept Selection and Further Developments

In this step, we focus on utilizing the information generated in the previous steps as a stimulus to further product development. The environmental impacts that were calculated in the previous step provide an overview of "hotspots." These are elements of product systems that contribute most to the overall environmental load. If, for instance, LCA shows that reusable bottles would result more emissions if not returned and refilled, then the suggested deposit system becomes a vital part. If a production process for the bottle material shows excessively high emissions of toxic substances, other materials can be considered. If a major flaw is detected at this stage for which no amendment is possible, a redirection of the development process is recommended which could reuse the lessons learnt from the first iteration.

The process of applying the framework does not stop here. An additional alternative can be taken into account and evaluated in similar ways. The proposed alternative can also be continuously improved by performing further iterations that serve to drive subsequent improvements. Furthermore, it is not necessary to complete a full iteration for every development project. Incremental changes, for example, can be developed without stakeholder involvement following only steps 3 to 5. A full iteration may however be beneficial for more radical eco-innovations where stakeholder inclusion is essential to the legitimacy and success of the project in question.

Repeated iterations can also assist in establishing and further developing partnerships with stakeholders that contribute with resources that are key to innovation. For this particular example, it may be useful to develop competences in materials that can potentially reduce the environmental impacts of reusable bottles. This could be achieved by developing new materials, by refining existing materials, or by finding new uses for materials when bottles can no longer be reused. To develop competences in these areas, it may be useful to collaborate with different types of partners, including bottle manufacturers, experts in materials science, and waste managers. The types of partners vary between projects, but the key is to focus on the types of resources (technical, financial or otherwise) that can assist in further reducing environmental impacts.

4.6 Concluding Remarks

Seeing sustainability as an opportunity rather than a constraint is key to its integration into daily engineering practices (Angell and Klassen 1999; Luttropp and Lagerstedt 2006; Maxwell and Van der Vorst 2003). This chapter aims to promote sustainability by (1) exploring ways to stimulate interactions with stakeholders that can assist in developing eco-innovations, (2) exploring ways to reconfigure QM tools such that they support both incremental and radical eco-innovation, and (3) examining ways to measure and quantify the environmental impacts of eco-innovation. Practitioners noted that a full iteration of our framework is needed for radical eco-innovations, but this is unnecessary for incremental eco-innovations where stakeholder preferences are already known. By promoting a flexible approach in applying our framework, we feel that it can support the development of both incremental and radical eco-innovations. In other words, while radical eco-innovations may require that practitioners perform all of our proposed practices, only practices three to six, for instance, may be required for incremental eco-innovations in cases where stakeholder needs are well understood.

Our proposed framework is novel in that it synthesizes ideas from three research fields to address existing research gaps within each of those fields. In addition to this conceptual novelty, we started to examine its practical utility and novelty by inviting practitioners to comment and provide feedback on our proposals. Generally, they felt that there is novelty in our approach, especially as regards the focus on interaction with stakeholders. Stakeholders, they argued, can both provide valuable information to product developers and ensure that radical eco-innovations are developed in an inclusive manner, thus providing a key source of legitimacy. Practitioners also confirmed our suspicions that LCA is typically a post hoc exercise and supported our idea that it should instead be an integral part of product development. Practitioners also welcomed the idea that sustainability be integrated via well-established tools such as QFD and Pugh concept selection.

Practitioners elaborated further by claiming that a broad and inclusive stakeholder approach is not essential for incremental innovations consisting of material substitutions in engine subcomponents, for instance. Instead, they argued that engineers require access to readily available information regarding the environmental impacts of different materials in order to make informed decisions. Practitioners also argued that such information should be presented in a monetized form such that environmental values are not "lost" among other criteria key to product development. Similarly, practitioners argued that although our framework works well in deriving a set of sustainability requirements that can boost eco-innovation, these criteria might not be prioritized as highly as other criteria such as cost requirements.

The main contribution of this chapter is a framework of practices, based on principles from three competence areas: EI, LCA, and QM. The framework supports a continuous and iterative application of life cycle thinking and an integration of sustainability considerations into the well-established QM toolbox. Existing QM practices are thus enhanced and used as an infrastructure for life cycle thinking and sustainability considerations. Our aim was to develop a framework that can assist in both developing and assessing eco-innovation in practical contexts. One limitation of our proposed framework is that it has not yet been subject to a comprehensive empirical evaluation. Further research is thus required to examine the utility of our approach. A second limitation is that our framework does not ensure that the most environmentally beneficial eco-innovations are selected for further development. By this, we mean that there is no guarantee that the environmental information generated via our framework will form the basis of decisions made in the concept selection stage, where other criteria related to cost and quality come into play. While our framework, with its stakeholder perspective, may assist in resolving this issue, future research must focus on how barriers to the development of eco-innovations in the concept selection stage can be overcome.

Acknowledgments This work was conducted within the Sustainable Production Initiative and the Production Area of Advance at Chalmers University of Technology. It was funded by the Swedish Governmental Agency for Innovation Systems (VINNOVA). The support is gratefully acknowledged.

References

Akao Y, Mazur GH (2003) The leading edge in QFD: past, present and future. Int J Qual Reliab Manage 20(1):20–35

Altshuller G (1999) The innovation algorithm: TRIZ, systematic innovation, and technical creativity. Technical Innovation Center, Worcester

Angell LC, Klassen RD (1999) Integrating environmental issues into the mainstream: an agenda for research in operations management. J Oper Manage 17:575–598

Baiman S, Rajan MV, Kanodia C (2002) The role of information and opportunism in the choice of buyer–supplier relationships/discussion. J Account Res 40(2):247–278

Baumann H, Tillman AM (2004) The hitch hiker's guide to LCA: an orientation in life cycle assessment methodology and application. Studentlitteratur, Lund

Baumann H, Boons F, Bragd A (2002) Mapping the green product development field: engineering, policy and business perspectives. J Clean Prod 10(5):409–425

Baumann H, Berlin J, Brunklaus B, Lindkvist M, Löfgren B, Tillman AM (2011) The usefulness of an actor's perspective in LCA. In: Finkbeiner M (ed) Towards life cycle sustainability management. Springer, Dordrecht, pp 77–85

Benoît-Norris C, Vickery-Niederman G, Valdivia S, Franze J, Traverso M, Ciroth A, Mazijn B (2011) Introducing the UNEP/SETAC methodological sheets for subcategories of social LCA. Int J Life Cycle Assess 16(7):682–690

Biemans W (1991) User and third-party involvement in developing medical equipment innovations. Technovation 11(3):163–182

Birch A, Hon K, Short T (2012) Structure and output mechanisms in Design for Environment (DfE) tools. J Clean Prod 35:50–58

Bocken NMP, Allwood JM, Willey AR, King JMH (2012) Development of a tool for rapidly assessing the implementation difficulty and emissions benefits of innovations. Technovation 32(1):19–31

Brunklaus B, Hildenbrand J, Sarasini S (2012) Eco-innovative measures in large Swedish companies: an inventory based on company reports. Vinnova, Stockholm

Burström von Malmborg F (2002) Environmental management systems, communicative action and organizational learning. Bus Strategy Environ 11(5):312–323

Carrillo-Hermosilla J, González PDR, Könnölä T (2009) Eco-innovation: when sustainability and competitiveness shake hands. Palgrave Macmillan, Hampshire

Dean JW, Bowen DE (1994) Management theory and total quality: improving research and practice through theory development. Acad Manage Rev 19(3):392–418

Delmas M, Toffel MW (2004) Stakeholders and environmental management practices: an institutional framework. Bus Strategy Environ 13(4):209–222

Dewald U, Truffer B (2011) Market formation in technological innovation systems – diffusion of photovoltaic applications in Germany. Ind Innov 18(3):285–300

Ferrer JB, Negny S, Robles GC, Le Lann JM (2012) Eco-innovative design method for process engineering. Comput Chem Eng 45:137–151

Fleischer G, Schmidt WP (1997) Iterative screening LCA in an eco-design tool. Int J Life Cycle Assess 2(1):20–24

Freeman RE (1984) Strategic management: a stakeholder approach. Pitman, Boston

Garvare R, Johansson P (2010) Management for sustainability – a stakeholder theory. Total Qual Manage 21(7):737–744

Geels FW (2004) From sectoral systems of innovation to socio-technical systems. Res Policy 33(6–7):897–920
Geels FW, Schot J (2007) Typology of sociotechnical transition pathways. Res Policy 36(3):399–417
Ghiya KK, Bahill AT, Chapman WL (1999) QFD: validating robustness. Qual Eng 11(4):593–611
Graedel TE (1998) Streamlined life-cycle assessment. Prentice Hall, Upper Saddle River
Griffin A, Hauser JR (1993) The voice of the customer. Mark Sci 12(1):1–27
Handfield R, Sroufe R, Walton SV (2005) Integrating environmental management and supply chain strategies. Bus Strategy Environ 14(1):1–19
Hart SL (1995) A natural-resource-based view of the firm. Acad Manage Rev 20(4):986–1014
Hauser JR, Clausing D (1988) The house of quality. Harv Bus Rev 66(3):63–73
Hellström T (2007) Dimensions of environmentally sustainable innovation: the structure of eco-innovation concepts. Sustain Dev 15(3):148–159
Hoffman AJ, Ventresca MJ (1999) The institutional framing of policy debates: economics versus the environment. Am Behav Sci 42(8):1368–1392
James P (1997) The sustainability circle: a new tool for product development and design. J Sustain Prod Des 1(2):52–57
Johansson G, Magnusson T (2006) Organising for environmental considerations in complex product development projects: implications from introducing a 'green' sub-project. J Clean Prod 14(15–16):1368–1376
Kano N, Seraku N, Takahashi F, Tsuji S (1984) Attractive quality and must-be quality. J Jpn Soc Qual Control (in Japanese) 14(2):39–48
Kesidou E, Demirel P (2012) On the drivers of eco-innovations: empirical evidence from the UK. Res Policy 41(5):862–870
Kitazawa S, Sarkis J (2000) The relationship between ISO14001 and continuous source reduction programs. Int J Oper Prod Manage 20(2):225–248
Levy DL, Egan D (2003) A neo-Gramscian approach to corporate political strategy: conflict and accommodation in the climate change negotiations. J Manage Stud 40(4):803–829
Liyanage S (1995) Breeding innovation clusters through collaborative research networks. Technovation 15(9):553–567
Luttropp C, Lagerstedt J (2006) EcoDesign and the ten golden rules: generic advice for merging environmental aspects into product development. J Clean Prod 14(15–16):1396–1408
Macinnis DJ (2011) A framework for conceptual contributions in marketing. J Mark 75(4):136–154
Markard J, Truffer B (2008a) Technological innovation systems and the multi-level perspective: towards an integrated framework. Res Policy 37(4):596–615
Markard J, Truffer B (2008b) Actor-oriented analysis of innovation systems: exploring micro–meso level linkages in the case of stationary fuel cells. Technol Anal Strateg Manage 20(4):443–464
Maxwell D, Van Der Vorst R (2003) Developing sustainable products and services. J Clean Prod 11(8):883–895
Meredith J (1993) Theory building through conceptual methods. Int J Oper Prod Manage 13(5):3–11
Mitchell RK, Agle BR, Wood DJ (1997) Toward a theory of stakeholder identification and salience: defining the principle of who and what really counts. Acad Manage Rev 22(4):853–888
Nehrt C (1996) Timing and intensity effects of environmental investments. Strategy Manage J 17(7):535–547
Okolo C, Pawlowski SD (2004) The Delphi method as a research tool: an example, design considerations and applications. Inf Manage 42:15–29
Oltra C (2011) Stakeholder perceptions of biofuels from microalgae. Energy Policy 39(3):1774–1781
Poudelet V, Chayer JA, Margni M, Pellerin R, Samson R (2012) A process-based approach to operationalize life cycle assessment through the development of an eco-design decision-support system. J Clean Prod 33:192–201
Pugh S (1991) Total design – integrated methods for successful product engineering. Addison-Wesley, Wokingham

Raharjo H (2013) On normalizing the relationship matrix in quality function deployment. Int J Qual Reliab Manage 30(6):647–661

Raharjo H, Brombacher AC, Xie M (2008) Dealing with subjectivity in early product design phase: a systematic approach to exploit QFD potentials. Comput Ind Eng 55(1):253–278

Rennings K (2000) Redefining innovation – eco-innovation research and the contribution from ecological economics. Ecol Econ 32(2):319–332

Rusinko CA (2005) Using quality management as a bridge to environmental sustainability in organizations. SAM Adv Manage J 70(4):54–60

Saaty TL (1980) The analytic hierarchy process. McGraw-Hill, New York

Sandén BA, Hillman KM (2011) A framework for analysis of multi-mode interaction among technologies with examples from the history of alternative transport fuels in Sweden. Res Policy 40(3):403–414

Sharma S, Vredenburg H (1998) Proactive corporate environmental strategy and the development of competitively valuable organizational capabilities. Strategy Manage J 19(8):729–753

Shewhart WA (1931) Economic control of quality of manufactured product. D. van Nostrand Company, New York

Sierzchula W, Bakker S, Maat K, van Wee B (2012) Technological diversity of emerging eco-innovations: a case study of the automobile industry. J Clean Prod 37:211–220

Silva DAL, Delai I, de Castro MAS, Ometto RO (2013) Quality tools applied to cleaner production program: a first approach towards a new methodology. J Clean Prod 47(4):174–187. doi:10.1016/j.jclepro.2012.10.026

Stevens RJL, Stevenson RJ (2012) Harnessing environmental sound technology for Chinese SMEs' environmental sustainable development. In: Proceedings of the 9th international conference on innovation and management. Eindhoven, The Netherlands, pp 315–320

Taguchi G (1986) Introduction to quality engineering: designing quality into products and processes. Asian Productivity Organization, Tokyo

Tischner U, Schmincke E, Rubik F, Prösler M (2000) How to do EcoDesign? A guide for environmentally and economically sound design. German Federal Environmental Agency, Berlin

Tsai MT, Chuang LM, Chao ST, Chang HP (2012) The effects assessment of firm environmental strategy and customer environmental conscious on green product development. Environ Monit Assess 184(7):4435–4447

Veugelers R (2012) Which policy instruments to induce clean innovating? Res Policy 41(10):1770–1778

Wilkinson A, Hill M, Gollan P (2001) The sustainability debate. Int J Oper Prod Manage 21(142):1492–1502

Yang Y, Holgaard JE, Remmen A (2012) What can triple helix frameworks offer to the analysis of eco-innovation dynamics? Theoretical and methodological considerations. Sci Public Policy 39(3):373–385

Chapter 5
Radical and Systematic Eco-innovation with TRIZ Methodology

Helena V.G. Navas

Abstract The main objective of this chapter is to support an implementation of systematic eco-innovation and radical eco-innovation with analytical tools and techniques of the Theory of Inventive Problem Solving (TRIZ) methodology. It also aims to increase opportunities for eco-innovative products and services while simultaneously enhancing the innovation capacity of organizations.

By applying the TRIZ techniques to the eco-innovation approach, competitiveness and innovation of a firm can be increased. Thereby, the development of clean production processes, waste recycling, and thus "environmentally friendly" products and services is supported, allowing enterprises to "green" their business, product, and management methods.

A survey was conducted regarding application opportunities for the most important pillars of the TRIZ methodology in an eco-innovation environment.

Some opportunities for TRIZ were analyzed in the domain of systematic eco-innovation, since TRIZ is seen as a scientific basis of systematic innovation.

It is difficult to find analytical tools that can truly provide support to radical innovation. TRIZ has successfully supported these activities; therefore, it was also proposed to extend this support to the activities of radical eco-innovation.

The study included the analysis of opportunities to use the most important TRIZ elements and techniques, namely, the levels of innovation, the contradictions, the analysis of resources and ideality, the scientific effects and databases, the inventive principles, and the contradiction matrix in environments of systematic and radical eco-innovation.

Keywords Eco-innovation • Systematic eco-innovation • Radical eco-innovation • Problem solving • TRIZ

H.V.G. Navas (✉)
UNIDEMI, Departamento de Engenharia Mecânica e Industrial, Faculdade de Ciências e Tecnologia, Universidade Nova de Lisboa, Quinta da Torre, 2829-516 Caparica, Portugal
e-mail: hvgn@fct.unl.pt

S. Azevedo et al. (eds.), *Eco-Innovation and the Development of Business Models*, Greening of Industry Networks Studies 2, DOI 10.1007/978-3-319-05077-5_5,

5.1 Introduction

The growing internationalization of enterprises requires the introduction of new technological innovation management practices with strong socio-environmental responsibility. All industrial programs must combine technological innovation with good practices, always based on environmental vision. The environmental impacts can have a strong influence on management practices in general and on innovation issues as well.

Eco-innovation implies innovation in products, processes, or business models that enables the company to achieve higher levels of environmental sustainability. Eco-innovation does not only demand the identification, implementation, and monitoring of new ideas aimed at improving the environmental performance of the organization but must also have an impact on the overall level of competitiveness.

Enterprises need to invest in systematic eco-innovation if they plan to win or at least survive. Innovation can no longer be seen as the product of occasional inspiration. It has to be transformed into a capacity, not a gift. Eco-innovation has to be learned and managed. Unexpected occurrences, inconsistencies, process requirements, changes in the market and industry, demographic change, and changes in perception or new knowledge can give rise to eco-innovation opportunities.

Systematic eco-innovation can be understood as a concept that includes the instruments necessary to develop the right inventions needed at that point in time and incorporate them into new products and processes.

Incremental eco-innovation is not always sufficient to prevent environmental impact of economic activities; thus, more radical eco-innovation initiatives are needed. Radical eco-innovation presupposes a profound shift in the use of resources. More radical forms of eco-innovation can cause a sustainable transition to be difficult to deploy (Hellström 2007).

Radical solutions are very important, especially considering the long-term gains. Traditional engineering and management practices can become insufficient and inefficient for the implementation of new scientific principles or for vast improvements of existing systems.

Organizations need innovation in the right dose and at the right moment. They need methodologies and analytical tools that can help implement radical changes and completely new techniques. The Theory of Inventive Problem Solving (TRIZ), brainstorming, collateral thinking, mind maps, and other methodologies can stimulate individual and collective creativity.

Also needed in order to become more competitive are new management paradigms. The environmental sustainability is a very pertinent issue in industrial management. Eco-initiatives allied to innovation and innovative technologies can improve the sustainability of organizations through low emissions to the nature and recycling strategies for products.

Some authors dedicate their studies to the evaluation of compatibility between TRIZ techniques and eco-innovation with, for example, the use of a TRIZ contradiction matrix with the eco-compass principle (Jones 2003), a collection of eco-innovation examples for all 40 TRIZ inventive principles (Chen 2003), or the elaboration of an eco-design guideline using the TRIZ Law of Evolution (Russo and Regazzoni 2008).

This chapter focuses on the integration of the major TRIZ analytical tools with a radical and systematic eco-innovation. It is a different approach from previous literature, which focuses on specificities of systematic and radical eco-innovation.

5.2 Theory of Inventive Problem Solving (TRIZ)

The Theory of Inventive Problem Solving, better known by its acronym (TRIZ), was developed by Genrich Altshuller in 1946 (Altshuller 1995). TRIZ is a theory that can assist any engineer in the inventing process.

The TRIZ methodology can be seen and used on several levels. At the highest level, the TRIZ can be seen as a science, as a philosophy, or as a way to be in life (a creative mode and a permanent search of continuous improvement). In more terms, the TRIZ can be seen as a set of analytical tools that assist both in the detection of contradictions on systems and in formulating and solving design problems through the elimination or mitigation of contradictions (Savransky 2000).

The TRIZ methodology is based on the following grounds:

- Technical systems
- Levels of innovation
- Law of ideality
- Contradictions

Every system that performs a technical function is a technical system. Any technical system can contain one or more subsystems. The hierarchy of technical systems can be complex with many interactions. When a technical system produces harmful or inadequate effects, the system needs to be improved. Technical systems emerge, ripen to maturity, and die (they are replaced with new technical systems). TRIZ systematizes solutions that can be used for different technical fields and activities.

In TRIZ, the problems are divided into local and global problems (Altshuller 1995). The problem is considered to be local when it can be mitigated or eliminated by modifying a subsystem, keeping the remaining unchanged. The problem is classified as global when it can be solved only by the development of a new system based on a different principle of operation.

Over the past decades, TRIZ has developed into a set of different practical tools that can be used collectively or individually for technical problem solving and failure analysis.

Generally, the TRIZ process is to define a specific problem, formalize it, identify the contradictions, find examples of how others have solved the contradiction or utilized the principles, and, finally, apply those general solutions to the particular problem.

Figure 5.1 shows the steps of the TRIZ process.

Fig. 5.1 Steps of the TRIZ's algorithm for problem solving (Fey and Rivin 1997)

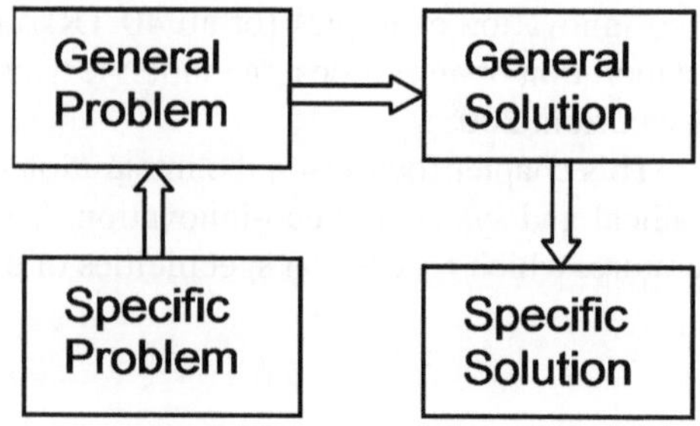

It is important to identify and to understand the contradiction that is causing the problem as soon as possible. TRIZ can help to identify contradictions and to formalize problems to be solved. The identification and the formalizing of problems is one of the most important and difficult tasks, with numerous impediments. The situation is often obscured.

The problem can be generalized by selecting one of the TRIZ tools. The generic solutions available within TRIZ can be of great benefit when choosing corrective actions.

The integral development of TRIZ consists of a set of concepts (Radeka 2007):

- Problem formulation system
- Physical and technical contradictions solving
- Concept of the ideal state of a design
- Analysis "substance field"
- Algorithm of Inventive Problem Solving (ARIZ)

Altshuller built a contradiction matrix, classifying the contradictions as follows (Altshuller 1995):

- Physical contradiction, which occurs when two mutually incompatible requirements refer to the same element of the system
- Technical contradiction, which occurs when the improvement of a particular attribute or characteristic of the system causes the deterioration of another attribute

The first step in the conflict solving process is drawing up a statement of the problem in order to reveal the contradictions contained in the system. Then, the parameters that affect and improve system performance are identified.

5.3 TRIZ Levels of Innovation for the Evaluation of Systematic and Radical Eco-innovation Initiatives

There are different methods to assess the levels of eco-innovation. Different methods take into account different aspects of eco-innovation initiatives. However, eco-innovation requires analysis on different levels; therefore, the evaluation systems must be organized accordingly. There is a need for approaches to capture complexities of a multilevel eco-innovation analysis.

The measurement of innovation is necessary; it contributes to the establishment of longer-term policies for eco-innovation.

Altshuller's analysis of a large number of patents reveals that the inventive value of different inventions is not equal. Altshuller systematized the solutions described in patent applications by dividing them into five levels (Altshuller 2001):

- Level 1: routine solutions using methods well known in their area of specialty. Level 1 is not highly innovative. This category constitutes about 30 % of the total.
- Level 2: small corrections in existing systems using methods known in the industry. This level makes up approximately 45 % of the total.
- Level 3: major improvements that solve contradictions in typical systems of a particular branch of industry. About 20 % of the total are on this level, which is where creative design solutions appear.
- Level 4: solutions based on the application of new scientific principles. They solve the problem by replacing the original technology with new technology. Nearly 4 % of the total are classified to be on this level.
- Level 5: innovative solutions based on scientific discoveries not previously explored. This level consists of less than 1 % of the total.

The five-level TRIZ classification can be used for analysis of eco-innovation initiatives. The same problem can be solved by solutions with different levels of innovation. The five-level classification can be applied for the evaluation of innovation level of different eco-innovation solutions, and it can be used during the development process of new proposals.

The development of a new solution can follow different procedures:

- Conventional improvement of existing system (levels 1 and 2)
- New forms but with existing principles of operation (levels 2 and 3)
- Creation of new system generation with new principles of operation (levels 4 and 5)

The creative solutions classified in levels 4 and 5 (and especially the solutions at level 5) can be considered radical.

For example, the need to reduce fuel consumption in automobiles has led manufacturers of engines and vehicles to introduce several new features and modifications. They do not possess the same level of innovation. The level of creativity can be determined according to the following scale that is adapting the TRIZ five levels of innovation:

Level 1 (compromising design) – travel with the windows rolled up because the aerodynamic performance improves and fuel consumption is lower.

Level 2 (resolving the technical contradiction) – install a speed limitation device, since the increase in speed increases the fuel consumption.

Level 3 (resolving the physical contradiction) – reconsider the design of the gearbox, as a good steady speed can help spend less fuel.

Level 4 (new technology) – change from traditional gasoline to LPG.
Level 5 (new phenomena) – rather than traditional internal combustion engines, switch to using electric or hybrid vehicles.

The five levels of innovation also can be useful for the prognosis of evolution of a specific system (Kraev 2006).

One could observe, for example, the evolution of diapers for babies. Earlier diapers were made of cloth and later evolved to disposable diapers. The evolution of disposable diapers is related to several criteria such as being healthier, more comfortable, cleaner, and more economical. In the future, the diapers must continue to evolve in order to be made of breathable, chemical-free, and biodegradable materials as well as being soft for skin contact and super absorbent.

5.4 Technical and Physical Contradictions and Eco-innovation

Technical and physical contradictions constitute one of the most important terms of the TRIZ vocabulary.

The TRIZ Axiom of Evolution reveals that, during the evolution of a technical system, improvement of any part of that system can lead to conflict with another part.

A system conflict or contradiction occurs when the improvement of certain attributes results in the deterioration of others. The typical conflicts are reliability/complexity, productivity/precision, strength/ductility, etc. A technical problem is defined by contradictions.

The traditional way of contradiction solving is through the search of possible compromise between contradicting factors, whereas TRIZ aims to remove contradictions and compromises. The inconsistencies are eliminated by modification of the entire system or by modification of one or more subsystems.

A traditional approach is based on trade-offs through preferable combinations of characteristics in conflict. TRIZ aspires to solve the contradictions by modification of systems to avoid deterioration of any characteristic in case of improvement of other characteristics.

Altshuller distinguished three types of contradictions (Altshuller 1986):

1. Administrative contradiction – contradiction between the needs and abilities
2. Technical contradiction – the classical engineering "trade-off," an inverse dependence between parameters/characteristics of a machine or technology
3. Physical or inherent contradiction – opposite/contradictory physical requirements of an object

An identification and analysis of contradictions should be included in any process of TRIZ inventive solving of problems. When a contradiction is identified, it becomes easier to find creative and effective solutions for the problem. A contradiction does not solve a problem, but it gives direction for a solution.

We can now use two ways of problem solution. One way is to resolve all contradictions by applying one of the appropriate TRIZ analytical tools for solving technical contradictions (e.g., 40 inventive principles, contradiction matrix, or others). The other way is to transform the technical contradictions into physical contradictions resolving all contradictions at the physical level (e.g., with databases of physical phenomena and effects).

All technical contradictions can be transformed into a corresponding physical contradiction.

Any physical contradiction can be resolved using one of the four principles (principles of physical contradiction solving):

- Separation of contradictory characteristics in time
- Separation of contradictory properties in space
- System transformation
- Phase transformation (physical and/or chemical transformation of substances)

Eco-innovative initiatives can contain some technical contradictions, such as the following:

- The goal of building vehicles that use less fuel conflicts with the engine power (lower horsepower).
- The goal of building vehicles that use less fuel conflicts with acceleration.
- The usage of electric vehicles as an alternative to internal combustion engines, but the distances between recharging are much smaller in electric vehicles.
- Heavier batteries needed as the distance range of the electric vehicle increases.

5.5 TRIZ Resource Analysis in an Eco-innovation Environment

The identification of available resources around a problem and the maximization of their use are important for finding cost-effective and environmentally friendly solutions. TRIZ demands that the analysis of resources take into account the negative as well as the positive resources in a system (Mann 2000).

The improvements must continue until the resources are fully utilized.

Resources can be grouped according to the following (Savransky 2000):

1. Natural or environmental resources
2. System resources
3. Functional resources
4. Substance resources
5. Energy and field resources
6. Time resources
7. Space resources
8. Information resources

Altshuller also grouped resources in the following categories:

1. Based on accessibility
   (a) Internal (limited to the main elements of the system)
   (b) External, including resources from the general environment and those which are specific for the given system
   (c) Resources from the supersystem or other accessible, inexpensive resources (including waste)
2. Based on readiness for utilization
   (a) Readily available resources
   (b) Derived (modified readily available resources)

The key for sustainability is resources productivity.

TRIZ analysis of resources can be useful in eco-innovation initiatives related to more efficient and responsible usage of resources, including energy use.

The gains in resource efficiency generally result from process improvements. Thus, sporadic initiatives of eco-innovation must evolve to the continuously planned and scheduled activities; the eco-innovation must become systematic.

The traditional incremental improvement of existing technologies is no longer sufficient. All economic activities need to radically increase the efficiency of resource utilization.

The application of the TRIZ analytical tools and techniques can be especially useful for the radical eco-innovation both in the generation of innovative and revolutionary solutions as well as in the resource analysis and forecast.

The resources can be used to solve problems according to the following workflow:

1. Formulate the problem.
2. Build the list of resources in the following order: internal, external, by-products, and complex resources.
3. Define what kind of resources is needed to solve the problem.
4. Estimate each of the existing resources and the effects of its use.
5. Propose how to use the resource.

For example, consider the following problem: it is necessary to increase the efficiency of a fireplace without any interior reconstruction (Kraev 2007). The list of resources is as follows: firewall brick, fire, hot-air conductivity system, fuel, convection air flow, ambient chilly air, atmospheric pressure, gravitational field, and geomagnetic field. It is necessary to increase the heat transport capacity.

The convection air current linked to the system and convection heat transfer are both useful, free resources that we can try to use.

Solution: It is known that the fireplace heats the air in the room by convection heat transfer. In fact, the air in the room will be heated considerably by placing a metal sheet with small air gaps on the front edges of the fireplace. In this way the air will heat much more quickly in the narrow space between the fireplace and the metal sheet. Hot air will come out of the top, while cold air will come out from the bottom of the metal sheet (Kraev 2007).

5.6 System Ideality for Radical and Systematic Eco-innovation

The law of ideality states that any technical system tends to reduce costs, energy wastes, and space and dimensional requirements as well as become more effective, more reliable, and simpler. Any technical system, during its lifetime, tends to become more ideal.

We can evaluate an inventive level of a technical system by its degree of ideality.

The ideality can be calculated as the ratio of a system's useful functions to its harmful functions with the formula:

$$\text{Ideality} = \text{useful functions} / \left(\text{harmful functions} + \text{cost}\right) \tag{5.1}$$

Useful functions include the following:

- Primary useful functions – the purpose for which the system was designed
- Secondary functions – other useful outputs
- Auxiliary functions – functions that support the primary useful functions, such as corrective functions, control functions, housing functions, transport functions, etc.

Harmful functions include all harmful factors associated to the system (e.g., costs, area that it occupies, emission of noises, expended energy, resources needed for system maintenance, etc.).

The level of ideality increases with the increase of useful functions and reduces with the increase of harmful functions.

There are three ways to increase an ideality of a technical system:

- Increasing the useful functions
- Reducing any harmful or expensive function
- A combination of the first two paths

According to the TRIZ methodology, an ideal system does not exist. An absolute ideality is impossible to achieve, but other relative levels of ideality are achievable. A real system can approximate to the ideal system by increasing the useful functions and eliminating the harmful functions through contradiction solving, more efficient use of resources, and the reduction of system complexity and number of components.

The ideality can be used both to improve existing systems and also for the creation of new technologies or new systems to fulfill specific functions.

The concept of increasing the degree of ideality is crucial for predicting the evolution of the system.

There are several concepts derived from the same concept of ideality, such as ideal final result, ideal final goal, ideal solution, ideal product, ideal process, etc.

Ideal final result (IFR) is the ultimate idealistic solution of a problem when the desired result is achieved by itself.

Systematic eco-innovation can be supported by TRIZ ideality, whereas the planned and continuous improvement can be made in terms of increasing the level of ideality of a given system.

Table 5.1 Ideality matrix for the camping stove (Navas 2013)

Parameter	1.	2.	3.	4.	5.	6.	7.	8.
1. Volume		+			–	–	–	
2. Weight	+				–	–	–	
3. Firing time					+			
4. Noise level								
5. Time required to boil water	–	–	+				–	+
6. Tank capacity	–	–	+		+		+	+
7. Burning time at maximum flame	–	–			–	–		–
8. Boiled water per unit of gas	–	–			+	+	–	

– Harmful iteration

+ Useful iteration

The initiatives of a radical eco-innovation can be seen as actions aimed at dramatically increasing the level of ideality. The ideal final result can be very useful, especially for a radical eco-innovation.

For example, consider the case study focused on increasing the energy efficiency of a camping stove.

The customer requirements were collected, pooled, and prepared by an affinity diagram, yielding the following list:

- Volume
- Weight
- Firing time
- Noise level
- Time required to boil water
- Tank capacity
- Burning time at maximum flame
- Boiled water per unit of gas

Table 5.1 contains the matrix of ideality built for the camping stove.

The ideality matrix helps identify the interactions between the technical requirements and distinguish the positive and negative effects of iterations. For example, weight reduction can lead to reduction in volume but may lead to reduction of the tank capacity.

The ideality matrix (Table 5.1) contains 11 positive (useful) and 19 negative (harmful) interactions. In this case, the level of ideality is

$$I = 11/19 \approx 0.579$$

Based on this analysis, measures can then be established to increase the level of ideality by increasing the useful functions, reducing any harmful or expensive function, or by combination of the first two paths.

5.7 Scientific Effects and the Application of Databases for Eco-innovation Problem Solving

TRIZ includes a database of scientific effects structured according to technological functions. According to TRIZ, the scientific effects are one of the principles for contradiction solving by transformation of an action or field to another with the application of physical, chemical, biological, geometric, or another phenomena.

Currently, there are over 8,000 known effects and different phenomena, and 400–500 of them are most applicable in the practice of engineering activity.

Special tables and descriptions of scientific phenomena exist, which give us the opportunity to define the required effect of an output action or function that should be performed according to the identified problem. Also, there are special software with databases of scientific and engineering phenomena. These programs allow effective selection based on the desired function. Some software provides access to more than 4,500 engineering and scientific effects, theorems, laws, and phenomena.

The use of scientific effects and phenomena helps to develop solutions at the highest level of innovation. Scientific effects can be used to solve problems outside the field where the original problem was found.

The databases of scientific effects can help all initiatives of eco-innovation, especially radical eco-innovation dispelling the fear of using new techniques but also avoiding repetition of wrong solutions.

For example, consider a study conducted in a plant that traditionally used toxic organic solvents. The main objective of the study was the elimination or drastic reduction of use of toxic solvents. The query databases helped tremendously in terms of both filtering solutions deemed to be inappropriate or potentially dangerous and also providing many ideas and suggestions for possible solutions. Solutions with lower levels of innovation (the toxic solvents replaced by less toxic solvents) were considered, and ideas for a radical change in the whole technological process of the company (total removing of the toxic solvents from the technological process) were analyzed.

5.8 Application of the Inventive Principles and Matrix of Contradictions

Altshuller (2001) found that, despite the great technological diversity, there are only 1,250 typical system conflicts. He also identified 39 engineering parameters or product attributes that engineers usually try to improve.

Table 5.2 presents the list of these parameters.

All 1,250 conflicts can be solved through the application of only 40 principles of invention (Altshuller 2001), often called Techniques for Overcoming System Conflicts, which are shown in Table 5.3.

Table 5.2 Engineering parameters according to TRIZ (Altshuller 2001)

No	Engineering parameter
1	Weight of moving object
2	Weight of nonmoving object
3	Length of moving object
4	Length of nonmoving object
5	Area of moving object
6	Area of nonmoving object
7	Volume of moving object
8	Volume of nonmoving object
9	Speed
10	Force
11	Tension, pressure
12	Shape
13	Stability of object
14	Strength
15	Durability of moving object
16	Durability of nonmoving object
17	Temperature
18	Brightness
19	Energy spent by moving object
20	Energy spent by nonmoving object
21	Power
22	Waste of energy
23	Waste of substance
24	Loss of information
25	Waste of time
26	Amount of substance
27	Reliability
28	Accuracy of measurement
29	Accuracy of manufacturing
30	Harmful factors acting on object
31	Harmful side effects
32	Manufacturability
33	Convenience of use
34	Repairability
35	Adaptability
36	Complexity of device
37	Complexity of control
38	Level of automation
39	Productivity

However, most of the principles of invention in Table 5.3 have a specific technical meaning introduced by Altshuller. For example, the principle of local quality involves (Terninko et al. 1998):

- Transitioning from a homogeneous structure of an object or outside environment/action to a heterogeneous structure

Table 5.3 Inventive principles of TRIZ

No	Inventive principle
1	Segmentation
2	Extraction
3	Local quality
4	Asymmetry
5	Combining
6	Universality
7	Nesting
8	Counterweight
9	Prior counteraction
10	Prior action
11	Cushion in advance
12	Equipotentiality
13	Inversion
14	Spheroidality
15	Dynamicity
16	Partial or overdone action
17	Moving to a new dimension
18	Mechanical vibration
19	Periodic action
20	Continuity of a useful action
21	Rushing through
22	Convert harm into benefit
23	Feedback
24	Mediator
25	Self-service
26	Copying
27	Exchange of expensive durable object by inexpensive short-lived object
28	Replacement of a mechanical system
29	Pneumatic or hydraulic construction
30	Flexible membranes or thin film
31	Use of porous material
32	Changing the color
33	Homogeneity
34	Rejecting and regenerating parts
35	Transformation of the physical and chemical states of an object
36	Phase transformation
37	Thermal expansion
38	Use of strong oxidizers
39	Inert environment
40	Composite materials

- Having different parts of the object carry out different functions
- Placing each part of the object under conditions most favorable for its operation

The inventive principles are simple analytical tools to solve technical contradictions and finally resolve the problem.

Table 5.4 Fragment of Altshuller's matrix of contradictions

Characteristic		Worsening characteristic	
		9 speed	21 power
Characteristic to be improved	19 energy spent by a moving object	8, 35	6, 19 37, 18

In practical activities, various methods are used to apply the 40 principles in the process of problem solving. The simplest method is to examine each contradiction and try to apply the principles of each of them or their combinations to solve the contradiction technique on the specific problem. Another method is the development of a technical contradiction and use of the contradiction matrix in order to determine the set of recommended principles to solve the problem (usually between two and four principles).

Altshuller built a contradiction matrix. The rows of the table of contradictions are populated with parameters whose adjustment improves the behavior of the system, and these intersect with the columns with parameters whose adjustment produces unwanted results. At the intersection are the numbers of invention principles that are suggested as being capable of solving the contradiction.

In the contradiction matrix, the rows and columns refer to Table 5.2. The numbers in the cells refer to Table 5.3.

For example, if the goal of a study is the reduction of fuel consumption of a vehicle, the modification can have a negative effect on the acceleration, speed, and power.

Table 5.4 contains the extract of the matrix of contradictions built into the case.

The conflict between the characteristics 19 and 9 can be solved by applying the principles of the invention 8 (counterweight) and 35 (transformation of the physical and chemical states of an object). The conflict between the characteristics 19 and 21 can be solved using the principles 6 (universality), 19 (periodic action), 37 (thermal expansion), and 18 (mechanical vibration).

5.9 Conclusions

The constant need for change is a current issue in industrial activities. Although organizations place priority on meeting their own objectives, ecological aspects must also be considered. Organizations need powerful and highly efficient analytical tools related to innovation and creativity. One of the most important factors for the success of industrial activities is the generation of ideas and innovation. The lack of creativity can lead to the failure of objectives. Creativity is crucial for competitiveness.

Some authors have tried to find common ground among some of the TRIZ techniques and eco-innovation. The novelty of this study is to survey the applicability of the key pillars of TRIZ to eco-innovation, especially systematic eco-innovation and radical eco-innovation.

The TRIZ methodology, with its strong theme of radical and systematic innovation, can contribute to accelerating the resolution of problems in the eco-innovation activities. The TRIZ analytical tools would be very useful for schematization of eco-innovation tasks, system analysis, identification, and formalization of contradictions and problematical situations and their solving processes.

TRIZ is a methodology especially suited for supporting the development of radical solutions where traditional techniques usually do not yield positive results.

Moreover, TRIZ is considered the scientific basis of systematic innovation. Thus, the extension of TRIZ for systematic eco-innovation and radical eco-innovation seems to be a logical step of evolution.

Acknowledgments The present author would like to thank the Faculty of Science and Technology of the New University of Lisbon (UNL) and the Portuguese Foundation for Science and Technology (FCT) through the Strategic Project no. PEst-OE/EME/UI0667/2011. Their support helps make our research work possible.

References

Altshuller GS (1986) To find an idea. Nauka, Novosibirsk (in Russian)

Altshuller GS (1995) Creativity as an exact science: the theory of the solution of inventive problems. Gordon and Breach Publishers, New York

Altshuller GS (2001) 40 principles: TRIZ keys to technical innovation. Technical Innovation Center, Worcester

Chen JL (2003) Eco-innovative examples for 40 TRIZ inventive principles. The TRIZ Journal, August

Fey VR, Rivin EI (1997) The science of innovation: a managerial overview of the TRIZ methodology. TRIZ Group, Southfield

Hellström T (2007) Dimensions of environmentally sustainable innovation: the structure of eco-innovation concepts. Sust Dev. Wiley InterScience 15:148–159

Jones E (2003) Eco-innovation: tools to facilitate early-stage workshop. Ph.D. thesis, Brunel University

Kraev V (2006) Kraev's Korner: levels of innovations. The TRIZ Journal, November

Kraev V (2007) Kraev's Korner: resource analysis. The TRIZ Journal, January

Mann DL (2000) The four pillars of TRIZ, invited paper at engineering design conference, Brunel, June

Navas HVG (2013) TRIZ: design problem solving with systematic innovation. In: Denis AC (ed) Advances in industrial design engineering. Rijeka, Croatia, InTech, pp 75–97. doi:10.5772/55979

Radeka K (2007) TRIZ for lean innovation: increase your ability to leverage innovation across the enterprise and beyond. Whittier Consulting Group, Inc. Camas, WA

Russo D, Regazzoni D (2008) TRIZ law of evolution as eco-innovative method. In: Proceedings of IDMME – virtual concept 2008, Beijing, China, 8–10 October

Savransky SD (2000) Engineering of creativity: introduction to TRIZ methodology of inventive problem solving. CRC Press, Boca Raton

Terninko J, Zusman A, Zlotin B (1998) Systematic innovation: an introduction to TRIZ (theory of inventing problem solving). St. Lucie Press, Boca Raton

Part II
Application: Surveys and Case Studies on Eco-innovation Deployment

Chapter 6
Eco-innovation on Manufacturing Industry: The Role of Sustainability on Innovation Processes

Patrícia A.A. da Silva, João C.O. Matias, Susana Garrido Azevedo, and Paulo N.B. Reis

Abstract New products and new services appear when the companies introduce eco-innovations, as well as new markets, "greener" products, and "cleaner" processes. In this context, it is recognized that the deployment of eco-innovation's processes by companies represents a decisive factor for the competitiveness of companies all over the world. Therefore, the present study aims to analyze the influence of the introduction of eco-innovation practices by the Portuguese manufacturing companies on their propensity to innovation. Simultaneously, the contribution of internal and external factors for the eco-innovation behavior of the Portuguese manufacturing industry is analyzed. For this purpose, a quantitative research is developed using a survey based on the most innovative companies. It is possible to conclude that the introduction of both environmental benefits inward the companies and the after-sales has influenced on the innovative capacity of the Portuguese manufacturing industry. Furthermore, some external and internal factors stimulate the introduction of eco-innovations.

Keywords Eco-innovation • Manufacturing sector • Community Innovation Survey

---

P.A.A. da Silva • J.C.O. Matias (✉) • P.N.B. Reis
Faculty of Engineering, Department of Electromechanical Engineering, University of Beira Interior, Covilhã, Portugal
e-mail: patriciaalipiosilva@gmail.com; matias@ubi.pt; preis@ubi.pt

S. Garrido Azevedo
Department of Management and Economics, University of Beira Interior, Covilhã, Portugal
e-mail: sazevedo@ubi.pt

S. Azevedo et al. (eds.), *Eco-Innovation and the Development of Business Models*, Greening of Industry Networks Studies 2, DOI 10.1007/978-3-319-05077-5_6,

6.1 Introduction

Given the growing human need to acquire new products, the economic development requires not only resources, but it also contributes to the unrestrained production of waste responsible for the planet degradation. In this context, sustainable development has been assumed as a guideline for the integration of economic, environmental, and social policies by the companies influencing their production processes and also their competitiveness. Thus, innovation processes toward sustainable development (eco-innovations) have received increasing attention during the past years. Radical eco-innovations are key enablers for improving a knowledge-based, resource-efficient, greener, and competitive European economy (Montalvo et al. 2011). However, the implementation of radical eco-innovations involves several factors and is affected by various barriers (Del Río et al. 2010).

This study provides the identification and empirical analyses of the main factors that influence the Portuguese manufacturing industry in terms of eco-innovation. The results present the environmental benefits recognized by the companies and also their correlation with the introduced innovation. In this context, according to the literature review, several research hypotheses are formulated in order to test and analyze the influence of some factors on the innovative activities and performance of Portuguese manufacturing industries. To empirically test the formulated hypothesis, secondary data was provided by the "Office of Planning, Strategy, Evaluation and International Relations of Ministry of Science, Technology, and Higher Education (GPEARI/MCTES)" – belonging to the "CIS (Portuguese Community Innovation Survey)" – and supervised by Eurostat. The CIS is a database where we can have access to the companies' innovation process in Portugal, in terms of product, process, and organizational innovation and marketing, as well as their determinants and weaknesses/capabilities (GPEARI 2010a, b).

This study intends to analyze the influence of the introduction of eco-innovation practices by manufacturing companies on their propensity to be innovative.

The chapter is structured as follows: after the introduction, the relevant literature on eco-innovation is present. After, the methodology used in this study is described and the results present and discussed. Finally, the main conclusions of the research are drawn.

6.2 Background

In the last decades, the growing of the economic activity has been accompanied by concerns about climate change, energy security, and scarcity of natural resources. On the other hand, environmental problems such as air pollution and decreasing of fossil oil resources are becoming pressing issues for international organizations, national governments, and consumers (Jansson 2011).

Recently, the OECD (2009) report suggests that the global greenhouse gas emissions are likely to increase by 70 % by 2050. This has contributed for the increased importance of the sustainable development concept, which gradually has emerged as a new development paradigm. Besides that initially the sustainable development was considered as a constraint, it has become an opportunity, as it combines three advantages: a response to climate and energy challenges, a way out of a crisis, and a potential growth path (Faucheux and Nicolaï 2011). Thus, several strategies have been suggested for building more sustainable companies. These strategies attempted to create long-term added value for consumers and employee by not only creating a "green" strategy but also taking into consideration every dimension of how a business operates in the social, cultural, and economic environment (Pratoom and Cheangphaisarn 2011).

In order to re-evaluate the relationships between organizations, technology, and society, many authors argue that the managers should have a strategic planning for the future considering however the environmental concerns (Ulhøi 2008; Stead and Stead 2008; Shrivastava 2008; Parnell 2008). Thus, innovation processes through sustainable development have received increasing attention for the past years. According to some authors (Bresciani and Oliveira 2007; Ulhøi 2008; Shrivastava 2008; Yang et al. 2010), when an environmental management policy is adopted, the companies increase their competitiveness through cost reduction, quality improvement, and implementation of new processes/products.

Moreover, derived from the increased globalization of the economic activity, the gradual integration of markets, and the evolution of consumer needs, there are new challenges to companies which increasingly strive them to innovate (Moreira et al. 2012). The increased efficiency in resources and energy use and the investment in a broad range of innovations to improve environmental performance could be an important driver for creating new industries and jobs (OECD 2009). In fact, since the 1980s, the increasing instability of the competitive environment, with shorter product and technological life cycles, has forced companies to reconsider their innovation strategies in order to widen their technology base (Faria et al. 2010). Taking the regulations to create new markets, as an example, they seem to be particularly relevant for eco-innovation in business contexts (Wagner and Llerena 2011). On the other hand, the industries are not just waiting for the governments to decree regulations, but a growing number of companies have adopted strategies that allow them to be ahead of the legislators. They take the initiative in designing their environmental and sustainability policies (Lindhqvist 2007). According to OECD (2009), although government regulations and standards have helped to reduce environmental impacts to a large extent, they are not the most efficient way to reduce emissions and do not offer enough incentives to innovate beyond end-of-pipe solutions. Therefore, sustainability arises as a process that brings social, economic, environmental, and technological changes through innovation, and it is increasingly recognized as a significant driver of economic growth (OECD 2009).

In this context arises the concept of eco-innovation as an innovation (Faucheux and Nicolaï 2011) that reduces environmental burdens and contributes to improve a situation according to given sustainable targets (Rennings 2000; Faucheux and Nicolaï 2011). For EIO (2010), the eco-innovation consists of the introduction of new or significantly improved products (good or service), processes, organizational changes, or marketing solutions that reduce the use of natural resources (including materials, energy, water, and land) and decrease the harmful substances across the whole life cycle. For Demirel and Kesidou (2011), eco-innovations have a central role in improving environmental technologies that measure, detect, and treat pollution from the source to the product's life cycle end, ensuring its life span has a minimal environmental impact. According to other authors, it is a process of developing new ideas, behavior, products, and processes that contribute to a reduction in environmental burdens or to ecologically specified sustainability targets (Rennings 2000; Hellström 2007; Del Río et al. 2010; Jansson et al. 2011; Wagner and Llerena 2011; Demirel and Kesidou 2011). Based on this definition, eco-innovation has a double externality feature (Wagner and Llerena 2011). However, some authors have referred specifically to eco-innovation as simply reducing environmental impacts through waste minimization (Hellström 2007).

The eco-innovation contributes to create eco-companies and might be defined as a subclass innovation, which improves the economic and environmental development (OECD 2009). Recently, the concept has been reformulated (Lobo 2010) to activities that involve equipment use, work, production, information networks, or products whose main goal is to retract, treat, reduce, prevent, or delete pollution or another environmental degradation caused by human activity.

As pointed out by Nuij (2001), an eco-innovation is the response from industry and the academic community to the challenge of sustainable product development that aims to develop new products and services that are not based on redesign or incremental changes to existing products but rather on providing the consumer with the function they require, in the most eco-efficient way. The incremental improvement is not enough, and industry must be restructured (OECD 2009) as well as technological products and systems (Hellström 2007). In this context, industry leaders and policymakers have looked at innovation as the key to make radical improvements in corporate environmental practices and performance. Radical innovations are those changes that lead to substantial improvements of products and processes that, however, do not necessarily lead to a systemic change (EIO 2010). On the other hand, researchers, as well as eco-innovation literature, have often expressed a need for radical and systemic technological changes to achieve demanding environmental sustainability goals (Del Río et al. 2010; OECD 2009).

According to Rennings (2000), incremental innovations can be characterized as continuous improvements of existing technological systems, while radical innovations are discontinuous. Radical eco-innovations have more substantial changes at the production system level, such as those involved in industrial ecology, including closed-loop systems in which waste becomes inputs for new processes (Hellström 2007; Del Río et al. 2010). Radical eco-innovations are also considered as key enablers for securing a knowledge-based, resource-efficient, greener, and competitive European economy (Montalvo et al. 2011). In other words, radical eco-innovation benefits companies by leading them to new products and services, as well as creating new markets, such as eco-products, for example, certified products. In addiction, the increasing scarcity of certain resources and the high prices of raw materials and energy may drive the need for more efficient processes and technologies also opening new markets (EIO 2010). However, according to Hellström (2007), most eco-innovations take place in the incremental mode so far.

Despite the different concepts and views about radical eco-innovations, OECD (2009) defends that eco-innovation as a whole contributes to the evolution in industry practices and industry has increasingly adopted cleaner production by reducing the amount of energy and materials used in the production process. They are considering the environmental impact throughout the product's life cycle and are integrating environmental strategies and practices into their own management systems. However, Jansson et al. (2011) say that relatively little attention has been paid to the adoption of innovations that are marketed as being more environmentally friendly than the alternatives on the market. According to Hellström (2007), most companies are hard put to integrate environmental concerns into their corporate strategy, and this is an impediment to the development of radical eco-innovation. On the other hand, the results from a survey of ten OECD governments on existing national strategies and policy initiatives show that an increasing number of countries now perceive environmental challenges not as a barrier to economic growth but as a new opportunity. Despite these facts, not all surveyed countries seem to have a specific strategy for eco-innovation (OECD 2009) which makes pertinent to explore the possible dimensions and ways for the introduction of eco-innovation, namely, in Portuguese manufacturing industry context. In this context, the following two research questions are proposed in this study:

- Which influence more the propensity of manufacturing companies to the eco-innovation? The introduction of eco-innovations in the company or the introduction of eco-innovation after-sales?
- Do internal and external factors contribute to the eco-innovation of Portuguese manufacturing industry? Which are the factors that most contribute to eco-innovation?

6.3 Methodology

Based on the literature review, four hypotheses are formulated for our empirical study. Our first interest is to analyze the main factors that are correlated with the entrepreneurial innovative capability of Portuguese manufacturing industry in terms of eco-innovation. There are few empirical studies based on surveys which analyze on one hand the environmental benefits adopted by the companies and on the other if they are adopted to obtain those benefits (Jansson 2011; Jansson et al. 2011; Wagner and Llerena 2011). Also, several authors (OECD 2009; Moreira et al. 2012) argue that sustainable development is an important influence on innovative processes and can be an important driver to help companies to be more competitive and survive in the market.

However, the database used in this study considers as eco-innovations the new or significantly improved product, process, method, concept, or policy with environmental benefits, not considering the economic and social dimensions of the sustainable development. Furthermore, several authors (Rennings 2000; Hellström 2007; OECD 2009; Del Río et al. 2010; Jansson et al. 2011; Wagner and Llerena 2011; Demirel and

Kesidou 2011) have considered the eco-innovation as the environmental improvements and the reduction of environmental burdens/impacts. Hence, we were forced to restrict this study to the environmental benefits introduced by the companies.

In this context, our purpose is to study the association between environmental benefits and the introduction of innovation in the companies. Thus, the innovation intensity is evaluated by the product and processes' innovation. A product innovation is the introduction of a good or service that is new or significantly improved in what concerns its characteristics or intended uses. This includes significant improvements in technical specifications, components and materials, incorporated software, user friendliness, or other functional characteristics (OECD 2005; GPEARI 2010a, b; EIO 2010). A process innovation is the implementation of a new or significantly improved production or delivery method. This includes significant changes in techniques, equipment, and/or software (OECD 2005; GPEARI 2010a, b). We will consider the innovation inside the company and the innovation after-sales. Thus, the following hypotheses are suggested:

H_1: *Companies that intend to introduce eco-innovations in the company present more propensity to innovation.*

H_2: *Companies that intend to introduce eco-innovation after-sales present more propensity to innovation.*

The third and fourth hypotheses intend to analyze the relationship between the companies' innovation process and the introduction of eco-innovations in response to some factors. These factors can be external, for example, the introduction of regulations, or internal, for example, the existence of regular procedures to identify and reduce environmental impacts. The literature review states that attending to the sustainable targets and the competitiveness of companies, there are several factors which could influence the introduction of eco-innovations in order to reduce environmental problems and contribute to improve the current situation. Hence, the following two hypotheses are proposed:

H_3: *Companies that intend to introduce eco-innovations in response to external factors present more propensity to innovation.*

H_4: *Companies that intend to introduce eco-innovations in response to internal factors present more propensity to innovation.*

6.3.1 Sample and Data Collection

In the present study, the secondary data was collected from the "Office of Planning, Strategy, Evaluation and International Relations of Ministry of Science, Technology, and Higher Education (GPEARI/MCTES)" – belonging to the 7th Community Innovation Survey (CIS 2008) – and supervised by the Eurostat. CIS 2008 was carried out in 31 countries, i.e., the 27 EU member states, 3 EU candidate countries (Croatia, Iceland, and Turkey), and Norway (EUROSTAT 2013). It is a data source where the activities and the innovation process are evaluated every 2 years, in terms

Table 6.1 Classification of manufacturing sectors according to NACE Rev. 3

NACE Rev. 3	Manufacturing industry	Share on total sample (%)
10–12	Food products and beverages, tobacco products	7.6
13–15	Textiles, clothing, and leather	10.9
16–18	Wood, paper, and printing industry	10.5
19–23	Oil, chemical and pharmaceutical industry, and nonmetallic mineral products	21.6
24–25	Metallurgical and metal products	20.8
26–30	Computers, electrical equipment, motor vehicles	16.9
31–33	Furniture and other manufactured goods	11.8

of product, process, and organizational innovation and marketing as well as their determinants and fragilities/capabilities (GPEARI 2010a, b). The collection took place between May 2009 and April 2010 although the reference periods relate to the years 2006–2008.

The data set includes all industrial trade and services in Portuguese companies with more than 10 workers, at the date of application of the questionnaires, and it consists of 6,593 observations, given an 83 % response rate (GPEARI 2010a, b). However, we limited the target sample of this study to Portuguese manufacturing companies, according to the classification of economic activities – NACE Rev. 3 (Nomenclature of Economic Activities in the European Community, third revision). Therefore, only 1,563 companies are considered in the sample. About 52.9 % of the companies in the sample are small companies having between 10 and 49 employees, about 35.3 % of the companies have between 50 and 249 employees and 11.7 % have more than 250 employees. Attending to the manufacturing sector, the most representative in the sample is the oil, chemical, and pharmaceutical industry and nonmetallic mineral products (NACE 19–23), followed by metallurgical and metal products (NACE 24–25). Table 6.1 gives an overview of the sample.

6.3.2 Measures

In this study, the dependent variable *total innovation* reflects a situation in which the companies innovate at both levels: the product and the process innovation level. This information is provided from the CIS 2008s statistical framework ("B. Product innovation" and "C. Process innovation"). The innovation introduced by new or significantly improved goods or manufacturing processes was obtained from these frameworks. This dimension is presented as a dichotomy variable which adopts value 0 if the company does not innovate at product or service level and value 1 for those that innovate at product and service level. We have considered just the innovation on goods and production methods due to sample characteristics.

The independent variables are represented in Table 6.2 which summarizes the variables and measures used in this study, and they are provided from CIS

Table 6.2 Hypothesis, independent variables definition, measures, and values

Hypotheses	Independent variables	Type of measures	Values
H_1	ECOMAT – reduction of material used per unit produced	Ordinal	0 = no
	ECOEN – reduction of energy used per unit produced		1 = yes
	ECOCO – CO_2 reduction produced by the companies		
	ECOSUB – substitution by less polluting or hazardous materials		
	ECOPOL – reduction of noise pollution, air, water, or soil inside the company		
	ECOREC – recycling of waste, water, or material		
H_2	ECOENU – reduction of energy consumption	Ordinal	0 = no
	ECOPOS – reduction of noise pollution, air, water, or soil after-sales		1 = yes
	ECOREA – improved product recycling after use		
H_3	ENREG – existing environmental regulations or fiscal duties (taxes/fees) on pollution	Ordinal	0 = no
	ENREGF – environmental regulations or taxes expected to be introduced in the future		1 = yes
	ENGRA – availability of support from the central government, subsidies, or other financial incentives for eco-innovation		
	ENDEM – current or expected demand for eco-innovations by customers/market		
	ENAGR – voluntary adoption of codes of conduct or participation in sectorial agreements for the implementation of good environmental practices		
H_4	ENVID – procedures to identify and reduce environmental impacts regularly	Ordinal	0 = no 1 = yes, implemented before January 2006 2 = yes, implemented or significantly improved after January 2006

2008s statistical framework related to innovation with environmental benefits. First, this statistical data collected from CIS is about innovations with environmental benefits introduced in the company and after-sales, and second, it is about external and internal environmental factors that contribute to the eco-innovation of Portuguese industry.

6.3.3 Data Analysis

Based on the literature review in the previous chapter, we checked that the innovation capacity of the companies is a complex phenomenon influenced by several factors, including sustainable factors that are in the basis of the eco-innovation concept. In this context, it is necessary to explore the relationship between sustainable development factors and its influence on the innovative capacity of Portuguese manufacturing companies.

The type of secondary data collected promotes essentially dichotomous variables, which are measured in qualitative ordinal scale. This justifies the type of statistical analysis used attending to the hypothesis formulated (Table 6.2).

In order to test whether there are differences statistically significant between dependent variable (total innovation – TOTAL_INOV) and a set of independent variables (*ECOMAT, ECOEN, ECOCO, ECOSUB, ECOPOL, ECOREC*, among others), the Pearson's chi-squared test also known as chi-squared test was used. In this test, the expected count of all cells is compared with the observed count to infer from the relationship between the variables. If the differences between these counts (expected and observed) are not statistically significant, the variables are independent, or otherwise, we must reject the hypothesis of independence (Howell 2007).

The Pearson's chi-square assumes that no cells have expected count less than 1 and no more than 20 % of the cells have expected count less than 5 observations. If the chi-square's assumptions are not guaranteed, the observed significance level may not be correct (Howell 2007; Tabachinick and Fidell 2007; Hair and Joseph 2010; Acton et al. 2009; Kinnear and Gray 2011).

However, the chi-squared test only gives information about the variables' independence, and it does not give any information on the correlation level. Thus, the next step is to verify the variables' association using the Spearman's rho, also called Spearman's correlation coefficient because we have ordinal variables (Howell 2007; Tabachinick and Fidell 2007; Pallant 2007; Hair and Joseph 2010; Acton et al. 2009; Kinnear and Gray 2011).

The correlation measure evaluates the intensity and the direction of the variables' relation. The intensity is the absolute value of the correlation coefficient, and the coefficient signal is the direction. The correlation coefficient varies between −1 and 1 (Hair and Joseph 2010). Values near −1 or 1 indicates there is a strong relationship between two variables, negative if it is near −1 and positive if it is near 1. The count nearest to zero presents less relation intensity (Tabachinick and Fidell 2007; Howell 2007; Gamst et al. 2008; Hair and Joseph 2010). According to Cohen, cited by Pallant (2007), the intensity of correlation measures is considered weak if $0.10 \ll \rho \ll 0.29$ or $-0.10 \ll \rho \ll -0.29$, moderate if $0.30 \ll \rho \ll 0.49$ or $-0.30 \ll \rho \ll -0.49$, and strong if $0.50 \ll \rho \ll 1$ or $-0.50 \ll \rho \ll -1$.

Both tests consider a significance level of 5 %, and sometimes, when possible, they consider a significance level of 1 %.

In the next step, both the independent tests of Pearson's chi-square and the Spearman's correlation coefficient are computed to verify the influence of sustainable factors in the Portuguese manufacturing industries' innovation.

To test the proposed hypotheses, the SPSS 20 – *Statistical Package for the Social Sciences 20.0* – is used.

6.4 Results

The results obtained after data analysis are present by hypothesis.

At this stage of the research, the hypotheses are tested using the Pearson's chi-squared test of independence. For testing the research H_1 The *companies which intend to introduce innovations with environmental benefits present more propensity to innovation* and after the conditions of applicability of the Pearson's chi-squared test are verified, it is found that there are statistical evidences to state that environmental benefits in the company are related to the introduction of manufacturing industry innovation at a significance level of 5 %. In other words, the environmental benefits influence the innovation of manufacturing industries (Table 6.3).

The results indicate that the percentage of companies which introduced product innovation or process innovation (43.8 %) is lesser than the percentage of cases which introduced both of them (56.3 %) (Table 6.4). Despite that the inferential statistical analysis shows that manufacturing industry' innovation is dependent of

Table 6.3 Pearson's chi-squared test for H_1

Variables	Pearson's chi-square	Sig.
ECOMAT	23.148	0.000
ECOEN	30.934	0.000
ECOCO	14.453	0.000
ECOSUB	29.780	0.000
ECOPOL	44.672	0.000
ECOREC	26.611	0.000
N total	1,563	

Table 6.4 Distribution of innovations with environmental benefits in the company according to the degree of innovation

	Product innovation or process innovation (%)	Product innovation and process innovation (%)	Total innovation (%)
ECOMAT	6.7	8.7	15.4
ECOEN	7.1	9.4	16.5
ECOCO	4.8	6.2	11.0
ECOSUB	7.1	9.3	16.4
ECOPOL	8.1	10.7	18.8
ECOREC	10.0	11.9	21.9
Total	43.8	56.2	100

Table 6.5 Spearman's rho (ρ) correlations for H_1

	TOTAL_INOV	ECOMAT	ECOEN	ECOCO	ECOSUB	ECOPOL	ECOREC
TOTAL_INOV	1	0.122	0.141	0.096	0.138	0.169	0.130
ECOMAT		1	*0.526*	0.398	0.320	0.339	0.320
ECOEN			1	0.487	0.348	0.417	0.353
ECOCO				1	0.431	0.453	0.282
ECOSUB					1	*0.534*	0.392
ECOPOL						1	*0.519*
ECOREC							1

Weak if $\rho = 0.10$ to 0.29; moderate if $\rho = 0.30$ to 0.49; strong if $\rho = 0.50$ to $\rho = 1$

all environmental benefits in the company, we verify that a major percentage of companies pay attention to *ECOREC* (21.9 %), following *ECOPOL (18.8 %) and then ECOEN* (16.5 %).

According to the chi-squared test results, the null hypothesis is rejected which makes possible to state that the introduction of environmental benefits of innovation in the companies is related to the propensity for business innovation. Being so, the variables' correlations are going to be tested using the Spearman's rho test. As verified in Table 6.5, there are statistical evidences to state that the sample presents a positive correlation between all variables, at a significance level of 5 and 1 %. However, some weak correlations are also identified ($\rho = 0.096$ and $\rho = 0.282$) between *total innovation* variable and variables related to the environmental benefits in the company (reduction of material used per unit produced; reduction of energy used per unit produced; CO_2 reduction produced by the companies; substitution by less polluting or hazardous materials; reduction of noise pollution, air, water, or soil; and recycling of waste, water, or material).

The analysis of the variables' correlation shows that there are moderate positive correlations between almost all of these variables in contrast to strong positive correlations between the variables ECOMAT and ECOEN, between the variables ECOSUB and ECOPOL, and between the variables ECOREC and ECOPOL.

Summing up, the introduced innovation by the manufacturing industries is positively influenced by the environment benefits in the company (p value = 0.000 for all the variables), although it is possible to verify a weak correlation, as shown in Table 6.5. A strong positive correlation between some environmental benefits is also verified. These facts require that new models should be formulated considering the independent variables, in order to study these variables' behavior and the influence on each other's behavior.

Considering the H_2 *Companies that intend to introduce innovations with environmental benefits in after-sales present more propensity to innovation* and after the conditions of applicability of the chi-squared test are verified, it was found that there are statistical evidences to state that environmental benefits resulting from the use of product after-sales are related to the introduction of manufacturing industry innovation, at a significance level of 5 %. That is, the environmental benefits resulting from the use of a product after-sales influence the innovation of manufacturing industries (Table 6.6).

Table 6.6 Pearson's chi-squared test for H_2

Variables	Pearson's chi-square	Sig.
ECOENU	25.452	0.000
ECOPOS	29.527	0.000
ECOREA	21.350	0.000
N	1,563	

Table 6.7 Distribution of innovations with environmental benefits resulting from the use of a product after-sales according to the degree of innovation

	Product innovation or process innovation (%)	Product innovation and process innovation (%)	Total innovation (%)
ECOENU	12.6	20.8	33.4
ECOPOS	12.6	20.8	33.4
ECOREA	12.6	20.8	33.4
Total	37.8	62.4	100

Table 6.8 Spearman's rho (ρ) correlations for H_2

	TOTAL_INOV	ECOENU	ECOPOS	ECOREA
TOTAL_INOV	1	0.128	0.138	0.117
ECOENU		1	*0.551*	0.475
ECOPOS			1	*0.562*
ECOREA				1

Weak if $\rho=0.10$ to 0.29; moderate if $\rho=0.30$ to 0.49; strong if $\rho=0.50$ to $\rho=1$

The percentage of companies that introduced product or process innovation (37.8 %) is lesser than the percentage of companies which introduced both innovations (62.4 %), but according to Table 6.7, they are equally distributed (33.4 %) by environmental benefits in after-sales.

According to Table 6.8, there are statistical evidences to state that the total samples present positive correlations between all the variables at a significance level of 5 and 1 %.

The *total innovation* variable and *environmental benefits resulting from the use of product after-sales* variable (reduction of energy consumption; reduction of noise pollution, air, water, or soil; and improved product recycling after use) present a positive but weak correlation (between $\rho=0.117$ and $\rho=0.138$), as shown in Table 6.8.

Considering the results, it is possible to state that there is a moderate and positive correlation between ECOREA and ECOENU variables (0.475) and there are strong positive correlations between ECOENU and ECOPOS variables (0.551) and between ECOREA and ECOPOS (0.562).

In this context, the introduced innovation by the manufacturing industries is positively influenced by the environmental benefits of using a product after-sales (p value=0.000 for all the variables), although the weak correlation, as shown in Table 6.8.

Table 6.9 Pearson's chi-squared test for H_3

Variables	Pearson's chi-square	Sig.
ENREG	20.100	0.000
ENREGF	5.376	0.020
ENGRA	6.063	0.014
ENDEM	15.825	0.000
ENAGR	19.333	0.000
N total	1,563	

Table 6.10 Distribution of eco-innovations in response to external factors according to the degree of innovation

	Product innovation or process innovation (%)	Product innovation and process innovation (%)	Total innovation (%)
ENREG	11.7	15.4	27.1
ENREGF	7.5	9.1	16.6
ENGRA	2.0	3.1	5.1
ENDEM	8.1	11.2	19.3
ENAGR	14.1	17.8	31.9
Total	43.4	56.6	100

Table 6.11 Spearman's rho (ρ) correlations for H_3

	TOTAL_INOV	ENREG	ENREGF	ENGRA	ENDEM	ENAGR
TOTAL_INOV	1	0.114	0.059	0.062	0.101	0.112
ENREG		1	0.463	0.181	0.308	0.326
ENREGF			1	0.308	0.383	0.307
ENGRA				1	0.306	0.190
ENDEM					1	0.378
ENAGR						1

Weak if $\rho=0.10$ to 0.29; moderate if $\rho=0.30$ to 0.49; strong if $\rho=0.50$ to $\rho=1$

Considering now the H_3 *Companies that intend to introduce eco-innovations in response to external factors present more propensity to innovation* and after the conditions of applicability of the chi-squared test are verified, the results of the test show that there are statistical evidences to affirm that the introduction of eco-innovations in response to external factors is related to the manufacturing industry innovation, at a significance level of 5 %. According to Table 6.9, all external factors influence the innovation of manufacturing industries.

According to Table 6.10, the percentage of companies which introduce product or process innovation (43.4 %) is lesser than the number of cases which introduce both innovations ($n=56.6$ %) (Table 6.9). The results show that companies pay more attention to certain external factors, such as *ENAGR* (31.8 %), following *ENREG* (27.2 %) and *ENDEM* (19.3 %).

Table 6.11 shows that there are statistical evidences to state that the total samples present a positive correlation between all the variables at a significance level of 1 % (and consequently 5 %) with the exception of two values representing correlations with a significance level of only 5 % (0.059 and 0.062).

Table 6.12 Chi-squared test for H_4

Variable	Pearson's chi-square	Sig.
ENVID	16.329	0.000
N total	1,563	

Table 6.13 Distribution of eco-innovations in response to internal factors according to the degree of innovation

ENVID	Product innovation or process innovation (%)	Product innovation and process innovation (%)	Total innovation (%)
No	22.8	31.0	53.8
Implemented before January 2006	5.6	12.2	17.8
Implemented or significantly improved after January 2006	9.3	19.1	28.4
Total	37.7	62.3	100

The results identify a positive weak correlation between *total innovation* and all variables related to the eco-innovations in response to external factors. As shown in Table 6.11, the results present coefficients between $\rho = 0.059$ and $\rho = 0.114$.

There is a positive moderate correlation between the variables which represent external factors, with the exception of ENGRA and ENREG variables (0.181) and between ENAGR and ENGRA (0.190) which have a positive weak correlation.

Therefore, we conclude that manufacturing industry innovation is positively influenced by the external factors despite the weak correlation (Table 6.11).

Finally, attending to the H_4 *Companies that intend to introduce eco-innovations in response to internal factors present more propensity to innovation* and after verifying the conditions of chi-squared test applicability, it was found that there are statistical evidences to state that introductions of eco-innovations in response to internal factor variables are related to the innovation of manufacturing industry, at a significance level of 5 %. Therefore, internal factors influence the innovation of manufacturing industries (Table 6.12).

According to Table 6.13, the percentage of companies which introduced product or process innovation (37.7 %) is lesser than the percentage of companies that introduced both innovations (62.3 %). However, most manufacturing industries do not have regular procedures to identify and reduce their environmental impacts (53.8 %), although there have been an increase in its implementation after January 2006 (28.4 % compared to 17.8 % before January 2006).

Table 6.14 shows statistical evidences to affirm that the sample presents a positive correlation between *total innovation* and *ENVID*, at a significance level of 5 and 1 %. The findings show a positive but weak correlation between these variables with a value of 0.096 (Table 6.14).

Table 6.14 Spearman's rho (ρ) correlations for H_4

	TOTAL_INOV	ENVID
TOTAL_INOV	1	0.096
ENVID		1

Weak if $\rho = 0.10$ to 0.29; moderate if $\rho = 0.30$ to 0.49; strong if $\rho = 0.50$ to $\rho = 1$

Summing up, the introduction of innovation by the manufacturing industries is positively influenced by the procedures to identify and reduce environmental impacts regularly (p value = 0.000) although there is a weak positive correlation (0.096).

6.5 Discussion of the Results

This study aims to analyze the influence of the introduction of eco-innovation practices by Portuguese manufacturing companies on their propensity to innovation. The second purpose was to analyze the contribution of internal and external factors for the eco-innovation.

According to several authors (Bresciani and Oliveira 2007; Ulhøi 2008; Shrivastava 2008; Yang et al. 2010), when an environmental management system is adopted, the companies increase their competitiveness and become more sustainable, influencing positively the manufacturing industry innovation behavior. However, there are always weak correlations.

Both benefits introduced, in the company or after-sales, are correlated with companies' innovation. But in the company, the most important sustainable factor is ECOREC (recycling of waste, water, or material) with 21.9 %, and the least important is ECOCO (CO_2 reduction produced by the companies) with 11.0 % responses. In after-sales, the sustainable benefits are equally distributed (33.4 %).

The external and internal factors are also correlated with companies' innovation. ENAGR (voluntary adoption of codes of conduct or participation in sectorial agreements for the implementation of good environmental practices) is the most important external factor for the companies in their eco-innovations with 31.8 %. These arguments are in line with the eco-innovation literature, where Lindhqvist (2007) argues that the industries take the initiative in designing the environmental and sustainability policies and are not just waiting for the governments to decree regulations, but a growing number of companies have adopted strategies that allow them to be ahead of the legislators. The second most important factor is ENREG (existing environmental regulations or fiscal duties (taxes/fees) on pollution) with 27.2 % and according to the literature, the regulations create lead markets, which seem to be particularly relevant for eco-innovation in business contexts (Wagner and Llerena 2011).

Despite our results that corroborate the literature review, it is confirmed that there are barriers which affect the adoption of eco-innovations (Del Río et al. 2010) and hence of radical eco-innovations. Moreover, this study shows how research on eco-innovation is still in its infancy, particularly concerning systemic eco-innovations which have greater potential for overall environmental improvements, but they are also highly complex, involving non-technological changes as stated in OECD (2009).

6.6 Conclusions

The aim of this study was to investigate the influence of the introduction of eco-innovation practices by Portuguese manufacturing companies on their propensity to be innovative. With the purpose of improving comprehension of eco-innovation in goods and processes and identifying the main determinants of eco-innovation in the manufacturing sector, several hypotheses for investigation were formulated based on the literature review. Furthermore, the results presented above contribute to the existing literature in various ways. Firstly, the data allowed us to analyze the eco-innovations introduced by the manufacturing industries and secondly test the relationship between innovative capability of Portuguese manufacturing industry and sustainable development. Finally, it allowed us to measure this correlation.

This investigation highlights four factors which stimulate and/or limit eco-innovative capability in manufacturing industry: environmental benefits in the company, environmental benefits in after-sales, external factors, and internal factors. It was taking these four factors into account that the various hypotheses tested empirically were formulated.

Though the literature reviewed and our findings clarify that innovative performance of Portuguese manufacturing industry is influenced by sustainable development factors, it has also been verified a positive weak correlation in all the empirical tests. According to the results obtained, some environmental benefits in the company and after-sales showed a positive strong correlation between themselves, and some variables have effects on other variables. Therefore, we need to formulate new models which include these variables in order to measure their behavior and how they influence each other.

The main contribution of this study for managers is the identification of the main factors that influence the entrepreneurial innovative capability of Portuguese manufacturing industry in terms of eco-innovation. According to this study, companies should invest on reduction of material and energy used/consumed; reduction of CO_2 produced by the companies; substitution by less polluting or hazardous materials; reduction of noise pollution, air, water, or soil inside the company and after-sales; recycling of waste, water, or material; improved product recycling after use; adoption of environmental regulations; adoption of codes of conduct or participation in sectorial agreements for the implementation of good environmental practices; and adoption of procedures to identify and reduce environmental impacts regularly factors as a way of being more innovative in environmental terms. The results present the environmental benefits adopted by the majority of the companies and if they are adopted to obtain those benefits.

According to this research, managers should invest in the promotion of eco-innovation in their companies introducing radical improvements in corporate environmental practices and performance, as well as improvements in products and in the process. With the identification of the main factors that influence the eco-innovation behavior, managers can easily and fastly adopt eco-innovative behaviors inward their companies.

The present work presents some limitations specially in terms of confidentiality. For example, the industries' location and the information about specific sections by NACE Rev. 3 are absent. In this context, specific information was excluded, and it was not possible to draw up a comparison of results with previous CIS so as to assess evolutionary tendencies in the area of innovation activities and expenditure. It was also very difficult to distinguish specific companies because all of them are grouped in a general NACE. However, the most important limitation we came across with was the impossibility to know if the companies have introduced radical or incremental innovations.

References

Acton C, Miller R, Fullerton D, Maltby J (2009) SPSS statistics for social scientists, 2nd edn. Palgrave Macmillan, Houndmills

Bresciani S, Oliveira N (2007) Corporate environmental strategy: a must in the new millennium. Int J Bus Environ 4:488–501

CIS 2008 (2008) Inquérito Comunitário à Inovação – CIS 2008. Gabinete de Planeamento, Estratégia, Avaliação e Relações Internacionais, Ministério da Ciência, Tecnologia e Ensino Superior, Lisboa

Del Río P, Carrillo-Hermosilla J, Konnola T (2010) Policy strategies to promote eco-innovation: an integrated framework. J Ind Ecol 14:541–557

Demirel P, Kesidou E (2011) Stimulating different types of eco-innovation in the UK: government policies and firm motivations. Ecol Econ 70:1546–1557

EIO – Eco-Innovation Observatory (2010) Methodological report

Eurostat European Commission (2013) Eurostat Metadata. http://epp.eurostat.ec.europa.eu/cache/ITY_SDDS/EN/inn_esms.htm#stat_pres. Accessed 4 June 2013

Faria P, Lima F, Santos R (2010) Cooperation in innovation activities: the importance of partners. Res Policy 39:1082–1092

Faucheux S, Nicolaï I (2011) IT for green and green IT: a proposed typology of eco-innovation. Ecol Econ 70:2020–2027

Gamst G, Meyers LS, Guarino AJ (2008) Analysis of variance designs: a conceptual and computational approach with SPSS and SAS. Cambridge University, Cambridge

GPEARI (2010a) Documento Metodológico CIS 2008 – Inquérito Comunitário à Inovação. Gabinete de Planeamento, Estratégia, Avaliação e Relações Internacionais, Direcção de Serviços de Informação Estatística em Ciência e Tecnologia, Lisboa

GPEARI (2010b) Sumários Estatísticos CIS 2008 – Inquérito Comunitário à Inovação. Gabinete de Planeamento, Estratégia, Avaliação e Relações Internacionais, Direcção de Serviços de Informação Estatística em Ciência e Tecnologia, Lisboa

Hair J, Joseph F (eds) (2010) Multivariate data analysis: a global perspective, 7th edn. Pearson Education, Upper Saddle River

Hellström T (2007) Dimensions of environmentally sustainable innovation: the structure of eco-innovation concepts. Sustain Dev 15:148–159

Howell D (ed) (2007) Statistical methods for psychology, 6th edn. Thomson Wadsworth, Pacific Grove, USA

Jansson J (2011) Consumer eco-innovation adoption: assessing attitudinal factors and perceived product characteristics. Bus Strategy Environ 20:192–210

Jansson J, Marell A, Nordlund A (2011) Exploring consumer adoption of a high involvement eco-innovation using value-belief-norm theory. J Consum Behav 10:51–60

Kinnear PR, Gray CD (2011) IBM SPSS statistics 18 made simple. Psychology Press, Hove/New York

Lindhqvist T (2007) Corporate management tools to address sustainability challenges, a reflection on the progress and difficulties experienced during the last decade. Environ Eng Manage J 6:351–355

Lobo A (2010) Eco-empresas e Eco-inovação em Portugal: Breve Análise Retrospectiva 1995–2008. Departamento de prospectiva e planeamento e relações internacionais, Ministério do Ambiente, do Ordenamento do Território, Lisboa

Montalvo C, Diaz-Lopez FJ, Brandes F (2011) Potential for eco-innovation in nine sectors of the European economy. Task 4, Horizontal Report 4, Europe INNOVA Sectoral Innovation Watch, for DG Enterprise and Industry, European Commission. Available at http://www.prepare-net.com/sites/default/files/eco-innovation_opportunities_in_nine_sectors_of_the_eu_economy.pdf

Moreira J, Silva MJ, Simões J, Sousa G (2012) Drivers of marketing innovation in Portuguese firms. Amfiteatru Econ 31:195–206

Nuij R (2001) Eco-innovation: helped or hindered by integrated product policy. J Sustain Prod Des 1:49–51

OECD – Organisation for Economic Co-operation and Development (2005) OSLO manual: proposed guidelines for collecting and interpreting technological innovation data, 3rd edn. Organisation for Economic Co-operation and Development, Paris

OECD – Organisation for Economic Co-operation and Development (2009) Sustainable manufacturing and eco innovation: towards a green economy. Policy Brief, available at www.oecd.org/sti/innovation/sustainablemanufacturing

Pallant J (ed) (2007) SPSS survival manual: a step by step guide to data analysis using SPSS for Windows, 3rd edn. Open University, Berkshire

Parnell JA (2008) Sustainable strategic management: construct, parameters, research directions. Int J Sustain Strateg Manage 1:35–45

Pratoom K, Cheangphaisarn P (2011) The impact of strategy for building sustainability on performance of software development business in Thailand. Asian J Bus Manage 3:32–39

Rennings K (2000) Redefining innovation – eco-innovation research and the contribution from ecological economics. Ecol Econ 32:319–332

Shrivastava P (2008) Sustainable organisational technology. Int J Sustain Strateg Manage 1:98–111

Stead JG, Stead WE (2008) Sustainable strategic management: an evolutionary perspective. Int J Sustain Strateg Manage 1:62–81

Tabachinick B, Fidell L (eds) (2007) Using multivariate statistics, 5th edn. Pearson Education, Upper Saddle River, USA

Ulhøi JP (2008) Supporting the development of environmentally sustainable technologies and products: the role of innovation, informal cooperation and governmental agency. Int J Environ Pollut 32:121–133

Wagner M, Llerena P (2011) Eco-innovation through integration, regulation and cooperation: comparative insights from case studies in three manufacturing sectors. Ind Innov 18:747–764

Yang CL, Lin S-P, Chan Y-H, Sheu C (2010) Mediated effect of environmental management on manufacturing competitiveness: an empirical study. Int J Prod Econ 123:210–220

Chapter 7
Portraying the Eco-innovative Landscape in Brazil: Determinants, Processes, and Results

Flavia Pereira de Carvalho

Abstract Eco-innovations are a crucial tool for firms to redefine the environmental impacts of their productive activities toward a new paradigm of sustainable development. Even though this is an important subject, still little is known about how eco-innovations take place in firms, especially in emerging economies like Brazil. The purpose of this chapter is to investigate the landscape of eco-innovations in Brazilian firms, concerning the characteristics of eco-innovators, as well as the determinants, results, types of innovation, and the existence of cooperative arrangements for the development of eco-innovations. To do so it presents the results of an unprecedented survey on eco-innovative activities in Brazilian firms carried out in 2012. The methodology is quantitative, descriptive, and explanatory, using precise measurement to provide a representative picture of eco-innovations in Brazil. The results show that firms are mostly driven by the opportunity to create new businesses with their eco-innovations. Moreover, it reveals that most eco-innovative firms in our sample conduct systematic, in-house R&D activities; most eco-innovative firms participate in cooperative arrangements for innovation especially with universities and research institutes; and most eco-innovations are organizational, with incremental impacts.

Keywords Eco-innovation • Sustainability • Brazilian firms • Determinants • Results

F.P. de Carvalho (✉)
Fundação Dom Cabral, Nova Lima, Brazil

UNU-MERIT, Maastricht, The Netherlands
e-mail: flaviacarvalho@fdc.org.br

S. Azevedo et al. (eds.), *Eco-Innovation and the Development of Business Models*, Greening of Industry Networks Studies 2, DOI 10.1007/978-3-319-05077-5_7,

7.1 Introduction

In the second half of the twentieth century, recognition of global-scale environmental threats gained traction in international discussions. Awareness increased with the recent scientific observation that "the rapid and accelerating technologically driven modification of our natural surroundings has changed them beyond the wildest Neolithic dreams" (Grey 1993: 464). These risks underscore the need for a more sustainable use of natural resources.

Experts have shed light on the environment's capacity to resist human patterns of development, questioning utilitarian frameworks that take natural resources for granted. Consequently, environmental responsibility contrasts with previous understandings of progress as an inevitable march forward, ensuring mastery over natural resources (Cohen 1997). Natural systems are not exogenous to human activities. Instead, these systems are complex and coevolving, directly affected by technological trajectories and social behavior.

The effect of technology on nature raises questions about whether past trends of prosperity can be broadened (or even sustained) in the future (Clark et al. 2005). Achieving greater harmony between economic development and nature requires a sea change in the impact businesses have on the environment – therefore demanding a new paradigm for production, consumption, and disposal. Changing the existing paradigm is difficult, however, since it demands alternative actions that benefit the economy, society, and the environment.

This chapter presents eco-innovations as the mechanism that allows the paradigm shift, from business as usual toward "a new path of economic development in which technological advances and social changes combine to reduce, by an order of magnitude, the environmental impacts of economic activity" (Jacobs 1997: 9). Eco-innovations are understood here as new products, processes, services, organizational processes, or business models that significantly reduce environmental damage.

Despite its importance, little is known about the subject, especially in countries with an emerging economy, such as Brazil. The present chapter comes with the purpose to fill this gap, presenting the results of a survey on eco-innovation in Brazilian firms. The main objective of the chapter is to draw a portrait of eco-innovation in Brazil, in order to understand their main types, determinants, results, the main characteristics of eco-innovators, and whether these eco-innovative firms are involved in cooperative arrangements.

The remainder of the chapter is organized as follows: next section brings the theoretical background, exploring the concept of eco-innovation, the importance of cooperative arrangements, and the literature describing the innovative landscape in Brazil. Section 7.3 sets the methodology and Sect. 7.4 brings the results of the survey. The chapter ends with some reflections on the results of the research, as well as its limitations and ways forward for future studies.

7.2 Theoretical Background

7.2.1 Eco-innovation[1]: Concepts, Determinants, Types, and Results

Innovation is an important tool for addressing environmental vulnerabilities and for enabling nature to better withstand man-made disruptions. The Stockholm Resilience Centre is one of a handful of organizations investigating "biophysical boundaries at the planetary scale within which humanity has the flexibility to choose a myriad of pathways for human well-being and development (Rockström et al. 2009: 6)." The organization's preliminary analysis found humanity has transgressed three boundaries: climate change, biodiversity loss, and the nitrogen cycle. Yet, the study concluded that the threshold for permanent environmental change was still uncertain. Research has not determined how long problems can persist before they impair nature's ability to restore itself (Rockström et al. 2009). Finding solutions to environmental threats requires new relationships between society and nature that ultimately will drive the economic actors toward a new model of production, consumption, and disposal.

Environmental challenges should not be viewed exclusively as constraints; they also can present opportunities for economic and social prosperity. Fostering a better relationship between industrial activities and natural resources creates win/win situations for both economic and environmental performance. These situations were described by Hart (1997) and also by Porter and van der Linde (1995).

Hart presented opportunities for firms to drive innovation and to build a growth trajectory through the internalization of environmental concerns. Similarly, Porter and van der Linde proposed (and confirmed) the hypothesis that environmental regulations foster efficiency and innovation. Regulations, according to them, are not impediments to economic activities, as commonly presented in political discourses, but rather opportunities to increase competitiveness. More recent works corroborated the importance of environmental regulations to eco-innovations in German (Horbach et al. 2012) and English companies (Demirel and Kesidou 2011).

During the 2000s, several studies supported technical solutions to environmental hazards and defined the concept of eco-innovation, which can be synthesized as:

> …the production, assimilation or exploitation of a product, production process, service or management or manner of doing business that is new for the organization (developing or adopting these) and that results, through the life cycle, in a reduction of environmental risks, pollution and other negative impacts of the use of resources (including the use of energy) compared to relevant alternatives. (Kemp and Pearson 2007: 7)

[1] Different names are used in literature to refer to eco-innovations: environmental innovations, green innovations, and sustainability-oriented innovations. All terms focus on methods that have a positive environmental impact.

Table 7.1 Taxonomy of eco-innovations

Innovation	Examples
Environmental technologies	Pollution control technology
	Clean technologies
	Green energy technology
Organizational innovations	Methods of pollution prevention
	Environmental management
	Management of the value chain
Innovation of products and services	Ecologically beneficial products
	Ecological and less resource-intensive
Green system innovations	More ecologically beneficial alternative system of consumption and production

Gains from eco-innovations refer to the environmental benefit attained, such as cleaner air or more available water, and the economic outcome for the firm. Innovations can be radical or incremental: the former includes discoveries that completely disrupt current activities, and the latter covers significant improvements to already existing products or processes. Both radical and incremental eco-innovations also fit into different categories, as described on Table 7.1.

The first category of eco-innovations comprises environmental technologies. These technologies may be remedial (end-of-pipe), reducing the effects of existing technology. They can also be clean alternatives to more polluting technologies. The second category covers organizational innovations that incorporate environmental issues, including strategies applied to production processes, infrastructure, and logistics that limit environmental damage. Internal audits, staff training, and waste and pollution prevention methods are examples of this category (Kemp and Foxon 2007).

Environmentally beneficial products and services, such as green certifications or biodegradable products, fit into the third category. Services include waste and pollution management, environmental consultation, and other activities that decrease the negative environmental impact of production methods. The last category contains green innovations of systems, which "involve a wide range of changes in technological production, knowledge, organization, institutions and infrastructures and possibly changes in the behavior of consumers" (Kemp and Foxon 2007: 9). The category covers alternative systems of production and consumption that are more environmentally beneficial than existing systems. Organic agriculture and renewable energy systems (like the system required for electric cars) are examples. These comprehensive innovations reach a wide range of actors and imply changes in broad sectors of economic activity.

The definition of eco-innovation prioritizes the environmental result over the motivation of the firm to innovate. In the same line, the literature usually does not discriminate between intentional and unintentional results. Discovering what drives firms to undertake eco-innovations is essential, however, if these are to be fostered systematically.

Table 7.2 Determination of eco-innovations

Regulations	Implementation of environment policies
	Anticipation of environment regulations
Firm-specific factors	Cost economics
	Better productivity
	Innovation in organizational management systems
	R&D activities
	Networks and cooperation
Market factors	Growing consumer awareness on environmental issues
	Expectation of increased participation in new market segments
Environmental concerns	Environmental concerns from companies that are "doing the right thing"

The literature states that four factors are the main drivers of eco-innovations (Belin et al. 2009; Horbach et al. 2012): regulations, market factors, firm-specific factors, and environmental concerns, presented in Table 7.2.

Environmental regulations induce firms to internalize the externalities created by their activities. Seen as such, these regulations impose extra costs for firms and constrain their competitiveness. Porter and van der Linde (1995), however, introduced a different perspective, claiming that environmental regulations force firms to innovate and improve their resource efficiency and productivity. Regulations would, as a result, increase turnover and profits (Porter and van der Linde 1995; Bernauer et al. 2006; Belin et al. 2009).

Market factors are the pressures consumers and competitors place on companies. Customer demands are proven drivers of firm behavior. Consumers are increasingly aware of – and concerned about – environmental threats and want to know what firms are doing about them. Several opportunities exist for the creation of products and services better suited to new market demands.

Firm-specific factors are mainly companies' technological and innovative capabilities, strategies, and key competencies as well as their goals for higher productivity, eco-efficiency, and other cost-reduction benefits (Bernauer et al. 2006). The literature generally refers to the existence of R&D activities and R&D expenditures as a *proxy* for technological and innovative capabilities of firms.

Pure environmental motivations are based on strong ethical values that compel firms to do the right thing, regardless of regulations or consumer pressure. These motivations are built on management values that encourage organizations to take their roles in society seriously (Bansal and Roth 2000).

Learning about the results of eco-innovations is equally important. There are economic effects and environmental effects resulting from eco-innovation activities. Studies on eco-innovation in the context of European firms have found diverse results, but the number of companies reporting positive economic results was higher than those reporting negative economic results (Arundel and Kemp 2009). At the end of the day, regardless of the environmental threats posed in their way, most firms do eco-innovate for economic reasons and economic return might be the decisive point regarding eco-innovation decisions.

7.2.2 *Cooperation for Eco-innovation*

The concept of cooperation involves a relation between the social players and the notion of shared effort, convergence of objectives, and coordination at different levels. Cooperation, therefore, is not only an exchange of information between the parties. It includes economic and business relations and is seen as indispensable in the aggregation of value and for development (Lins 2009). Cooperative arrangements for innovation have been studied in their various forms: socio-technical networks, innovation networks, innovation systems, triple helix, strategic alliances, open innovation, among other approaches (Ferro 2010).

Numerous studies provide evidence of the role of external sources of knowledge for the companies as important inputs for innovation in general (Teece 1992; Veugelers 1997; Hagedoorn 2002). The growing speed of technological change, the greater complexity of technical and scientific knowledge, and the shortening of technological life cycles have significantly increased the costs of innovation, along with the speed of technology obsolescence (Hagedoorn 2002). It has therefore become unlikely that only one company or group will possess all the knowledge necessary for the development of a technology (Benfratello and Sembenelli 2002; Chesbrough 2003; Cortês et al. 2005). Such transformations make cooperative partnerships for innovation a more frequent arrangement, allowing greater access to knowledge along with cost reductions.

Studies on eco-innovations in developed countries suggest that partnerships for innovation are a possible positive factor for their occurrence (Belin et al. 2009). This argument is based on the fact that new regulatory instruments and also new technological trajectories put in question the current processes and the knowledge base of the company. Access to external sources (suppliers, consultants, research institutes, universities) is a means for the company to have contact with the new capabilities needed to eco-innovate (Belin et al. 2009; Hart 1997). Cooperative arrangements for eco-innovation also deal with the uncertainty factor, especially in the case of basic science or new fields of knowledge (Barbieri et al. 2010). These can be attenuated through innovation partnerships.

Another factor is also cited as favorable to the establishment of partnerships for eco-innovation – the fact that environmental issues involve a wide range of interested parties that go beyond the ambit of the market, such as the local community and activist groups (Hall and Vredenburg 2003), which means to say, questions connected to the environment create complex interactions between the company and the economic, social, and institutional environment in which it is inserted. As a consequence, eco-innovation cannot be understood as an isolated decision of the company (Hall and Vredenburg 2003), there being not infrequently pressure exerted by other players and even their desire to participate in the process of searching for environmentally positive alternatives (Hart 1997; Barbieri et al. 2010).

In resume, eco-innovations seem to demand to a larger extent on external sources of knowledge and information in order to take place (Belin et al. 2009). Studies point to a greater dependence on research undertaken by public institutions, which can be explained by the greater need to access relatively recent technologies for the development of environmental innovation (Belin et al. 2009). Eco-innovation thus appears as a process in which knowledge and resources are distributed among the various players interconnected (Kemp and Foxon 2007).

7.2.3 Innovation and Eco-innovation in Brazil

To contextualize innovation in Brazil, this section brings some statistics on the general innovative behavior of Brazilian firms. The data are based on results from the fourth edition of the Brazilian Innovation Survey (PINTEC), conducted in 2008 by the Brazilian Statistics Office (IBGE).

From a universe of more than 100,000 companies, the survey found an innovation rate of 38.6 %, which means that around 40,000 firms stated they implemented some innovation in product, process, or organizational method. Innovation in Brazilian firms is still modest when compared to rates provided by similar surveys from advanced countries. In Germany, innovation rates exceeded 75 %, and the average innovation rate of the EU-27 countries was approximately 52 % (Eurostat 2012).

Figures on R&D activities for innovation are also very humble: only 11.6 % of innovative firms had conducted these activities. Out of the R$54.1 billion Brazilian firms spent on innovative activities in 2008, only R$15.3 billion was on R&D, representing 28 % of all innovation expenditures (IBGE 2010). In 2008, the gross expenditures on R&D in Brazil as a whole were slightly above 1 % of GDP. In contrast, Germany spent 2.69 %, and Denmark spent 2.85 % (Eurostat 2012).

Among the innovative firms, only 10.4 % of PINTEC's respondents have engaged in cooperation for innovation (IBGE 2010). From this small group, 63 % attribute significant importance to relations with suppliers, 45.8 % to cooperation with clients or consumers, and 32.4 % to relations with universities and research institutes. These numbers suggest that Brazilian companies are not very keen on seeking external knowledge or cooperating for innovation.

PINTEC had no direct questions on eco-innovations. It did inquire about the results of firms' innovations, however, and received responses that supported four categories of environmental results: reduction in the use of water, energy, and inputs and a positive environmental impact overall. Brazilian firms were not excelling in these aspects, either. Only 12.2 % of Brazilian firms reported a reduction in water consumption, 24.1 % lowered energy consumption, and 26.2 % decreased materials consumption. In addition, 33.1 % of industrial companies reported a positive environmental effect from the innovation (IBGE 2010) (Fig. 7.1).

A previous study conducted in 2012 used data from PINTEC to determine aspects of firms' behavior that increase the probability an eco-innovation will be implemented (Arruda et al. 2012).

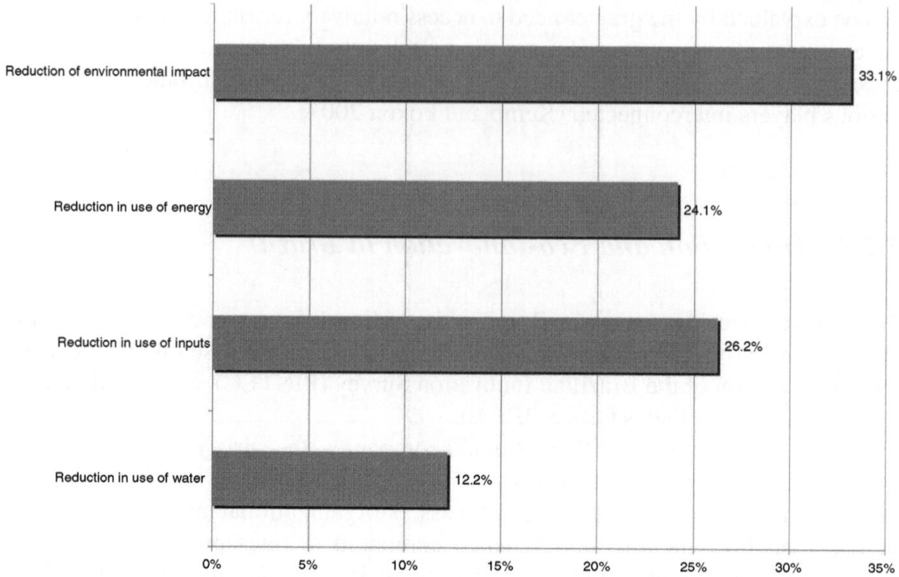

Fig. 7.1 Main environmental impacts from innovations: results from PINTEC

The results indicated that systematic R&D activities contributed to eco-innovation. Engagement in cooperative arrangements – the use of external sources of knowledge for innovation – was equally positive. Patent requests or grants also increased the occurrence of eco-innovation. The study concluded that firms that use government programs to foster innovation also had a higher propensity for eco-innovation.

These results signalize that eco-innovation tends to occur in environments where management has a strong commitment to knowledge generation, research, development, and innovation. Firms that lack systematic efforts and resources directed to the innovative process and that adopt innovations created elsewhere are less likely to generate positive environmental impacts.

In the next section, we present our methodology and a description of the survey findings.

7.3 Methodology

The empirical work is quantitative, using precise measurement to provide a representative picture of eco-innovations in Brazil and better understand the relationships between the key variables set by the theoretical background.

The research purpose is thus both descriptive and explanatory. Due to the objective of portraying a general representation of eco-innovations in Brazilian firms, this work provides an accurate description through the analysis of statistical indicators, obtained through a cross-sectional survey that targeted a specific population. Its explanatory nature, on the other hand, derives from its intent of overlapping results of key variables and explaining their relationships (Gil 1999).

Therefore, a research that combines descriptive and explanatory lenses allows understanding the characteristics of certain groups we gathered data from, determining the proportion of our sample that behaves in a certain way and explains the relationships between key variables that were raised by the literature on eco-innovation.

In order to gather data to draw a portrait of eco-innovation in Brazil, we ran a survey with Brazilian firms. The population of this study refers to all Brazilian firms, of which a sample of 98 responded to the questionnaire. Our unit of analysis are eco-innovations carried out by Brazilian enterprises. The main variables used in the survey and described in the chapter are presented on Table 7.3.

Table 7.3 Variables in the survey

Variables	Definition
Determinants	
Regulations	Eco-innovation is driven by regulations, either existing or anticipatory
Market factors	Eco-innovation is driven by market factors, such as consumer pressure and market demand for greener products
Firm – specific	Eco-innovation is driven by firm-specific factors, such as the search of cost economies, productivity, efficiency
Environmental concerns	Eco-innovation is driven by the management's decision to "do the right thing"
Firm characteristics	
Size	Size of the firm, measure by net revenues
Sector	Firm sector
R&D	If the firm reported having done R&D activities in the past year
Types of eco-innovation	
Product	If the eco-innovation is a product or service
Process	If the eco-innovation is a production process
Organizational	If the eco-innovation was an organizational change
Technology	If the eco-innovation is a technology
Environmental initiatives	
Life cycle analysis programme	The company has an LCA programme
Annual sustainability report	The company releases a sustainability report
Environmental certification (ISO 14001)	The company has ISO 14001 certification
Environmental training programme	The company has as environmental training programme
Water use reduction/treatment programme	The company has a water consumption reduction programme
Energy reduction programme	The company has energy consumption reduction programme

(continued)

Table 7.3 (continued)

Variables	Definition
Environmental management programme	The company has an environmental management program (other than ISO)
Team responsible for environmental issues	The company has a team responsible for environmental issues
Waste reduction programme	The company has a waste reduction programme
Cooperation partners	
Universities/research institutes	The company cooperated with universities/research for the eco-innovation
Clients/consumers	The company cooperated with clients/consumers for the eco-innovation
Competitors	The company cooperated with competitors for the eco-innovation
Other firms in the group	The company cooperated with other firms in the group for the eco-innovation
Consultancy	The company cooperated with consultancies for the eco-innovation
Suppliers	The company cooperated with suppliers for the eco-innovation
NGOs	The company cooperated with NGOs for the eco-innovation

7.3.1 Sample and Data Collection

We base our study on an unprecedented survey, conducted from May through July 2012, with Brazilian companies in manufacturing and services sectors. The study employed a stratified random sample with companies of different sizes. Questionnaires were submitted electronically. Survey Monkey® was used to design the questionnaire and manage the responses; 98 companies replied. The sample was within the 95 % reliability interval with a significance of 10 %, which was expected for populations between 50 and 100,000, the approximate number of large companies operating in Brazil (Rea and Parker 2005).

The survey design was based on extensive research on earlier surveys and also on PINTEC's survey when similar questions were asked. The purpose of the survey was to provide a broader understanding of eco-innovation in Brazilian firms and to answer two main questions:

- What are the determinants, types, and results of eco-innovation in Brazilian firms?
- What are the characteristics of eco-innovative firms in Brazil?

The questionnaire was divided into five sections: (A) general company data; (B) general information on the innovative activities of the company; (C) objectives of eco-innovations; (D) processes of eco-innovations; and (E) results of eco-innovations.

Section A contained questions on capital ownership, revenues, and share of revenues obtained from exports. Firms were also asked about the importance of environmental issues for their business as well as about the existence of environmental

Table 7.4 Distribution of the sample by sectors

Sectors	Number of firms
Services	25
Machinery, equipment and technology	14
Mining and steel	13
Chemicals and pharmaceuticals	6
Pulp and paper	4
Petroleum, gas and energy	10
Food and drink	8
Vehicles and parts	7
Construction	4
Textiles and clothing	4
Others	3

management initiatives (ISO 14001, sustainability report, among others). Section B included questions on companies' innovation activities, such as the amount of resources invested in innovation, expenditures on R&D, and the sources of funds intended for innovation. Section C examined the nature and occurrence of eco-innovations more specifically. Section D focused on the processes involved in eco-innovation activities, particularly, the sources of external knowledge used and whether cooperative arrangements for innovation had taken place. Finally, Section E inquired about the results of eco-innovations and about possible barriers that would prevent eco-innovations from being successfully implemented.

7.3.2 Characterization of the Respondents

We received 98 valid questionnaires from firms representing various economic sectors. Most of the sample (58 %) was composed of large firms with more than 500 employees. Domestic firms comprised the majority of the sample (61 %); 22.4 % of the companies were owned by foreign capital, and 16.6 % of them had mixed capital ownership. The size of the firms in the sample was also evident in their revenues: 73.5 % of the firms had revenues higher than R$20 million (approximately US$10 million). Table 7.4 presents the sectoral distribution of firms.

7.4 Results

7.4.1 Characteristics of Eco-innovators, Innovative Efforts, and Cooperation for Eco-innovation

Seventy-two percent of the 98 firms that responded to the survey reported the implementation of general innovations: a high percentage compared to the innovation rate of firms in PINTEC (38.6 %). Compared with the innovation rate obtained by PINTEC for firms with more than 500 employees, however, this rate was similar. As

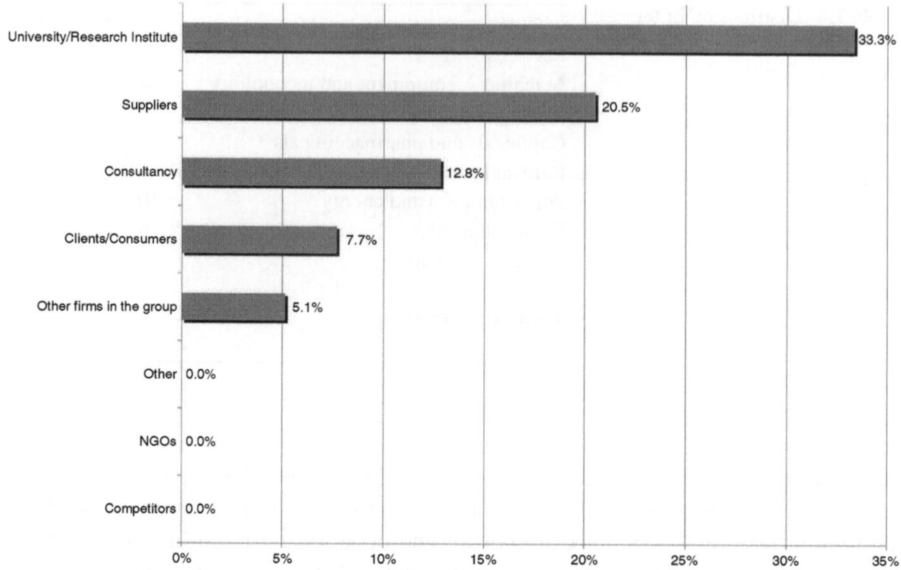

Fig. 7.2 Partner in cooperative arrangements for eco-innovation

to the degree of novelty, 42 % of our sample firms reported innovations that were new to their companies or to the national market.

When we focused specifically on environmental impacts, we found that 63 % of innovative firms engaged in eco-innovations in 2011, or 45 % of the total sample. Again, the share was impressive, given eco-innovative results shown in the PINTEC sample, which did not exceed 30 % of innovative firms. A study investigating eco-innovation in UK firms found a similar share of eco-innovative firms in that country to that found by our survey (Demirel and Kesidou 2011). A simple explanation for the higher number was that firms interested in eco-innovation were more likely to answer our questionnaire. Conversely, PINTEC's survey questioned a much larger sample of Brazilian firms, since it was sponsored and conducted by Brazil's federal government and had a compulsory character.

Moreover, 75 % of innovative firms reported undertaking R&D activities – of which 65.4 % refers to continuous R&D efforts. This information corroborated the idea that companies with internal R&D activities obtain better results in science-based innovation (Pavitt 1984) and tend to generate innovations with higher impact. The high percentage of firms eco-innovating and conducting R&D was not coincidental. Earlier studies established a relationship between the presence of R&D activities and the occurrence of eco-innovations in Brazilian (Young and Lustosa 2007), German, and French firms (Belin et al. 2011). It is reasonable to expect that strong, targeted R&D efforts would engender eco-innovations.

Differently from PINTEC, we found that 86 % of the eco-innovative firms in our sample established cooperative arrangements to access other knowledge sources. Firms named universities and other research institutions as their main partners, followed by suppliers (Fig. 7.2). The literature on eco-innovations maintains that firms have a lot to gain from collaborative arrangements (Belin et al. 2011; Hart

1997). Having a variety of actors and stakeholders involved in the innovation process diversifies knowledge and reduces risks, costs, and uncertainty.

Previous studies indicated that cooperation made eco-innovation more frequent (Belin et al. 2011; Arruda et al. 2012). Since eco-innovations are often linked to game-changing knowledge – and, consequently, subjected to greater complexity and uncertainty – these innovations are keener on benefiting from cooperative arrangements. External partnerships widen the range of available knowledge and capabilities, including both tangible and intangible resources. As innovations do not necessarily derive from new knowledge, partnerships increase the pool of resources available that can be combined in different ways to generate a novelty.

It is also interesting to observe that NGOs are not taken as important partners for innovation in Brazil. While Brazil has approximately 300,000 active NGOs, the United States has more than 1.5 million[2] and India 3.3 million.[3] Moreover, only 0.8 % of all Brazilian NGOs are related to environmental preservation,[4] which might help understand their small contribution to the eco-innovative process.

7.4.2 Types of Eco-innovations

Acquiring more information on existing eco-innovations was also important, in order to determine what categories of innovation – such as processes, products, or technologies – were targeted most frequently, along with their degree of novelty.

Organizational innovations accounted for the largest share at 36 %, mostly from the introduction of waste recycling, water reuse, or initiatives to reduce energy and materials consumption. Most times organizational innovations are not new to the world, though they represent different and more attractive ways to get things done by the enterprise.

New technologies had the second largest share, encompassing 33 % of the eco-innovations implemented by the sample firms. We were surprised by this nonintuitive result, since new-technology generation tends to be more radical and science-based than process innovation, which corresponded to 17 % of the surveyed companies. Finally, new products were the less representative type of eco-innovation comprising 14 % of the total (Fig. 7.3).

7.4.3 Determinants of Eco-innovations

Exploring the determinants of eco-innovations produced some interesting results (Fig. 7.4). New-business creation was, by far, the main determinant: 23 % of firms indicated it was their key driver to eco-innovate. It seems surprising that a

[2] http://www.humanrights.gov/2012/01/12/fact-sheet-non-governmental-organizations-ngos-in-the-united-states/

[3] http://southasia.oneworld.net/news/india-more-ngos-than-schools-and-health-centres#.UeSxtkI0Ib4

[4] http://abong.org.br/ongs.php

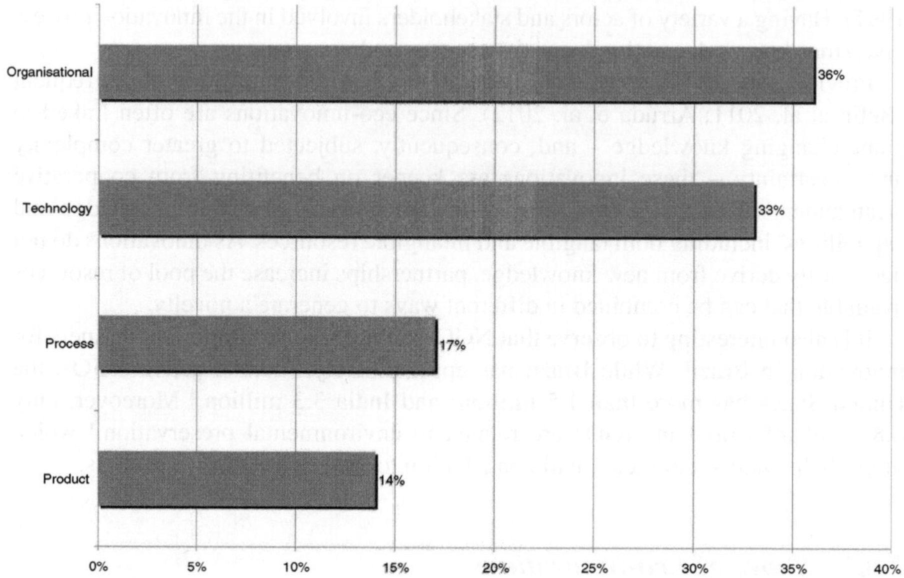

Fig. 7.3 Types of eco-innovation

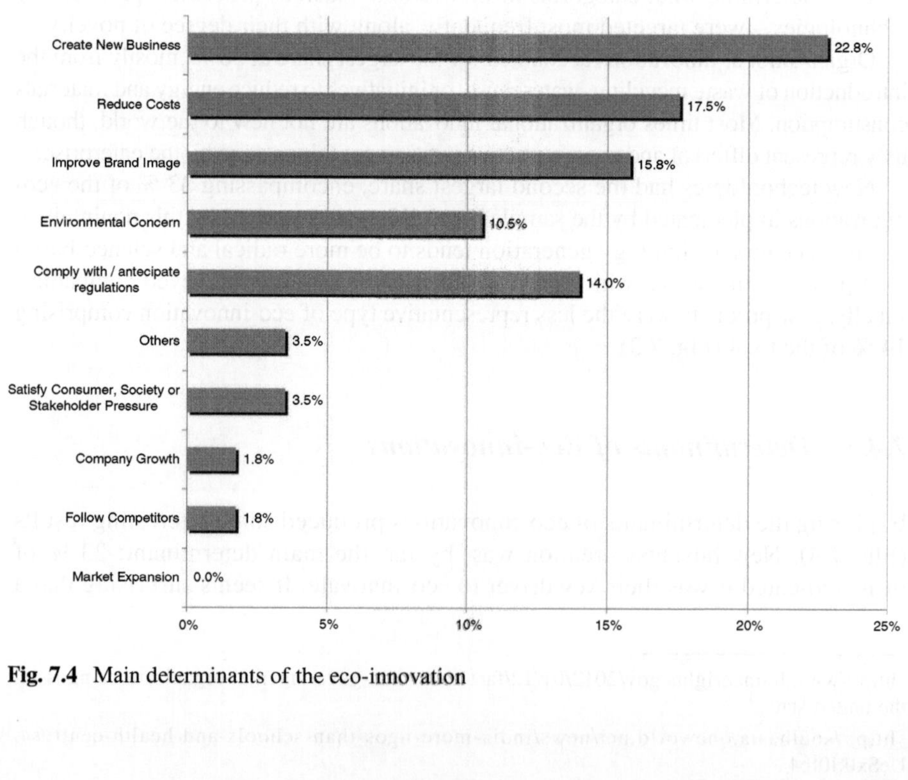

Fig. 7.4 Main determinants of the eco-innovation

number of firms were eco-innovating to create new businesses, which means they are generating products and technologies that can help markets to become greener, or that they are adding environmentally beneficial products and technologies to their portfolios. Justification for new-business creation as the key driver of eco-innovation was supported by the perception that environmental strategies are niche activities that are more knowledge-intensive than general innovations. This group of firms was of special interest because the technologies or products they disseminate throughout the economy could enable other firms to employ more environmentally friendly production methods. In other words, these firms' core businesses are eco-innovations and, consequently, they approach environmental concerns strategically.

Cost reduction was the second most important determinant: 17 % of the sample firms stated it was the reason for their eco-innovation. Actually, we expected cost reduction to be the main determinant, a trend that would have conformed with the incremental approach to general innovation more frequently employed by Brazilian firms. Cost reduction was followed by brand reputation and image improvement, for 16 % of the respondents. These qualities were critical to businesses, especially large firms in industries that degrade the environment and in an economy with increasing prices for environmental resources.

Environmental concerns, in turn, were the main determinant for only 10 % of eco-innovative firms. Reductions in waste production and energy consumption were the most mentioned environmental concerns. It is curious and somewhat worrisome that environmental concerns did not motivate many firms to take action. The behavior of the majority of the firms demonstrated that they are not adopting a proactive approach to important environmental issues or that their desire to eco-innovate is based mostly on traditional business goals, such as longevity, competitiveness, and profit. A reactive approach to environmental concerns, however, made it less likely for environmental strategies to become part of firms' corporate culture.

Fourteen percent of companies listed regulatory compliance as their main determinant. Brazilian environmental law is one of the most rigid in the world, and the low percentage of companies listing compliance to regulations as the main determinant was surprising. The regulatory push was a critical factor for eco-innovation in EU firms (Belin et al. 2011; Demirel and Kesidou 2011). Moreover, literature has discussed the role of regulations as drivers of firms' competitive advantage by using innovation to comply with rules and decrease environmental threats (Porter and van der Linde 1995). The results from our questionnaire indicated that Brazilian regulations are either difficult to translate into innovation opportunities or that Brazilian environmental laws are not tough enough to put firms onto an eco-innovative path. Possibly, firms in our sample were eco-innovating beyond compliance, using environmental opportunities as a way to differentiate their companies in the marketplace.

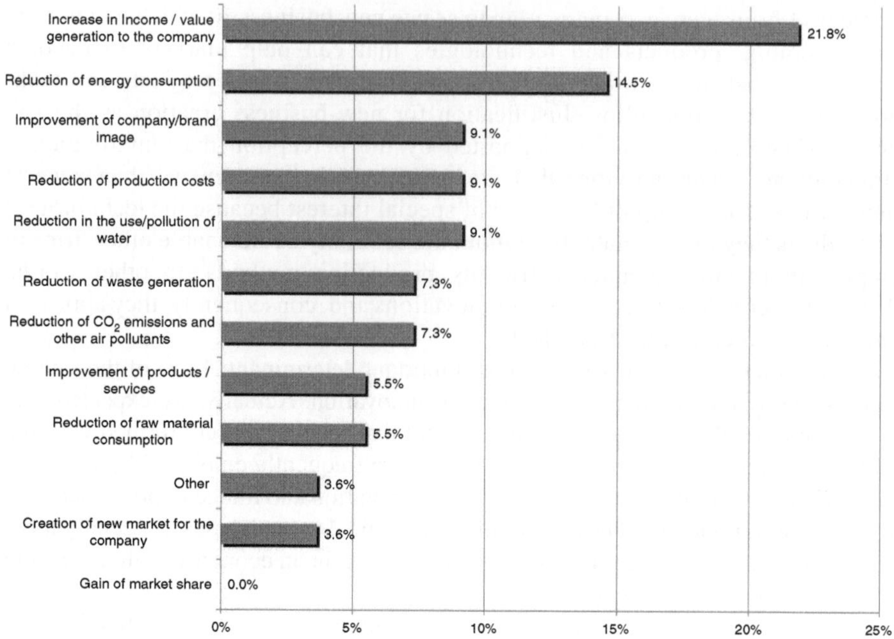

Fig. 7.5 Main results of the eco-innovations

7.4.4 Results of Eco-innovations

Firms were also asked to report the results obtained from their eco-innovation efforts (Fig. 7.5). The most important result firms mentioned was value creation: the generation of additional income through the sale of eco-innovative products, services, or technologies. The second result obtained was a reduction in energy consumption, cited as the main result by approximately 15 % of our sample firms. Other important results included an enhanced reputation and decreased production costs.

Interestingly, most firms in our sample reported being engaged in extracting value from their eco-innovation, and for that reason they tend to conduct initiatives that are more prone to generate competitive gains. Therefore, outcomes, which are usually associated with cost reduction – such as improving efficiency in the use of materials – are not among the highest results to this sample.

In sum, two distinct groups of eco-innovative firms emerged from our sample. The first group employed a reactive and limited approach to innovation, and we can correctly assume this group obtained limited benefits from their eco-innovations. These firms tend to eco-innovate as a response to regulatory demands or to pressures from competitors or clients, possibly by adapting their production processes toward more efficient resource use. A second group took advantage of the tremendous opportunities for innovation that environmental issues generated. These firms are making high profits by serving emerging market demands and by creating new market niches with environmentally responsible products or technologies.

7.5 Concluding Remarks

The purpose of this chapter was to draw a portrait of eco-innovations in Brazil. Since no earlier study had consistently examined the drivers, methods, dimensions, and impacts of eco-innovations in Brazil, we aimed to fill this gap in the existing literature, focusing our attention on some key variables: characteristics of eco-innovative firms; the determinants, types, and results of eco-innovations; and the existence of cooperative arrangements for eco-innovation.

Our results highlight the existence of two groups according to their approach to eco-innovations. The largest group reactively responds to pressures from their main stakeholders, adapting their production processes to these new demands or as means of optimizing their use of natural resources (such as energy). The smallest group performs radical innovations. Despite the low number of firms in this category, some patterns can be observed: they are relatively young, small, and often emerged as spin-offs from universities. This suggests that opportunities do exist in Brazil, but not many firms are seeking radical impacts with eco-innovations. The ones that do pursue radical paths tend to be more science-based and focus on specific technological niches.

The findings reveal that creating new market segments is an important driver of eco-innovation in the sample firms: a result we corroborate with theoretical observations that green innovations should be viewed as opportunities, rather than constraints to economic activity. More specifically, these opportunities can fulfill existing consumer demands for environmentally friendly products and services or create new market segments. The latter is potentially disruptive, since it relies more heavily on novel technological trajectories, business models, products, and services.

We should reiterate that although regulatory policies are not a main driver of eco-innovation for Brazilian firms, these policies have the potential to create markets for environmental solutions and technologies, such as waste management – an organizational innovation often mentioned by respondents. Future governmental policies could shape businesses' environmental impacts and stimulate further eco-innovations.

This observation supports the Porter hypothesis and the perception that environmental regulations can enhance firms' competitiveness by forcing them to invest in energy efficiency and waste reduction and to pay more attention to their products' life cycles. Findings in Brazil are similar to findings from countries such as England and Germany, which disprove the widespread notion that weak environmental laws have a positive effect on countries' competitiveness. Yes, weak laws attract multinational companies unwilling to cope with strict environmental standards, but strong environmental legislation promotes growth by stimulating innovative behavior.

The evidence that eco-innovative firms conduct systematic, in-house R&D activities supports the prevailing theoretical description that internal R&D achieves better results in science-based innovation (Pavitt 1984), such as initiatives that involve environmental stewardship. Our study shows, though, that market pressures also are essential determinants of eco-innovation. In other words, as pointed out by Freeman (1979) for general innovation, eco-innovations also result from complex and nonlinear dynamics that are influenced by both market and technological aspects.

More than 85 % of eco-innovative firms participate in cooperative arrangements for innovation, a statistic that reflects the increased complexity of the knowledge required for eco-innovation. As suggested by Belin et al. (2009), innovations with positive environmental outcomes seem to demand to a larger extent on external sources of knowledge and information in order to take place. Public universities and research institutes are the main partners, suggesting that eco-innovation in Brazil might be science-intensive and that Brazilian universities are national benchmarks in environmentally related studies. This is consistent with evidences raised by Belin et al. (2009) suggesting greater dependence on research undertaken by public institutions. Suppliers are also important partners to eco-innovate, signalizing that eco-innovations have the potential of stimulating changes in the value chain of eco-innovators.

Regarding the nature of Brazilian eco-innovations, the research observed that most of them are organizational. This means that most firms are improving their initiatives in order to cut costs by optimizing the use of natural resources (such as energy and water) and minimizing environmentally hazardous outputs (as waste and emissions). These types of eco-innovation are consistent with the characteristics of the major group within our sample: firms generating incremental results. New technologies had the second largest share, due to the existence of a group that performs radical innovation, seizing competitive gains through exploring latent opportunities. These two types of eco-innovations are also consistent with their most cited results: market factors, such as value creation, and firm-specific factors, as reduction in use of energy.

Finally, we want to reemphasize that our study was limited to 98 companies. Although our sample included enterprises of different sizes and from different sectors, it is not representative of the entire Brazilian business landscape. We believe that the sample helps us to increase our understanding of eco-innovation determinants and profiles of eco-innovators, but it does not allow us to generalize our findings or to apply them to all Brazilian firms. Nevertheless, studies mapping key variables related to eco-innovations are essential to subsidize public policies aiming at fostering a more innovative and environmentally beneficial landscape, also stimulating academic debates on an important topic to address issues on sustainable development. Our work analyzes descriptive results from the sample companies; further quantitative studies would enable us to better comprehend the variables and their interconnections.

References

Arruda C, Carvalho F, Ferreira G, Dutra H (2012) Cooperação e Inovações Ambientais: Uma Análise de Empresas Brasileiras a partir da PINTEC. Paper presented at the XXXVI EnANPAD, Rio de Janeiro

Arundel A, Kemp R (2009) Measuring eco-innovation. UNU-MERIT working paper series, no. 17. UNU-MERIT, Maastricht

Bansal P, Roth K (2000) Why companies go green: a model of ecological responsiveness. Acad Manage J 43(4):717–736

Barbieri J, Vasconcelos I, Andreassi T, Vasconcelos F (2010) Inovação e Sustentabilidade: Novos Modelos e Proposições. RAE, São Paulo 50(2):146–154

Belin J, Horbach J, Oltra V (2009) Determinants and specificities of eco-innovations – an econometric analysis for France and Germany based on the Community Innovation Survey. In: DIME workshop on environmental innovation, industrial dynamics and entrepreneurship, Utrecht University

Belin J, Horbach J, Oltra V (2011) Determinants and specificities of eco-innovations: an econometric analysis for the French and German Industry based on the community innovation survey. DIME (Dynamics of Institutions and Markets in Europe), working paper no. 10

Benfratello L, Sembenelli A (2002) Research joint ventures and firm level performance. Res Policy 31:493–507

Bernauer T, Engels S, Kammerer D, Seijas J (2006) Explaining green innovation. Center for Comparative and International Studies (CIS), ETH Zurich, University of Zurich, working paper no. 17

Chesbrough H (2003) Open innovation: the new imperative for creating and profiting from technology. Harvard Business School Press, Boston

Clark W, Crutzen P, Schellnhuber H (2005) Science for global sustainability: toward a new paradigm. Center for International Development (CID), Harvard University, working paper no. 120, March, 2005

Cohen M (1997) Risk society and ecological modernization: alternative visions for post-industrial nations. Futures 29(2):105–119

Cortês M, Pinho M, Fernandes A, Smolka R, Barreto A (2005) Cooperação em empresas de base tecnológica: uma primeira avaliação baseada numa pesquisa abrangente. São Paulo em Perspectiva 10(I):85–94

Demirel P, Kesidou E (2011) Stimulating different types of eco-innovation in the UK: Government policies and firm motivations. Ecol Econ 70(8):1546–1558

Eurostat (2012) Science, technology and innovation in Europe. Available at: http://epp.eurostat.ec.europa.eu/statistics_explained/index.php/Main_Page

Ferro AFP (2010) Gestão da inovação aberta: práticas e competências em P&D Colaborativa. Tese (Doutorado em Política Científica e Tecnológica) Programa de Pós-Graduação em Política Científica e Tecnológica, Instituto de Geociências Universidade Estadual de Campinas

Freeman C (1979) The determinants of innovation. Futures 11(3):206–215

Gil A (1999) Métodos e técnicas de pesquisa social, 5th edn. Atlas, São Paulo

Grey W (1993) Anthropocentrism and deep ecology. Aust J Philos 71(4):463–475

Hagedoorn J (2002) Inter-firm R&D partnerships: an overview of major trends and patterns since 1960. Res Policy 31:477–492

Hall J, Vredenburg H (2003) The challenges of innovating for sustainable development. Sloan Manage Rev 45(1):61–68

Hart S (1997) Beyond greening: strategies for a sustainable world. Harv Bus Rev 75:66–76

Horbach J, Rammer C, Rennings K (2012) Determinants of eco-innovations by type of environmental impact: the role of regulatory push/pull, technology push and market pull. Ecol Econ 78:112–122

IBGE (2010) Pesquisa de Inovação Tecnológica – PINTEC 2008. IBGE, Rio de Janeiro

Jacobs M (1997) The quality of life: social goods and the politics of consumption. In: Jacobs M (ed) Greening the millennium? The new politics of the environment. Blackwell Publishers, Oxford, pp 47–61

Kemp R, Foxon T (2007) Typology of eco-innovation. MEI (Measuring Eco-Innovation Project), Maastricht, working paper

Kemp R, Pearson P (2007) Final Report MEI (Measuring Eco-Innovation Project), deliverable 15

Lins S (2009) Sinergia Fator de Sucesso nas Realizações Humanas. Campus-Elsevier, Rio de Janeiro

Pavitt K (1984) Sectoral patterns of technical change: toward a taxonomy and a theory. Sci Policy Res Univ Sussex 13:343–374

Porter M, van der Linde C (1995) Toward a new conception of the environment-competitiveness relationship. J Econ Perspect 9:97–118

Rea L, Parker R (2005) Designing and conducting survey research: A comprehensive guide. Jossey-Bass, San Francisco

Rockström J, Steffen W, Noone K, Persson Å, Chapin FS III, Lambin E, Lenton T, Scheffer M, Folke C, Schellnhuber H, Nykvist B, de Wit C, Hughes T, van der Leeuw S, Rodhe H, Sörlin S, Snyder P, Costanza R, Svedin U, Falkenmark M, Karlberg L, Corell R, Fabry V, Hansen J, Walker B, Liverman D, Richardson K, Crutzen P, Foley J (2009) Planetary boundaries: exploring the safe operating space for humanity. Ecol Soc 14(2):32

Teece D (1992) Competition, cooperation, and innovation: organizational arrangements for regimes of rapid technological progress. J Econ Behav Organ 18(1):1–25

Veugelers R (1997) Internal R & D expenditures and external technology sourcing. Res Policy. Elsevier, 26(3):303–315

Young C, Lustosa M (2007) Meio Ambiente e Competitividade na Indústria Brasileira. In: Revista de Economia Contemporânea, vol 5. IE-UFRJ (Institute of Economics), Rio de Janeiro, pp 231–259

Chapter 8
Contextual Factors as Drivers of Eco-innovation Strategies

The Definition of an Organizational Taxonomy in the Brazilian Cellulose, Paper, and Paper Products Industry

Marlete Beatriz Maçaneiro and Sieglinde Kindl da Cunha

Abstract This chapter sought to identify an organizational taxonomy through the reactive and proactive eco-innovation strategies adopted by firms in the Brazilian cellulose, paper, and paper products sector. The taxonomy was the result of a cluster analysis and an analysis of the frequency of internal and external contextual factors in the organization in each cluster. The quantitative empirical approach was used, through a cross-sectional survey. Data collection was conducted using a questionnaire answered by 117 firms in the sector. The data were mostly analyzed using analysis of variance (ANOVA) and Tukey's test. The analysis was conducted using an *eco-innovation strategies* construct composed of 13 variables, with a later analysis of the impact of 28 variables composing six structures of contextual factors. Among the main results was the definition of taxonomy for the firms in the sample that was identified through their eco-innovation strategies. They were defined as reactive, indifferent, proactive, or eco-innovative organizations. This taxonomy was defined from the eco-innovation strategies and their contextual factors, which we consider to be an advance in the literature on this theme.

Keywords Eco-innovation • Environmental strategies • Organizational taxonomy • Eco-innovation drivers

---

M.B. Maçaneiro (✉)
State University Midwest – UNICENTRO, Guarapuava, PR, Brazil
e-mail: marlete.beatriz@yahoo.com.br

S.K. da Cunha
Environmental Management, Positivo University, Curitiba, PR, Brazil
e-mail: skcunha21@gmail.com

S. Azevedo et al. (eds.), *Eco-Innovation and the Development of Business Models*, Greening of Industry Networks Studies 2, DOI 10.1007/978-3-319-05077-5_8,

8.1 Introduction

Many aspects of environmental sustainability have been debated in academia in recent decades in a wide range of fields of knowledge, with concerns being raised over the environment. Meanwhile, the theme of innovation has focused strictly on economic concerns, such as competitiveness and demand pressures. There have been obstacles to incorporating the inherent processes of these phenomena as environmental management in the context of innovation and innovation management supported by the assumptions of the environmental field.

This line of study deals with eco-innovations, which are innovations that stress sustainable development. Throughout their lifecycle, they focus on reducing environmental risk, pollution, and other negative impacts that stem from the use of resources compared with alternative uses (Arundel and Kemp 2009; Rennings 1998).

Authors who study this theme point out that there have been relatively few studies and actions that focus on the overlapping of innovation and environmental sustainability. This results in theoretical, methodological, and political uncertainty in terms of implementation and management (Andersen 2006, 2008; Andrade 2004; Arundel and Kemp 2009; Baumgarten 2008; Maçaneiro and Cunha 2010; Maçaneiro 2012). Therefore, this field is classified as one that has not been widely explored, especially in Brazil, but which has been attracting growing attention in the international literature, especially in the European Union and the United States.

Blackburn (2008) claims that environmental management programs are not always given due consideration by firms and are often not included in their essential strategies. The matter is considered more as a way out in reaction to activists, the media, and environmental regulatory agencies. When this happens, firms perceive environmental management as a risk, in terms of cost, that can also harm their reputation, sales, and growth. Other authors have corroborated this point of view, including Foxon and Andersen (2009), Lustosa (1999), Nidumolu et al. (2009), Romeiro and Salles Filho (1996), and Young et al. (2009). They point out that firms believe that the additional costs of handling environmental matters will be high and that these costs are a threat to their survival, making them less competitive. In many cases, this issue is not handled as a proactive business strategy but rather as a reactive strategic action.

However, environmental matters should be viewed by firms as a stimulus for innovation and technological, economic, and competitive opportunities, which are proactive strategies. Managerial knowledge and attitudes to technological changes and environmental concerns should be encouraged by environmental regulations. In other words, regulations should spur firms to innovate, and firms should see this pressure as a chance to improve productivity and competitiveness (Ansanelli 2003; Ashford 2000; Porter and van der Linde 1995).

In addition to regulations, other contextual factors have an impact on eco-innovation strategies and consequently on the environmental performance of organizations. These factors are linked to the internal and external environment of firms, including government incentives and the effects of a firm's local reputation. They also include how it is viewed as part of its sector and in terms of market conditions that affect the organization's image. There are also internal organizational factors. Taking internal and internal factors into account is crucial for a firm's competitiveness because it enables it to define realistic proactive strategies. According to Menguc et al. (2010, p. 280), "Both the internal and external perspectives [...] are complementary and capture the extent of a firm's social performance and responsiveness." Nevertheless, Sharma et al. (2007, p. 269) claim that "even with the evidence accumulated over the last decade that proactive environmental strategies are likely to be accompanied by improved financial performance, we still lack a well-developed understanding of why only some firms in an industry implement such strategies."

In this sense, it is important to analyze the drivers of eco-innovation strategies and the factors that compose them in order to verify the relevance of these drivers in the definition of organizational strategies. It is also important to define an organizational taxonomy in terms of the adoption of eco-innovation strategies, which remains a gap in the literature. To view these issues, we decided that this study would be conducted at firms in the cellulose, paper, and paper products sector in Brazil. The activities of this sector are potentially highly polluting and use natural resources, as stated in the law that enacted the Brazilian National Environmental Policy (Brasil 1981).

According to Schmidheiny (1992), global demand for industrial wood rose gradually with industrialization in developed and developing countries. This led to a tendency to use reconstituted wood (particleboard and pressed wood) and wood fiber products instead of solid wood. In the paper industry, there was also concern over environmental issues. In the European Community, 50 % of the paper used in the 1990s was already made of recycled fiber. According to Miles and Covin (2000), the American forest products industry has shown an interest in environmental activities due to factors that affect its marketing and financial performance. This industry has developed major initiatives that encourage responsible environmental management. This has improved the reputation of the industry in terms of credibility, integrity, reliability, and responsibility.

In the case of Brazil, according to Juvenal and Mattos (2002, p. 1), this industry is internationally competitive and has an "[...] extremely advanced technological base that is capable of ensuring constantly increasing productivity." The Brazilian Association of Cellulose and Paper (Bracelpa 2012) points out that the sector is among the largest producers of cellulose and paper in the world. In 2010, Brazilian production was 14,164,369 t of cellulose, putting Brazil in fourth place in the world, behind only the United States, China, and Canada. In terms of paper, Brazil was in tenth place, with a total of 9,843,747 t in 2010.

The technological development of this sector in Brazil grew in leaps and bounds beginning in the 1950s, with investments in improving technologies and mature processes. According to Barbeli (2008, p. 108), "the production of paper and cellulose was constituted in a contextualized activity in a highly competitive and widely globalized market, whose production plants resort to relatively consolidated technological processes." Therefore, this industry developed sophisticated technology related to forests and achieved certified quality levels. Juvenal and Mattos (2002) highlight that the Brazilian cellulose and paper industry is currently supplied exclusively by planted forests, meaning high industrial yields and returns and guaranteeing low costs.

In this sense, eco-innovation in the sector has been constituted as an essential factor to its development due to the competitiveness of the sector, demands from clients, and strong competitors. Furthermore, when it comes to environmental aspects, as this is a Brazilian sectors that mostly deals in exports, this helps it to adapt to international environmental standards as a strategic factor (Serôa da Motta 1993). Nevertheless, the sector emits an excessive amount of nitrogen oxide, in addition to a high rate of organic loading and sulfur dioxide. This means that in this sector "[…] showing positive characteristics regarding environmental impact has become an increasingly important competitive element" (Romeiro and Salles Filho 1996, p. 106). According to Juvenal and Mattos (2002, p. 18), the Brazilian cellulose, paper, and paper products sector "[…] has incorporated the most rigorous standards. In addition to the adaptation of industrial units, paper recycling has reached a level of approximately 45 %."

These mean that this sector is an important subject to study and highly relevant to the development of the country. Therefore, the purpose of this chapter is to identify an organizational taxonomy from the proactive and reactive eco-innovation strategies adopted by firms in the Brazilian cellulose, paper, and paper products sector, resulting in a cluster analysis and an analysis of the impact of internal and external contextual factors on the organizations in each cluster.

8.2 Reactive and Proactive Eco-innovation Strategies in Environmental Management

The concept of eco-innovation is relatively new, stemming from recent debates and concerns over environmental impacts. The term eco-innovation itself was used for the first time by Fussler and James in their book *Driving Eco-Innovation*, published in 1996. It is defined as the production, application, or exploitation of a good, service, production process, organizational structure, management, or business method that is new to a firm or user. The results, during its lifecycle, are a reduction in environmental risks, pollution, and the negative impacts of the use of resources in comparison with their corresponding alternatives (Kemp and Foxon 2007; Arundel and Kemp 2009; Rennings 1998). According to Könnölä et al. (2008, p. 3), "While it is namely environmental impacts that define eco-innovation, economic and social impacts play a crucial role in its development and application and hence determine its diffusion path and contribution to competitiveness and overall sustainability."

Reid and Miedzinski (2008) highlight that eco-innovation can be considered as every type of innovation that leads to the use of fewer resources and less energy in the extraction of material and the manufacture, distribution, reuse, and recycling of a product[1] and its disposal, as long as resources are less intensely used in terms of a

[1] "Reuse means reusing materials and conserving their original properties or characteristics even after they have been used for an identical or similar use, as is the case of returnable packaging. Recycling is the transformation of residuals into new raw materials, involving the collection, processing and trade of residuals" (Barbieri 2002, p. 43).

product's lifecycle. Furthermore, Rennings (1998, p. 5) claims that "Eco-innovations can be developed by firms or non-profit organizations, they can be traded on markets or not, their nature can be technological, organizational, social or institutional." Technologies can be either curative or preventive, with the former repairing damage to the environment and referred to as end-of-pipe solutions.[2] The latter attempt to avoid damage and are known as cleaner production solutions[3] and are part of a broader approach. Organizational eco-innovations are those changes in management instruments (eco-audits) and innovations in services (management of energy demand and management of the transport of residuals). These require new infrastructure and changes to the system that go beyond changes to a given technology. Social eco-innovations, meanwhile, are expressions of sustainable consumption patterns, which have attracted increasing attention and are considered as changes in the values of people and their lifestyles, shifting towards sustainability. Finally, institutional eco-innovations are innovative institutional responses to problems of sustainability, such as local networks and agencies, global governance, and international trade.

Arundel et al. (2003) add that eco-innovation can be considered technical when it comes to new equipment, products, and production processes and organizational when it comes to structural changes within the organization to institute new habits, routines, and guidelines for the use of tools and environmental programs or it can be used as a business strategy. "Successful environmental innovation may often require both technical and organizational change" (Arundel et al. 2003, p. 325). From the moment when eco-innovations become integral parts of the corporate strategy of firms, solutions cease to be end-of-pipe and shift towards becoming preventive.

When curative technologies are used in an industrial system, damage is reduced but costs are high because equipment to control pollution is unproductive and may not reduce social costs (Hart 1995; Barbieri 2002). Nevertheless, technologies that prevent pollution are innovations that lower pollution levels and result in improved quality, performance, safety, and lower costs, with products having a higher resale value and being more likely to be recycled or reused.

In short, in this study eco-innovation is understood as an *innovation that consists of changes and improvement in environmental performance within the dynamic of the greening of products, processes, business strategies, markets technologies, and innovation systems. In this context it is defined by its contribution to the reduction of environmental impact of products, services, and organizational processes.*

It is in this context of eco-innovation that firms adopt different ranges of strategies to handle environmental issues, such as reactive and proactive strategies. Barbieri (2007), Sharma (2000), and Sharma et al. (1999) define reactive strategies as results

[2] End-of-pipe technologies are solutions that aim only to control pollution that has already been produced and come into play at the end of the production process, with no other substantial change in the lifecycle of the product (Barbieri 2002; Lustosa 2003).

[3] "Cleaner production is a wider approach to protecting the environment as it operates throughout all the phases of the manufacturing process and lifecycle of a product, including its use at home and in the workplace" (Barbieri 2002, p. 40).

in the form of actions for complying with regulations, i.e., they are actions that are externally imposed through environmental legislation. These strategies are nothing more than making sure that firms comply with legislation by controlling pollution. They invest in corrective technologies to resolve problems at the end of the production process, i.e., end-of-pipe technologies. This does not require a firm to develop skills or abilities for producing new technologies or new environmental processes.

Firms that employ reactive strategies do not view environmental management as a priority. They invest only to comply with environment regulations. These are viewed as an institutional restriction and even a threat in the form of additional costs. They are not viewed as an opportunity to improve managerial practices. Furthermore, the involvement of the top management in this aspect of the business is merely sporadic, as their environmental actions are confined to the causes of pollution (Barbieri 2007; Buysse and Verbeke 2003).

> From a business point of view, this approach means higher production costs that do not add any value to the product. They are also hard to reduce because they are legal requirements. [...] If these costs are added to the price of the end product, this type of solution is also of no interest to consumers. Understanding environmental concerns as an additional cost for firms and consumers is one of the most deeply ingrained business paradigms and is an obstacle to businesses becoming more active in the search for a solution to these problems. [...] From a business viewpoint, solutions that only seek to control pollution are fundamental, but insufficient. (Barbieri 2007, pp. 121–122)

On the other hand, proactive strategies are voluntary actions that reduce the environmental impact of operations, creating a competitive advantage by adopting eco-innovative technologies. These innovations are defined as environmental strategies for the *prevention of pollution* or *voluntary strategies*, which require the acquisition of new technologies, involving greater learning and the development of competitive organizational skills. In this case, the actors view environmental issues as an opportunity for competitive gains (Aragón-Correa 1998; Barbieri 2007; Sharma et al. 1999).

"In addition to monitoring and preventing pollution, the firm seeks market advantages and to neutralize threats resulting from existing environmental issues or those that might occur in the future" (Barbieri 2007, p. 125). Thus, firms construct chains of sustainable values by creating environmentally friendly products and services. They concentrate on reducing the consumption of nonrenewable and renewable resources, extending this idea from the factory and offices to the value chain. They also come to understand the concerns of consumers and check their sources of raw materials and distribution in partnership with nongovernmental organizations. The organizations acquire the necessary skills to understand how renewable and nonrenewable resources affect the ecosystems of business and industry, combining business models, technologies, and regulations in different industries. This ends up reducing costs and even creating new sustainable business models, leading to new forms of distribution and providing value to clients, which will all change the competitive base. This is where paradigm shifts occur in relation to environmental protection, resulting in the adoption of eco-innovations (Nidumolu et al. 2009). "The prevention of pollution increases a firm's productivity because fewer pollutants at the source mean resources saved, which enables the production of more goods and services with fewer inputs" (Barbieri 2007, p. 122).

The consequences of opting or not opting to manage eco-innovation mean a place among the competition or remaining in or leaving the market. Nowadays, amidst the aspects linked to the management of eco-innovation, firms should cease to consider environmental protection a legal requirement punishable by fines and sanctions. In this new scenario, this issue has become part of the organizational strategies for identifying threats and opportunities (Donaire 2007). For this purpose, organizational processes are needed that require a diversity of resources and considerable efforts on the part of entrepreneurs to make them sustainable.

8.3 Methodological Procedures

The study used a quantitative empirical approach, employing a cross-sectional survey conducted at 117 firms in the Brazilian cellulose, paper, and paper products sector. The data collection instrument was a self-administered computerized questionnaire (Hair et al. 2005), which was forwarded by the Internet using the Qualtrics® system. It was modeled in the form of an opinion poll, with the questions inquiring as to the perception of the respondents using a balanced 5-point scale with a neutral option, in the same style as the Likert scale. The content validity of the questionnaire was pretested by three specialist professors. It was also pretested with three managers in charge of environmental management of firms in the sector and two university professors in the fields of strategy and sustainability. The data were collected between July and October 2012.

The sampling for the research was defined as non-probabilistic (not random), in that the probability of the elements of a population being chosen is not the same. Therefore, the sample was defined by willingness to participate, in that certain people were invited to complete the questionnaire but could decide whether or not they were willing to take part in the study (Cooper and Schindler 2011). From a total population of 3,147 companies in the sector (IBGE 2010), we had access to the e-mail addresses and telephone numbers of 672 companies from all over Brazil. The data collection instrument was forwarded to these companies, with a response rate of 17.4 %.

Prior to the data analysis, the initial step was to validate and cleanse the data to eliminate possible flaws and distortions resulting from errors made during the completion of the questionnaire and verify missing values in the responses and outliers. For this purpose, an analysis was made of each variable per construct, using a Boxplot. No outliers were detected among the responses and neither were there missing values within the questions as the electronic system in use does not allow respondents to leave questions unanswered. What did happen was that some questionnaires were not finalized, with the respondents abandoning them and leaving them incomplete. These questionnaires were discarded, leaving a total of 117 valid questionnaires.

Table 8.1 Representativeness of sample companies (Data from field research, 2012, and IBGE 2010)

Region	Companies included in the PIA		No. of sample companies	Residual
	No. of existing	No. of expected		
Southeast	1,828	68.0	70	2.0
South	881	32.8	39	6.2
Northeast	291	10.8	4	–6.8
North	42	1.6	2	0.4
Midwest	105	3.9	2	–1.9
Total	3,147		117	

Statistical significance according to the chi-square test ($p > 0.05$) = 0.159

To verify the national representativeness of this composition, we analyzed the participation of companies in the *Pesquisa Industrial Anual* (PIA) of the IBGE (2010). Specifically, we performed a chi-square (goodness of fit) test to assess whether the frequency of the sample by Brazilian state is statistically representative of the overall population (Maroco 2003). The results are presented in Table 8.1.

The PIA comprised 3,147 companies, which resulted in the expected numbers after performing the chi-square (goodness of fit) test (see Table 8.1). Thus, it was possible to compare the number of sample firms by region and to infer that the sample was representative, since we obtained a significance level above 0.05, indicating no statistically significant differences (Hair et al. 2005).

In terms of size, the majority of responding organizations (45 %) are medium-sized, according to the Sebrae (2012) criterion. Another significant proportions are small (34 %), followed by large (17 %) and microenterprises (4 %). The characterization of sample firms can be summarized as follows: most are of Brazilian origin (97 %) and they are controlled by domestic capital exclusively (80 %). The mean time in existence is 36 years, and they operate in the domestic market only (63 %). Survey respondents mostly work in managerial positions (66 %) and have worked for the firm for a mean of approximately 10 years.

The respondents to whom the instrument was forwarded were responsible for the environmental management area/sector/division or similar department at their firms and were either managers, directors, or owners. Of the respondents, 30 % are directly employed in positions related to the environmental area of the firm, 36 % hold administrative positions, 20 % hold positions in quality and R&D, and 14 % work in production. In general, of the total number of respondents, 66 % hold directors' or managerial positions in these different areas of their organizations, which increases the quality of the data collected for this study. Furthermore, the respondents have been at their companies for an average of 10 years, which can be considered a significant period of time for the respondents to garner the necessary knowledge regarding the firm to complete the questionnaire.

The main techniques used for the data analysis were cluster analysis techniques, analysis of variance (ANOVA), and Tukey's test. These analyses were conducted from a construct denominated *eco-innovation strategies*, composed of 13 variables, with the later analysis of the incidences of 28 variables that made up six constructs of contextual factors. These sets and variables will be presented in the following section.

8.4 Composition of the Constructs and the Constitutive and Operational Definitions of the Variables of the Study

The *eco-innovation strategies* construct was divided into *proactive strategies* and *reactive strategies*. For the constitutive definition of this construct, it was considered that "firms with a reactive strategy attach high importance to government regulation, but only in a static sense, as an almost mechanistic and daily routine-driven response to new regulatory requirements" (Buysse and Verbeke 2003, p. 460). These are expensive and normally unproductive processes that reflect a reactive and selective posture regarding environmental issues, basically with end-of-pipe technologies. The firm "[...] centers its attention on the negative effects of its products and processes through point solutions. [...] seeking to control pollution without significantly altering the processes and products that produce it [...]" (Barbieri 2007, p. 118).

Proactive strategies, on the other hand, are voluntary actions that seek to lower pollution and other environmental impacts of a firm's operations with the support of the upper management and at the same time create a competitive advantage by adopting innovative environmental technologies. These actions are defined as *prevention of pollution* or *voluntary strategies*. They require the installation of new technologies and involve constant learning and the development of competitive organizational skills and improved total quality of the organization. These strategies are seen in a competitive light, and the term is used to describe innovating activities for the prevention of pollution (Buysse and Verbeke 2003; Hart 1995; Menguc et al. 2010; Sharma 2000; Sharma et al. 1999).

For the operational purposes of this study, the development of eco-innovative strategies was determined through variables that composed the *eco-innovation strategies* construct, based on the literature, as shown in Table 8.2. Thirteen variables were included in this question, using a 5-point Likert scale, that varied from *I totally disagree* to *I totally agree*. The mean of the first five variables composed the reactive strategies construct and the mean of the others composed the proactive strategies construct.

The literature contains a series of factors that can affect the formulation and type of environmental strategy adopted by organizations, which impact each reality differently. In this study, six constructs of contextual factors for the organizations were used, including internal and external factors – (1) external factors: *environmental regulation*, the *use of environmental and innovative incentives*, and *reputational effects* and (2) internal factors: *top management support*, *technological competence*, and *environmental formalization*. It should be emphasized that to conduct this study, these factors were defined in constitutive and operational terms for the composition of the research variables, based on the literature under study.

Table 8.2 Variables employed to measure the *eco-innovation strategies* construct

	Variable	*Question*: Using the following options, evaluate the degree of development in your firm concerning environmental strategy
Reactive strategies	Var01	The firm is only concerned with pollution at the end of the production process, using remediation techniques such as the decontamination of damaged soil
	Var02	The firm only acquires end-of-pipe technologies to deal with pollution before it enters the environment, such as sewage treatment plants, electrostatic precipitators, incinerators, and air pollution control equipment
	Var03	The firm invests in environmental actions only to comply with environmental legislation
	Var04	The firm invests in environmental technology and actions only to solve problems with activists and the media
	Var05	The firm views environmental management as an additional cost that can harm the growth of business
Proactive strategies	Var06	The firm uses marketing resources to handle environmental management
	Var07	The firm develops environmental strategies in its administrative work (recycling paper, use of recycled materials, reduced use of material, etc.)
	Var08	The firm uses environmental action in production (minimizing residuals, use of renewable energy, reuse of water, safe treatment and disposal of dangerous residuals, reduced CO_2 production, reuse of raw material, etc.)
	Var09	The firm conducts regular environmental audits
	Var10	The firm conducts an environmental analysis of the lifecycle of its products
	Var11	The firm has partnerships/agreements with other firms/institutions for environmental actions
	Var12	The firm has or makes possible environmental training programs for its managers and employees
	Var13	The firm has a system to prevent environmental accidents

Source: Prepared by the authors, based on Aragón-Correa (1998), Barbieri (2007), Blackburn (2008), Buysse and Verbeke (2003), Caracuel et al. (2011), Donaire (2007), Foxon and Andersen (2009), Fussler and James (1996), Nidumolu et al. (2009), Sharma (2000), Sharma et al. (2007), and Sharma et al. (1999)

The *environmental regulation* instruments are those defined as legal norms regarding environmental performance, such as command and control. There are also economic instruments, which affect costs and consumption, in addition to self-regulation by firms or industrial sectors (Schmidheiny 1992). More specifically, government actions concerning environmental issues are structured in different ways in terms of regulations of command and control, incentives, and subsidies (Kanerva et al. 2009).

To measure this construct, four variables were used, as shown in Table 8.3. A 5-point balanced scale with a neutral option was used, with the degree of relevance varying from very small to very large. The mean of the first two variables composed

Table 8.3 Variables employed to measure the *environmental regulation* construct

	Variable	*Question*: Using the options below, evaluate the degree of relevance in your firm of environmental regulations/legislation for each of the following variables
Cost/threat	Var14	Acquisition of technology to control pollution at the end of the production process
	Var15	Increased cost through fiscal and/or administrative sanctions due to responsibility for environmental damage, resulting in a threat to business growth
Opportunity	Var16	The development or acquisition of new products/processes/innovative technologies to prevent pollution, involving constant learning and developing organizational skills
	Var17	Regulation helps to guide the firm and helps it innovate, learn, and change its practices, with pressure to do so being seen as an improvement in productivity and competitiveness

Source: Prepared by the authors, based on Almeida (2010), Ansanelli (2003), Ashford (2000), Blackburn (2008), Buysse and Verbeke (2003), Hart (1995), Nidumolu et al. (2009), Porter and van der Linde (1995), Sharma (2000), and Sharma et al. (1999)

Table 8.4 Variables employed to measure the *use of environmental and innovative incentives* construct

Variable	*Question*: Using the options below, evaluate the degree of relevance of resources effectively obtained by the firm for environmental and innovative purposes
Var18	Nonrepayable government grants
Var19	Long-term government funding with special interest rates below the financial market rates (repayable)
Var20	Government support for the use of risk capital
Var21	Tax benefits for innovation and/or ecological products
Var22	International funding from investment funds for financing from international organisms and agencies

Source: Prepared by the authors, based on Finep (2012)

the *regulation viewed as a cost/threat* construct, and the mean of the other two composed *regulation viewed as an opportunity*. The mean of these constructs was translated to the *environmental regulation* construct.

In contextual factor *use of environmental and innovative incentives*, these incentives were considered as being defined in economic terms, supporting the incorporation of innovative and environmental technology in the organizations, and stimuli/restrictions on the private appropriability of the benefits of innovation, as measures to support technological innovation linked to the development, diffusion, and efficient use of new technologies and as incentives for support (Cassiolato and Lastres 2000; Dosi 1988).

To verify whether the firms in the study had ever contemplated any incentive for environmental innovation, they were asked the question in Table 8.4, with five variables and a 5-point scale with a neutral option, varying from *very small* to *very large*.

Table 8.5 Variables employed to measure the *reputational effects* construct

Variable	*Question*: In the options below, evaluate the degree of relevance of each factor/agent on the firm's actions to improve its image where environmental issues are concerned
Var23	Relationship with the supply chain (suppliers)
Var24	Conscious end consumers, industrial clients, and public clients
Var25	Relationships with environmental NGOs, business associations, media, or members of movements that aim to improve the environment or raise the environmental awareness of society
Var26	Environmental performance of competitors
Var27	Demand from investors to maintain profitability
Var28	Image in the eyes of more environmentally aware collaborators

Source: Prepared by the authors, based on Barbieri (2007), Buysse and Verbeke (2003), Camara and Passos (2005), Carrillo-Hermosilla et al. (2009), Donaire (2007), Nidumolu et al. (2009), and Passos (2003)

For the constitutive definition of the *reputational effects* construct, the work of Miles and Covin (2000) was referenced. To these authors, the reputation of a firm is translated into the perception of its most important stakeholders such as the owners, society, and the local and even international community, including current and future generations, clients, employees, suppliers and strategic partners, government agencies, banks, and other creditors and NGOs. Furthermore, Carrillo-Hermosilla et al. (2009) highlight that the main external factors pertaining to matters of effect of reputation that can have an influence are information and relationships with the supply chain and other stakeholders such as end consumers and public clients, environmental performance of competition to maintain competitiveness, relationships with business associations and NGOs that are sources of indirect and direct pressure regarding the development and adoption of eco-innovations, and social awareness, since civilian society can influence the adoption of environmental measures.

Reputational effects were measured using the question in Table 8.5, with six variables using a balanced 5-point scale with a neutral option ranging from *very small* to *very large*.

In the *top management support* construct, the study considered that "[…] leaderships subject to a more high-pressure external context with greater demands and opportunities concerning environmental issues tend to perceive more frequently that the environment plays a relevant role in firm business" (Souza 2004, p. 251). In this sense, key behaviors by the top managers include aspects such as capacity for management, authority, and influence to allocate adequate resources to environmental issues and define long-term environmental programs and policies. They also include obtaining the commitment and involvement of employees, rewarding them on environmental issues, including the involvement of stakeholders, the publication of environmental performance on the firm's image, and the support of interested parties (Berry and Rondinelli 1998).

The *top management support* was measured in the question posed in Table 8.6, which was formulated with four variables. A 5-point balanced scale was used, with a neutral option, which varied from a *very small* to a *very large* degree of relevance.

Table 8.6 Variables employed to measure the *top management support* construct

Variable	*Question*: Using the options below, evaluate the degree of relevance of the top management when defining the following variables
Var29	The upper management believes it is fundamental to handle environmental issues and have environmental programs and policies
Var30	The leaders of this organization have a policy to reward employees for environmental improvements
Var31	Organizational resources are earmarked for environmental initiatives
Var32	The leaders see the environment as highly strategic

Source: Prepared by the authors, based on Berry and Rondinelli (1998), Camara and Passos (2005), Donaire (1996), Menguc et al. (2010), Passos (2003), and Souza (2004)

Table 8.7 Variables employed to measure the *technological competence* construct

Variable	*Question*: Use the options below to evaluate to what extent the firm answers the following descriptions
Var33	The firm is considered the first to introduce new technologies and new products in the sectors
Var34	The firm has human resources to develop eco-innovations
Var35	It has the conditions to install and adapt to the adoption of new environmental technologies
Var36	The firm is engaged in collaborating with other institutions/organizations, forging strategic relationships and alliances

Source: Prepared by the authors, based on Carrillo-Hermosilla et al. (2009), and Menguc et al. (2010)

The *technological competence* construct was based on the idea that this is one of the factors that form the bases for adopting a proactive environmental strategy. This is because it enables high organizational skills such as learning, continuous innovation, and experimentation (Menguc et al. 2010), i.e., technological competence means the capacity to absorb, mainly as the result of innovation through investments in R&D (Cohen and Levinthal 1990).

Technological competence was measured by the question in Table 8.7, including four variables with the use of a 5-point Likert scale, ranging from *I totally disagree* to *I totally agree*.

The last construct was defined as *environmental formalization*, in which the impacting factors are those related to internal organizational structures for the adoption of organizational innovations to support eco-innovation (Kemp and Arundel 1998; Carrillo-Hermosilla et al. 2009). One of the important matters is the inclusion of functions, activities, authority, and specific responsibilities for handling environmental issues, enabling the dissemination of ideas to members of the organization at all levels and forming a formal firm commitment (Donaire 1996, 2007).

Environmental formalization was measured by the question in Table 8.8, formulated with five variables, using a 5-point Likert scale, with responses varying between *I totally disagree* and *I totally agree*.

With these constructs having been defined, the next step was data collection and analysis, as presented in the following section.

Table 8.8 Variables employed to measuring the *environmental formalization* construct

Variable	*Question*: Using the options below, evaluate to what extent environmental management is formalized at your firm
Var37	At the firm, environmental policy is clearly documented in the mission statement
Var38	The firm has a specific position/function/sector for environmental issues in its administrative sphere
Var39	The firm trades products with an ecological brand using environmental labeling
Var40	The firm has ISO 14000 environmental management and/or FSC (Forest Stewardship Council) and/or Total Quality Environmental Management (TQEM) certification
Var41	The firm has implemented some form of environmental management system

Source: Prepared by the authors, based on Almeida (2010), Barbieri (2007), Camara and Passos (2005), Carrillo-Hermosilla et al. (2009), Donaire (1996, 2007), Kemp and Arundel (1998), Lau and Ragothaman (1997), and Passos (2003)

8.5 Cluster Analysis and Definition of Organizational Taxonomy

To identify which eco-innovation strategies are used by the firms in the study and group them in accordance with their common characteristics (creating taxonomy), the multivariate cluster analysis technique was used. This analysis was initially employed to measure the clusters of firms for the *eco-innovation strategies* construct, for inclusion in a taxonomy. The first subsection will present the cluster analysis itself. This will be followed by the cluster analysis according to contextual factors.

8.5.1 Cluster Analysis

According to Malhotra (2006, p. 573, emphasis in the original), cluster analysis is used when "[…] there is not *a priori* information about the group or cluster membership for any of the objects. Groups or clusters are suggested by the data, not defined *a priori*." The type of method used in this study was the hierarchical agglomerative clustering using Ward's variance method. This test was initially used to verify the number of clusters indicated in which four groups were visualized by the dendrogram. In this case, the hierarchical test was conducted and the error bar graph was constructed, as shown in Fig. 8.1, into which were inserted the taxonomies created for each group.

The graph shows a taxonomy of the firms in the study in accordance with their clusters. Cluster 1 contains the *proactive organizations* (46 firms – 39 %). These have higher scores for proactive strategies and lower scores for reactive strategies. For the purposes of this study, this cluster perceives environmental management as a corporate strategy, but still in its early stages, developing some actions to minimize the environmental impacts of their products/processes.

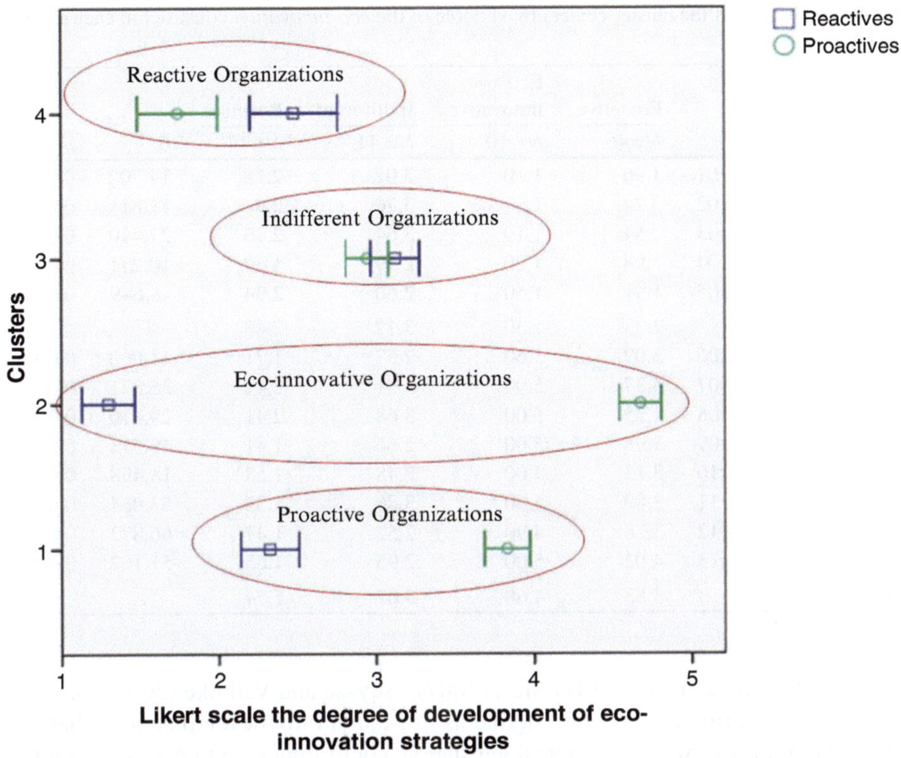

Fig. 8.1 Cluster analysis and definition of taxonomy through the *eco-innovation strategies* construct

However, in cluster 2 there are what this study considers to be *eco-innovative organizations* (10 firms – 9 %). These are firms with high scores for proactive strategies and very low scores for reactive ones. They do not include and totally disagree with reactive strategies and include a high degree of proactive strategies. In this study, these firms are considered eco-innovators, emphasizing sustainable development resulting in reduced environmental impact, pollution, and other negative impacts of the use of natural resources. They can be linked to the firms that Aragón-Correa (1998), Barbieri (2007), Sharma (2000), and Sharma et al. (1999) define as those that develop voluntary actions to prevent environmental impacts, creating a competitive advantage by adopting eco-innovative technologies.

Cluster 3 of this analysis contains *indifferent organizations* (44 firms – 38 %). These firms have no precisely defined proactive or reactive strategies. In other words, they do not take a stance in either respect regarding environmental actions, opting to assume a neutral position on the scale.

Cluster 4 contains the *reactive organizations* (17 firms – 14 %), which are those with higher scores for reactive strategies and lower scores for proactive strategies. In these firms, environmental management is not viewed as an organizational strategy but exists only to comply with the minimum requirements set by legislation. They can

Table 8.9 Means of the cluster centers by variable of the *eco-innovation* construct in each cluster and ANOVA test

Variable		Proactive $N=46$	Eco-innovative $N=10$	Indifferent $N=44$	Reactive $N=17$	F	p value
Reactive strategies	Var01	1.96	1.10	3.02	2.18	14.702	0.000*
	Var02	3.24	1.60	3.70	2.94	11.811	0.000*
	Var03	2.54	1.10	3.64	2.76	27.410	0.000*
	Var04	1.98	1.20	2.59	1.59	10.271	0.000*
	Var05	1.91	1.50	2.66	2.94	8.649	0.000*
Mean		2.33	1.30	3.12	2.48		
Proactive strategies	Var06	3.09	3.60	2.57	1.71	11.060	0.000*
	Var07	4.37	5.00	3.91	2.41	25.871	0.000*
	Var08	4.35	5.00	3.68	2.41	29.410	0.000*
	Var09	3.98	5.00	2.66	1.41	49.518	0.000*
	Var10	3.15	4.00	2.48	1.53	18.498	0.000*
	Var11	3.89	4.90	3.80	1.35	53.064	0.000*
	Var12	3.78	4.90	2.52	1.47	66.872	0.000*
	Var13	4.02	5.00	2.95	1.65	51.102	0.000*
Mean		3.83	4.68	3.07	1.74		

*p value < 0.05

be viewed as those firms that Barbieri (2007), Buysse and Verbeke (2003), Sharma (2000), and Sharma et al. (1999) define as those that only develop actions that are externally imposed by environmental legislation, controlling pollution with corrective technologies and with only the sporadic involvement of the upper management.

To prove this cluster profile and also provide some other characteristics of the firms, an analysis of the cluster centers by variables, also known as centroids, was conducted. "The centroids represent the mean values of the objects contained on the cluster of each of the variables. The centroids enable a description of each cluster, assigning them a name or label" (Malhotra 2006, p. 581). Furthermore, the ANOVA test was conducted which "[…] informs us if three or more population means are equal […]" (Field 2009, p. 299), i.e., it is a test that verifies whether the groups stem from populations with equal means. These values are shown in Table 8.9.

Table 8.9 confirms the taxonomy by the scores obtained in the constructs for eco-innovation strategies in each of the groups. In the proactive organizations cluster, the highest mean of the cluster center is found in var07, followed by var08. This shows that these firms develop environmental actions in their administrative work and in production but develop the other proactive actions to a lesser degree. Furthermore, they take into account the acquisition of end-of-pipe technologies, which would not be advisable for firms seeking to become more proactive. Meanwhile, the eco-innovative organizations have higher means in almost all the variables of the construct for proactive strategies. Practically, all the means were higher than four on the scale except var06, which indicates that these firms are more in favor of eco-innovation. On the other hand, in the case of the reactive strategies, the means varied between points 1 and 2 on the scale (*I totally disagree* and *I disagree*). The graph

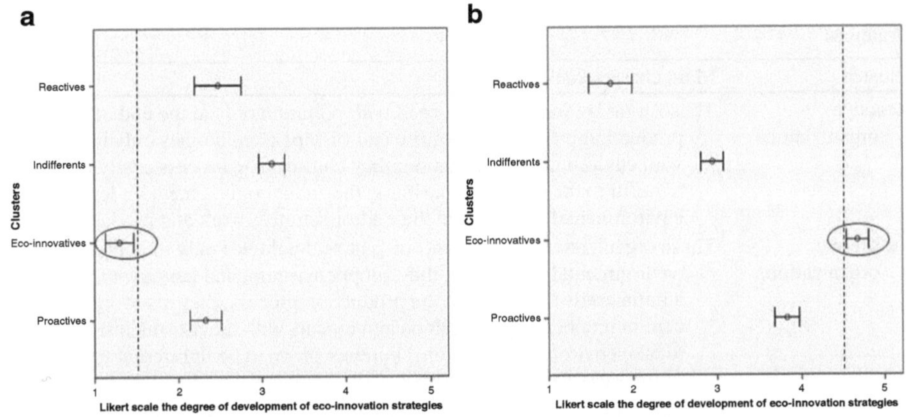

Fig. 8.2 Differences between the means of reactive and proactive strategies in the groups of firms. (**a**) Reactive strategies. (**b**) Proactive strategies

also shows that the medians of the indifferent organizations, both in terms of reactive and proactive strategies, had practically the same means in a neutral position on the scale. In the reactive organizations cluster, the highest means are found in the reactive strategies construct, especially var02 and var05. This shows that these firms are only concerned with acquiring end-of-pipe technologies and view environmental management as an additional cost that can have a negative effect on the growth of their business. Therefore, this analysis confirms the taxonomy identified in this study through the mean scores for each variable in each cluster center.

Regarding the analysis of variance (ANOVA), it is necessary to verify the F index and the statistical significance (*p* value) in Table 8.9. For the F index, when there is no "[…] difference between the means of populations, the index should be close to 1. If the means of population are not equal, the numerator should show this difference and the F index should be greater than 1" (Cooper and Schindler 2011, p. 496). The statistical significance was considered according to the confidence interval (p-value $\leq \alpha$), with a significant of $\alpha = 0.05$ as a parameter, this being the most commonly used (Cooper and Schindler 2011). There are statistically significant differences among the means, shown by the high F values and all the significance levels below 0.05.

However, it is also necessary to verify exactly in which groups there are differences. For this purpose, post hoc follow-up tests were conducted, using Tukey's test because it is widely employed in management studies and is generally more powerful than other such tests (Field 2009; Hair et al. 2005). Error bar graphs, shown in Fig. 8.2, were prepared for each construct.

The graphs in Fig. 8.2 regarding *reactive innovation strategies* show that cluster 2 (eco-innovative organizations) is different from the others, with its reactions being *I totally disagree* and *I disagree*. Concerning the *proactive eco-innovation strategies*, all the clusters differ from one another. However, cluster 2 differs from the others in terms of performance in this dimension in that the variables had higher scores among the three groups. Therefore, the eco-innovative organizations are different in

Table 8.10 Summary of the characteristics of each cluster as a result of their eco-innovation strategies

Cluster	Main characteristics
Reactive organizations	These organizations are concerned with pollution only at the end of the production process and acquire end-of-pipe technologies only to comply with environmental legislation; they consider the process costly and harmful to the growth of business; they adopt only a medium level of environmental measures in their administrative work and production
Indifferent organizations	These organizations acquire end-of-pipe technologies only to comply with environmental legislation; they adopt environmental measures in the administrative sectors and the production process; they also forge a certain number of partnerships/agreements with other firms/institutions to take environmental actions; but they prove to be indifferent to most of the variables in this study
Proactive organizations	These organizations take environmental measures in their administrative work and also in production and conduct periodic environmental audits; they form partnerships with other firms/institutions for environmental purposes and have a system for preventing environmental accidents
Eco-innovative organizations	These organizations take environmental actions at the corporate level and also in production, holding regular environmental audits and adopting environmental analysis processes regarding the lifecycle of products; they form partnerships and make agreements with other firms/institutions for environmental purposes and provide environmental training programs for their collaborators; they also have systems to prevent environmental accidents

relation to these strategies, with the reactive organizations with scores below the mean and the proactive organizations above the mean.

In summary, we weave the main characteristics of each typology defined in this taxonomy which can be seen in Table 8.10.

Having obtained the results of the clusters regarding eco-innovation strategies, the next topic will analyze at the behavior of these groups in relation to their contextual factors.

8.5.2 *Analysis of the Relationship Between the Clusters and Contextual Factors*

To verify the existence of relationships between the organizational respondents and contextual factors, an analysis was also conducted of the cluster centers by variables and analysis of variance using ANOVA (Field 2009). For this analysis, the groups defined in the cluster analysis were used, with the values shown in Table 8.11, in addition to the means by cluster and by construct.

In general, in the means of the constructs, the proactive and eco-innovative organizations have higher means than those in the indifferent and reactive clusters. In the means of the *use of environmental and innovative incentives* construct, the eco-innovative organizations have a higher means than the others, which may reflect

Table 8.11 Analysis of the differences between the mean values of the impact of contextual factors on each group of firms

Contextual factors	Proactive $N=46$	Eco-innovative $N=10$	Indifferent $N=44$	Reactive $N=17$	F	*p* value
Environmental regulation	3.85	4.45	3.06	3.12	12.449	0.000*
Use of environmental and innovative incentives	1.89	2.90	1.75	1.32	6.559	0.000*
Reputational effects	3.57	4.28	2.78	2.52	20.286	0.000*
Top management support	3.51	4.30	2.35	1.88	35.577	0.000*
Technological competence	3.51	4.18	2.91	2.12	30.657	0.000*
Environmental formalization	3.88	4.80	2.80	1.67	59.841	0.000*

*p value <0.05

a result that stems from greater government incentive to innovate, which justifies this definition once again. Furthermore, the eco-innovative organizations always have higher means in all the constructs, while the reactive organizations always have the lowest means. The analysis of variance (ANOVA) also shows statistically significant differences between the means, maintaining high F values and desirable levels of significance.

Nevertheless, it is still not possible to know exactly in which groups there are differences, which can be verified with the follow-up tests and Tukey's test (Field 2009; Hair et al. 2005). In the error bar graphs for each construct of factors, the differences between the means can also be seen, as shown in Fig. 8.3.

The graphs show that cluster 2 (*eco-innovative organizations*) differed from cluster 3 (*reactive organizations*) and cluster 4 (*indifferent organizations*) in every construct. Cluster 2 does not differ from cluster 1 (*proactive organizations*) in use of *environmental and innovative incentives*. Therefore, cluster 2 was the one that differed most from the others and also had the highest mean for impact of contextual factors.

There are no significant differences between clusters 1, 3, and 4 for *environmental regulation* and *use of environmental and innovative incentives*. In *reputational effects* and *top management support*, there are no important differences between clusters 3 and 4. In *technological competence* and *environmental formalization*, all the clusters differ from each other.

The result to be considered is that in all the constructs, cluster performance appears in order of importance for environmental issues: (1) eco-innovative organizations, (2) proactive organizations, (3) indifferent organizations, and (4) reactive organizations. This performance can be viewed in the graph in Fig. 8.4.

This graph clearly shows the difference between the clusters, and some considerations can be inferred concerning the firms. Cluster 1, *proactive organizations*, is made up of 46 firms and is the largest cluster in the study. It shows a performance between the neutral point on the scale and four, except in the case *use of environmental and innovative incentives* construct, which came below the two-point level. Cluster 2, *eco-innovative organizations*, has ten firms (the smallest cluster in the study). Its results are higher than all the others in contextual factors, with a highlight

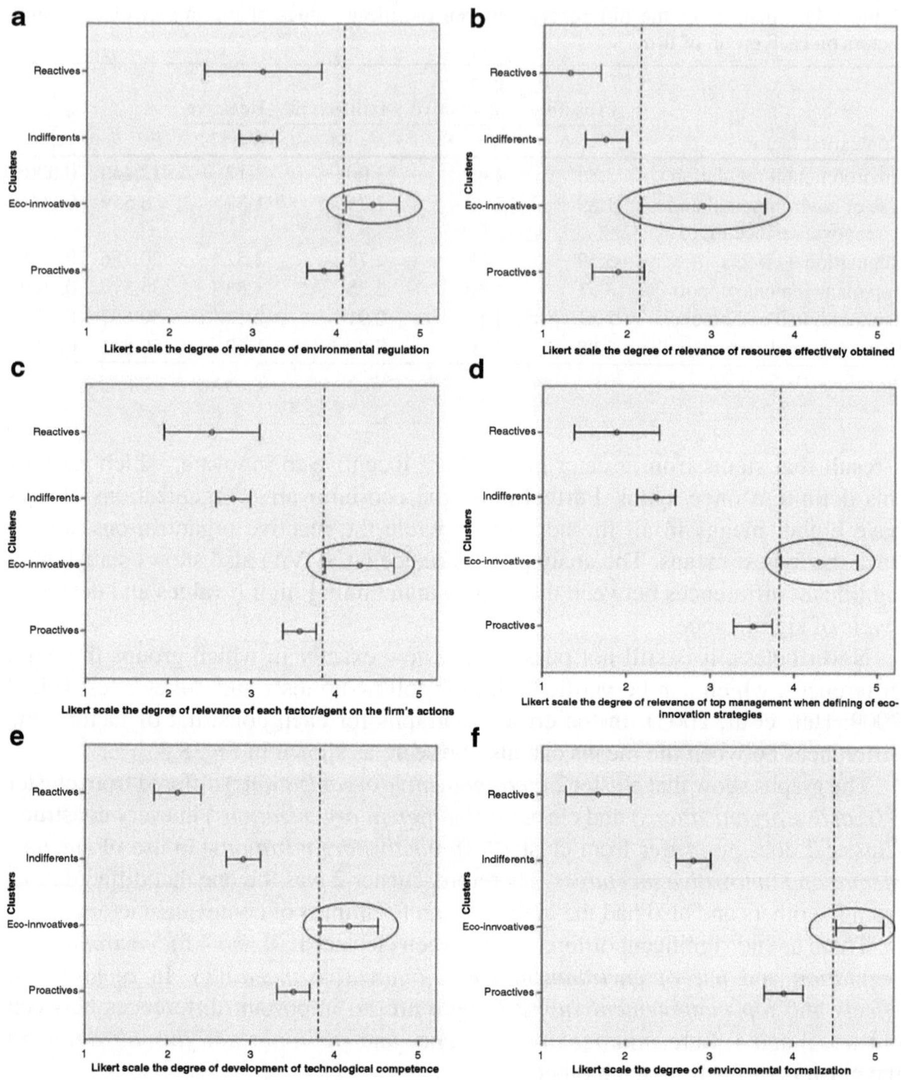

Fig. 8.3 Differences between the mean impacts of contextual factors in the groups of firms. (**a**) Environmental regulation. (**b**) Use of environmental and innovative incentives. (**c**) Reputational effects. (**d**) Top management support. (**e**) Technological competence. (**f**) Environmental formalization

for *environmental formalization*, in which these firms averaged close to five points. Cluster 3, composed of 44 *indifferent organizations*, has a mean result on the scale, which is demonstrative of its denomination. The last in the means of contextual factors is the cluster of *reactive organizations*, with 17 firms, with the lowest means in all the constructs.

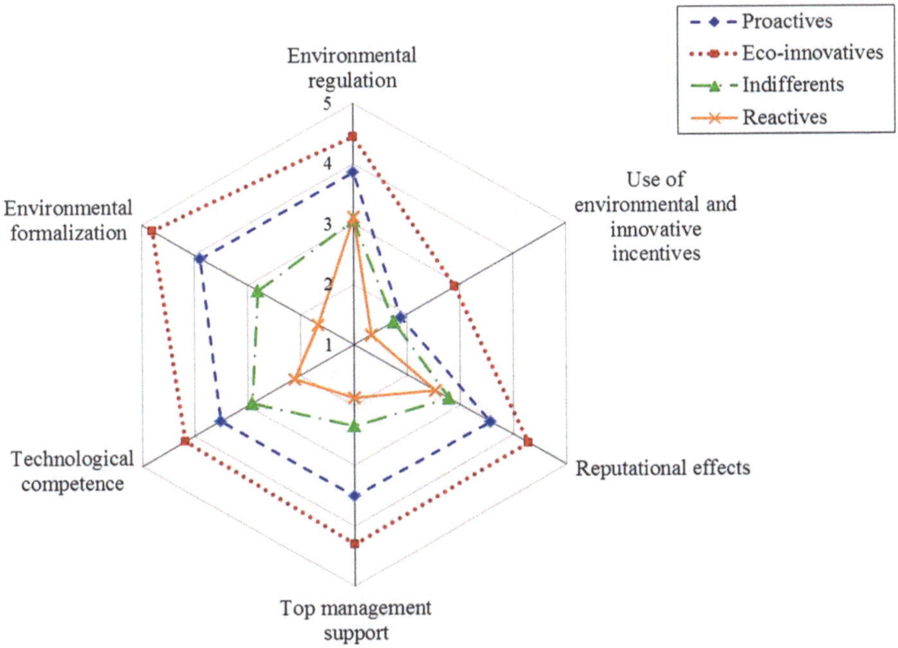

Fig. 8.4 Cluster performance of contextual factors

Thus, taking into account all these differences in mean values, it can be proven that once again this classification is fitting for the groups that were formed in terms of the characteristics of the sampled firms.

8.6 Final Considerations

This study stemmed from a review of the literature that showed that there is a need for studies that simultaneously look at the internal and external perspectives of an eco-innovation strategy and its effects. Therefore, this study aimed to identify an organizational taxonomy by examining the proactive and reactive eco-innovation strategies adopted by firms in the Brazilian cellulose, paper, and paper products sector. Furthermore, it sought to analyze the impact of internal and external contextual factors of organizations in each group of firms.

To achieve these goals, the study initially sought a theoretical basis that handled eco-innovation strategies and the formulation of proactive and reactive strategies for environmental issues, which were defined in constitutive and operational terms. Eco-innovation strategies are formulated according to changes in the context in which each organization is embedded, and the external and internal factors were important to this study because they have an impact on the decisions pertaining to the strategies in question.

When the factors were analyzed, the external factors were initially defined in constitutive and operational terms, starting with environmental regulation. It is worth mentioning that this should lead the firm to innovate and the firm should view this pressure as an opportunity to improve productivity and gain a competitive edge. In addition to regulation, other external contextual factors were also analyzed as having an impact on eco-innovation strategies and consequently on the environmental performance of organizations. These factors are linked to government environmental and innovative incentives and the effects of the firm's reputation in its context, which will in turn have an impact on the firm's image. These factors should be managed jointly and interactively between obstacles and drivers by adopting eco-innovation practices for firms to maintain a favorable real image regarding environmental concerns.

The internal factors that lead to eco-innovation management were also analyzed. These are mainly linked to support from the upper management, with the skills and capacity to absorb technology and formalize the environmental issues in their internal context. The more deeply rooted the environmental and innovative culture of a firm, the more effectively the internalization of strategies for adopting proactive eco-innovation strategies occurs.

Therefore, the objectives of this study can be considered as having been achieved through a cluster analysis and the identification of an organizational taxonomy in which the firms that participated in the study were classified as *reactive organizations*, *indifferent organizations*, *proactive organizations*, and *eco-innovative organizations*. This taxonomy was proved both by the analysis of these clusters with reactive and proactive eco-innovation strategies and the analysis of the contextual factors.

In the *reactive organizations*, eco-innovation is not considered an organizational strategy, and the companies strive to comply with the minimum requirements of environmental legislation. They develop actions to control pollution with corrective technologies, and the involvement of the top management is only sporadic. The *indifferent organizations* have no certain definition of their strategies, which lie between proactive and reactive. These companies do not take a significant strategic stance one way or another concerning environmental actions, with their actions lying between the proactive and reactive level. The *proactive organizations* are those who already perceive the importance of environmental management as a corporate strategy, albeit an incipient one, and develop some actions to minimize the environmental impact of their products/processes. Finally, the *eco-innovative organizations* do not consider reactivity as a strategy but instead take a high degree of proactive strategies into consideration. For this reason, they place importance on sustainable development, resulting in reduced environmental risks, lower levels of pollution, and other negative impacts resulting from the use of natural resource. They take voluntary actions to prevent environmental impacts, creating a competitive advantage through the adoption of eco-innovative technologies.

We consider this result as an advance in the literature on this theme because there was a gap to be filled. Furthermore, we can consider this study as an important contribution to the advancement of knowledge in the field of eco-innovation strategies and also the definition of external and internal factors that influence the adoption of these strategies.

It should be emphasized that the originality lies in the fact that this is a specially perceived and applied empirical study, providing information on the management of eco-innovation regarding its drivers and effects and enabling in-depth analyses to bridge gaps in the literature as noted by Kemp and Arundel (1998). More specifically, this study contributes to both existing theory and the management of eco-innovation in organizations. With these conclusions, this study can serve as a guide for the innovative focus of environmental management in industries in this and other sectors. It also makes a contribution to the research in this field.

Finally, the limitations of the study are those of sampling surveys, where it is rarely possible to determine the degree of accuracy of the findings and the fact that the questionnaires are completed by only one person at each firm. Therefore, the intentions of the respondent and the manner in which the responses are given cannot be gauged. However, to ensure aspects of validity and reliability, a content validity was applied, the questionnaire was pretested, and other precautions were taken regarding the methodology of the study.

In terms of suggestions for future studies, there are important lines of research that have yet to be explored regarding eco-innovation. These themes are related to the impact of the adoption of eco-innovation strategies on the social, environmental, and economic performance of firms and eco-entrepreneurship, i.e., the creation of ecological technology-based firms. Other themes are sources of information and technology transfer in the context of eco-innovation and the definition of government policies and the styles of environmental styles that are most adequate for innovation, with comparative studies of the effects of policies on environmental innovation, and an empirical international analysis regarding the specific characteristics of eco-innovation in different national systems and the environmental innovative capacity of different countries.

Acknowledgments This study was supported by Conselho Nacional de Desenvolvimento Científico e Tecnológico (CNPq).

References

Andersen MM (2006) Eco-innovation indicators. European Environment Agency, Copenhagen, February. http://130.226.56.153/rispubl/art/2007_115_report.pdf. Accessed 24 June 2010

Andersen MM (2008) Eco-innovation – towards a taxonomy and a theory. Annals of the 25th DRUID conference – entrepreneurship and innovation – organizations, institutions, systems and regions, June, Copenhagen

Ansanelli SLM (2003) Mudança institucional, política ambiental e inovação tecnológica: caminho para o desenvolvimento econômico sustentável? Annals of the 8th Encontro Nacional de Economia Política, Florianópolis

Aragón-Correa JA (1998) Strategic proactivity and firm approach to the natural environment. Acad Manage J 41(5):556–567

Arundel A, Kemp R (2009) Measuring eco-innovation. UNU-MERIT working paper series. http://www.merit.unu.edu/publications/wppdf/2009/wp2009-017.pdf. Accessed 16 June 2010

Arundel A, Kemp R, Parto S (2003) Indicators for environmental innovation: what and how to measure. In: Annandale D, Phillimore J, Marinova D (eds) International handbook on environment and technology management. Edward Elgar, Cheltenham, pp 324–339

Ashford NA (2000) An innovation-based strategy for a sustainable environment. In: Hemmelskamp J, Rennings K, Leone F, ZEW Economic Studies (eds) Innovation-oriented environmental regulation: theoretical approach and empirical analysis. Springer, New York/Heidelberg, pp 67–107. http://18.7.29.232/bitstream/handle/1721.1/1590/Potsdam.pdf?sequence=1. Accessed 29 Jan 2012

Barbeli MC (2008) Indústria de papel e celulose: estado da arte das tecnologias de co-geração de energia. Revista de Ciências Exatas e Tecnologia 3(3):107–122

Barbieri JC (2002) Desenvolvimento e meio ambiente: as estratégias de mudanças da Agenda 21, 5a edn. Vozes, Petrópolis

Barbieri JC (2007) Gestão ambiental empresarial: conceitos, modelos e instrumentos, 2a edn. Saraiva, São Paulo

Baumgarten M (2008) Ciência, tecnologia e desenvolvimento – redes e inovação social. Parcerias Estratégicas 26:102–123

Berry MA, Rondinelli DA (1998) Proactive environmental management: a new industrial revolution. Acad Manage Exec 12(2):38–50

Blackburn WR (2008) The sustainability handbook. Environmental Law Institute, Washington, DC

Brasil (1981) Lei nº 6.938, de 31 de agosto de 1981. Dispõe sobre a Política Nacional do Meio Ambiente, seus fins e mecanismos de formulação e aplicação, e dá outras providências. https://www.planalto.gov.br/ccivil_03/leis/l6938.htm. Accessed 12 July 2011

Bracelpa – Associação Brasileira de Celulose e Papel (2012) Dados do setor. http://www.bracelpa.org.br/bra2/sites/default/files/estatisticas/booklet.pdf. Accessed 29 Oct 2012

Buysse K, Verbeke A (2003) Proactive environmental strategies: a stakeholder management perspective. Strateg Manage J 24(5):453–470

Caracuel JA, Torres MÁE, Torres NEH, Salazar MDV (2011) La influencia de la diversificación y experiencia internacional en la estrategia medioambiental proactiva de las empresas. Investigaciones Europeas de Direccion y Economia de la Empresa 17(1):75–91. doi:10.1016/S1135-2523(12)60045-8

Carrillo-Hermosilla J, del Rio González P, Könnölä T (2009) Barriers to eco-innovation. In: Carrillo-Hermosilla J, del Rio González P, Könnölä T (eds) Eco-innovation: when sustainability and competitiveness shake hands. Palgrave Macmillan, New York, pp 28–50

Cassiolato JE, Lastres HMM (2000) Sistemas de inovação: políticas e perspectivas. Parcerias Estratégicas 8:237–255

Cohen WM, Levinthal DA (1990) Absorptive capacity: a new perspective on learning and innovation. Adm Sci Q 35(1):128–152

Cooper DR, Schindler PS (2011) Métodos de pesquisa em administração, 10a edn. Bookman, Porto Alegre

da Camara MRG, Passos LAN (2005) Inovação, competitividade ambiental e clusters na indústria química: um estudo das empresas da ABIQUIM. Annals of the 11th Seminário Latino-Iberoamericano de Gestión Tecnológica - ALTEC – Innovación Tecnológica, Cooperación y Desarrollo, Salvador

de Almeida FAS (2010) Influências das políticas ambientais no desempenho empresarial econômico e socioambiental: um estudo do setor de leite e derivados de Goiás. 205 f. Tese (Doutorado em Administração) – Universidade de São Paulo, São Paulo

de Andrade THN (2004) Inovação tecnológica e meio ambiente: a construção de novos enfoques. Ambiente & Sociedade VII(1):89–106. doi:10.1590/S1414-753X2004000100006

de Souza RS (2004) Fatores de formação e desenvolvimento das estratégias ambientais nas empresas. 2004. 283 f. Tese (Doutorado em Administração) – Universidade Federal do Rio Grande do Sul, Porto Alegre

Donaire D (1996) A internalização da gestão ambiental na empresa. Revista de Administração 31(1):44–51

Donaire D (2007) Gestão ambiental na empresa, 2a edn. Atlas, São Paulo

Dosi G (1988) Institutions and markets in a dynamic world. Manch Sch 56(2):119–146. doi:10.1111/j.1467-9957.1988.tb01323.x

Field A (2009) Descobrindo a estatística usando o SPSS, 2a edn. Artmed, Porto Alegre
Finep – Financiadora de Estudos e Projetos (2012) Modalidades de financiamento. http://www.finep.gov.br/pagina.asp?pag=20.06. Accessed 28 Oct 2012
Foxon T, Andersen MM (2009) The greening of innovation systems for eco-innovation – towards an evolutionary climate mitigation policy. Annals of the DRUID summer conference – innovation, strategy and knowledge, June, Copenhagen
Fussler C, James P (1996) Driving eco-innovation: a breakthrough discipline for innovation and sustainability. Pitman Publishing, London
Hair JF Jr, Babin B, Money AH, Samouel P (2005) Fundamentos de métodos de pesquisa em administração. Bookman, Porto Alegre
Hart S (1995) A natural–resource–based view of the firm. Acad Manage Rev 20(4):986–1014. doi:10.5465/AMR.1995.9512280033
IBGE – Instituto Brasileiro de Geografia e Estatística (2010) Pesquisa Industrial: empresa. http://www.ibge.gov.br/home/estatistica/economia/industria/pia/empresas/2010/defaultttabpdf.shtm. Accessed 24 Sept 2012
Juvenal TL, Mattos RLG (2002) O setor de celulose e papel. In: São Paulo EM, Kalache Filho J (Orgs.) BNDES 50 anos: histórias setoriais. BNDES, Rio de Janeiro, pp 1–21
Kanerva M, Arundel A, Kemp R (2009) Environmental innovation: using qualitative models to identify indicator for policy. United Nations University – UNU-MERIT, Working papers series. http://arno.unimaas.nl/show.cgi?fid=17863. Accessed 29 Jan 2012
Kemp R, Arundel A (1998) Survey indicators for environmental innovation. IDEA report, STEP Group, Oslo. http://www.step.no/old/Projectarea/IDEA/Idea8.pdf. Accessed 2 Aug 2010
Kemp R, Foxon TJ (2007) Tipology of eco-innovation. MEI project: Measuring Eco-Innovation. European Commission, August. http://www.merit.unu.edu/MEI/deliverables/MEI%20D2%20Typology%20of%20eco-innovation.pdf. Accessed 3 June 2010
Könnölä T, Carrillo-Hermosilla J, del Río Gonzalez P (2008) Dashboard of eco-innovation. Annals of the DIME international conference – innovation, sustainability and policy, Sept. University Montesquieu Bordeaux IV, France
Lau RSM, Ragothaman S (1997) Strategic issues of environmental management. S D Bus Rev 56(2):1–7
Lustosa MCJ (1999) Inovação e meio ambiente no enfoque evolucionista: o caso das empresas paulistas. Annals of the 37th Encontro Nacional de Economia (ANPEC), Belém, pp 1177–1194
Lustosa MCJ (2003) Industrialização, meio ambiente, inovação e competitividade. In: May PH, Lustosa MCJ, da Vinha V (Org.) Economia do meio ambiente: teoria e prática. Elsevier, Rio de Janeiro, pp 155–172
Maçaneiro MB (2012) Fatores contextuais e a adoção de estratégias de ecoinovação em empresas industriais brasileiras do setor de celulose, papel e produtos de papel. 2012. 237 f. Tese (Doutorado em Administração) – Universidade Federal do Paraná, Curitiba
Maçaneiro MB, da Cunha SK (2010) Eco-inovação: um quadro de referência para pesquisas futuras. Annals of the 26th Simpósio de Gestão da Inovação Tecnológica, Vitória
Malhotra NK (2006) Pesquisa de marketing: uma orientação aplicada, 4a edn. Bookman, Porto Alegre
Maroco J (2003) Análise estatística com utilização do SPSS, 2a edn. Sílabo, Lisboa
Menguc B, Auh S, Ozanne L (2010) The interactive effect of internal and external factors on a proactive environmental strategy and its influence on a firm's performance. J Bus Ethics 94:279–298. doi:10.1007/s10551-009-0264-0
Miles MP, Covin JG (2000) Environmental marketing: a source of reputational, competitive, and financial advantage. J Bus Ethics 23(3):299–311
Nidumolu R, Prahalad CK, Rangaswami MR (2009) Why sustainability is now the key driver of innovation. Harv Bus Rev 87:56–64
Passos LAN (2003) Gestão ambiental e competitividade: um estudo do setor químico brasileiro. 2003. 166 f. Dissertação (Mestrado em Gestão de Negócios) – Universidade Estadual de Maringá, Londrina
Porter M, van der Linde C (1995) Toward a new conception of the environment–competitiveness relationship. J Econ Perspect 9(4):97–118

Reid A, Miedzinski M (2008) Eco-innovation, Final Report for Sectoral Innovation Watch. Technopolis Group, Brussels. http://www.technopolis-group.com/resources/downloads/661_report_final.pdf. Accessed 29 June 2010

Rennings K (1998) Towards a theory and policy of eco-innovation – neoclassical and (Co-) evolutionary perspectives. Discussion paper no 98-24. Mannheim, Centre for European Economic Research (ZEW). ftp://ftp.zew.de/pub/zew-docs/dp/dp2498.pdf. Accessed 15 Apr 2010

Romeiro AR, Salles Filho S (1996) Dinâmica de inovações sob restrição ambiental. In: Romeiro AR, Reydon BP, Leonardi MLA (eds) Economia do meio ambiente: teoria, políticas e a gestão de espaços regionais. UNICAMP, Campinas

Schmidheiny S (1992) Mudando o rumo: uma perspectiva empresarial global sobre desenvolvimento e meio ambiente. Ed. da FGV, Rio de Janeiro

Sebrae – Serviço Brasileiro de Apoio às Micro e Pequenas Empresas (2012) Critérios e conceitos para classificação de empresas. http://www.sebrae.com.br/momento/quero-abrir-um-negocio/integra_bia?ident_unico=97. Accessed 26 Feb 2012

Serôa da Motta R (1993) Política de controle ambiental e competitividade – estudo da competitividade da indústria brasileira. IE/UNICAMP-IE/UFRJ-FDC-FUNCEX, Campinas

Sharma S (2000) Managerial interpretations and organizational context as predictors of corporate choice of environmental strategy. Acad Manage J 43:681–697. doi:10.2307/1556361. Briarcliff Manor, Academy of Management

Sharma S, Pablo AL, Vredenburg H (1999) Corporate environmental responsiveness strategies: the importance of issue interpretation and organizational context. J Appl Behav Sci 35(1):87–108. doi:10.1177/0021886399351008

Sharma S, Aragón-Correa JA, Rueda-Manzanares A (2007) The contingent influence of organizational capabilities on proactive environmental strategy in the service sector: an analysis of North American and European Ski Resorts. Can J Adm Sci 24(4):268–283. doi:10.1002/cjas.35

Young CEF, Podcameni MGB, Mac-Knight V, Oliveira AS (2009) Determinants of environmental innovation in the Brazilian Industry. Annals of the 4th Congreso de la Asociación Latinoamericana y del Caribe de Economistas Ambientales y de Recursos Naturales. UNA – Universidad Nacional Costa Rica, Heredia. http://www.ie.ufrj.br/gema/pdfs/DETERMINANTS%20OF%20ENVIRONMENTAL%20INNOVATION%20IN%20THE%20BRAZILIAN%20INDUSTRY.pdf. Accessed 27 June 2011

Chapter 9
Conceptualizing Industry Efforts to Eco-innovate Among Large Swedish Companies

Steven Sarasini, Jutta Hildenbrand, and Birgit Brunklaus

Abstract The term "eco-innovation" is of interest to policymakers and industrial practitioners that seek to marry environmental protection with economic development. Sweden has made some headway in that it has an international reputation for leadership on environmental issues and for creating policies that seek to boost eco-innovation in key industries. However, examining industrial efforts to eco-innovate is complicated by the fact that eco-innovation is poorly defined. The varying definitions and typologies currently in circulation pose risks to the field of eco-innovation research. In this chapter, we aim to consolidate existing conceptualizations by adapting an existing typology of eco-innovation. We then apply this typology to examine eco-innovation in large Swedish companies. The study finds that large Swedish companies focus the majority of their eco-innovative efforts on internal measures related to product and process changes. However, the companies in our sample are less adept at collaborating with suppliers, users, and other external partners that can boost eco-innovation. The study concludes by discussing the utility of our typology and by deriving recommendations for policymakers based on our findings.

Keywords Eco-innovation • Dimensions • Typology • Sweden

9.1 Introduction

Eco-innovation (EI) is a relatively new concept that refers to innovations with positive environmental impacts. Numerous governments have embraced EI as a means to resolve environmental problems as part of a wider program of ecological modernization

S. Sarasini (✉) • J. Hildenbrand • B. Brunklaus
Department of Energy and Environment, Division of Environmental Systems Analysis, Chalmers University of Technology, Gothenburg, Sweden
e-mail: steven.sarasini@chalmers.se; jutta.hildenbrand@chalmers.se; birgitb@chalmers.se

S. Azevedo et al. (eds.), *Eco-Innovation and the Development of Business Models*, Greening of Industry Networks Studies 2, DOI 10.1007/978-3-319-05077-5_9,

(Hajer 1995). Ecological modernists argue that environmental problems can be resolved via policy-led socio-technical change (Mol and Sonnenfeld 2000). Scholars have argued that countries can act as environmental pioneers by adopting policies that promote first-mover strategies and lead markets for EIs, setting an example and creating new export opportunities (Porter and van der Linde 1995; Huber 2008). Examining EI from a national perspective can thus assist policymakers and practitioners in their pursuit of such leadership.

Sweden is regarded as an environmental pioneer, having taken the lead on issues such as climate change (Kronsell 1997). In the 1990s, the Swedish government embraced ecological modernization and implemented a range of policies that aim to strengthen Sweden's environmental reputation (Lundqvist 2000; Uba 2010). EI is critical to this approach and is seen as the link between environmental protection and sustainable economic prosperity (Sarasini 2009). Examining Swedish industry efforts to tackle environmental problems via EI is instructive for at least two reasons. First, it is important to examine how well Swedish industry has succeeded in integrating environmental protection into the economy. Second, the specificities of Swedish efforts to eco-innovate may be of interest to practitioners who wish to follow a similar path.

The concept of EI is not well defined. Some scholars have proffered definitions (e.g., James 1997; Rennings 2000) and others have proposed typologies of EI as part of an inductive approach to understanding the concept (e.g., Hellström 2007). The lack of a cohesive understanding is problematic. Identifying and characterizing efforts to resolve ecological problems via EI is a first step in understanding opportunities and overcoming barriers to such efforts. The lack of a robust conceptualization jeopardizes the practical utility of research efforts in this area. Similarly, it is crucial that practitioners and policymakers share an understanding of EI when seeking to derive innovative opportunities from ecological problems and when seeking to learn from best practices. This is especially so if one considers that a range of radical EIs is needed to overcome some ecological problems (Huesemann 2003). A cohesive conceptualization is needed to enable EI research to have a significant and meaningful impact.

We aim to address this research gap by developing an improved conceptualization of the term EI. We build on previous attempts to characterize EI via typologies that describe the varying dimensions of EI. We review the literature on EI for definitions and existing typologies (Sect. 9.2). In Sect. 9.3 we present our methods and modify an existing framework for examining EI (Carrillo-Hermosilla et al. 2009). In Sect. 9.4 we use the framework to examine EIs in large Swedish companies. In Sect. 9.5 we discuss the utility of the framework and implications for practitioners and policymakers.

9.2 What Is EI?

EI is a relatively new term that is used interchangeably with others such as environmental technology and eco-efficiency (Hellström 2007). Environmental technologies are "technologies whose use is less environmentally harmful than relevant alternatives" (Kemp and Foxon 2007: 2). EI is broader in that it is not limited to technology. James (1997: 53) defines EI as "new products and processes which provide customer and business value but significantly decrease environmental impacts." Carrillo-Hermosilla et al. (2009) define EI as "technological change in production processes and products" and "change in the behaviour of individual users or organisations" that improves environmental performance. Rennings (2000) defines EI as:

> ...all measures of relevant actors (firms, politicians, unions, associations, churches, private households) which: develop new ideas, behavior, products and processes, apply or introduce them; and which contribute to a reduction of environmental burdens or to ecologically specified sustainability targets.

The two latter definitions deviate from traditional conceptions of innovation as composed of new products and processes (Schumpeter 1934; Edquist 2001). This deviation is based on the idea that different types of changes (technological, social, and institutional) are required to resolve ecological problems (Hellström 2007). Rennings (2000) states: "EIs can be developed by firms or nonprofit organizations, they can be traded on markets or not, their nature can be technological, organizational, social or institutional."

The nontechnical dimensions of EI set it apart from concepts such as environmental technology. Some scholars focus on these dimensions to emphasize the importance of EI from micro to macro perspectives. This, however, sometimes results in the loss of the environmental dimensions of EI. Andersen (2008: 5), for instance, argues that "The concept is closely related to competitiveness and makes no claim on the "greenness" of various innovations." Similarly, Kemp and Foxon (2007) argue that EI is defined by the *intention* to reduce environmental impacts:

> EI is the production, assimilation or exploitation of a novelty in products, production processes, services or in management and business methods, which aims, throughout its life cycle, to prevent or substantially reduce environmental risk, pollution and other negative impacts of resources use (including energy use). Novelty and environmental aim are the two distinguishing features (p. 5).

These conceptual inconsistencies pose problems. First, the field of EI research risks discord in that scholars may pursue different and possibly conflicting avenues of research. Second, policymakers require clear and coherent guidelines and information on which to base decisions. Third, industry practitioners can benefit from monitoring industry trends in EI as part of business intelligence activities. Again these types of activities may suffer if EI is poorly defined.

9.2.1 Conceptualizing EI Using Typologies

Another way to conceptualize EI is to derive typologies that capture the various dimensions of industry efforts to reduce environmental burdens. An accurate reading of what industry seeks to achieve using a comprehensive typology is the first step in addressing barriers to EI. Industrial actors play fundamental roles as regards ecological problems, and it is within industry that EIs must be developed and adopted to achieve a sustainable society.

Various typologies of EI have been put forward. Hellström (2007) argues that existing conceptualizations are biased towards incremental changes. Hellström shows empirically that whilst the majority of EIs are incremental, there are several radical examples of EI. Hellström distinguishes between products, processes, and sources of supply. He also distinguishes between component and architectural innovations, measured on an incremental-radical scale.

Andersen (2005) distinguishes between "add-on" EIs, "integrated" EIs, eco-efficient technological or organizational system innovations, and general-purpose eco-efficient innovations. In another proposed typology that focuses on "green IT," Faucheux and Nicolai (2011) distinguish between product and process dimensions of EI, between technological and organizational dimensions and a separate dimension that focuses on user acceptance.

International organizations such as the European Union (EU) and the Organization for Economic Co-operation and Development (OECD) have also commissioned studies on EI. An EU-funded project included a typology that focuses on environmental technologies, organizational innovations, product and service innovations and "green system innovations" (Kemp and Foxon 2007). A report by Technopolis distinguishes between "curative" or "end-of-pipe" technologies, product and service EIs, organizational EIs, and marketing EIs (Reid and Miedzinski 2008). A report from the OECD (2011) adds three further dimensions (targets, mechanisms, and impacts). The report labels organizational, marketing, and institutional changes as "targets" for EI.

9.2.2 Deriving an Appropriate EI Typology

The variety described above depends to some extent on the different perspectives from which one can examine EI (e.g., from micro to macroscales). It also illustrates the incoherent manner in which EI is presently conceptualized. In order to examine EI within large Swedish companies, we follow an inductive logic where we seek to refine an existing typology in a manner that befits our empirical material.

Carrillo-Hermosilla et al. (2009) outline a typology which describes dimensions of EI from a company perspective. We chose this typology because we feel it is the most comprehensive and coherent description of the eco-innovative activities pursued by companies. The typology distinguishes between four dimensions of EI.

Design dimensions refer to modifications to products or processes in terms of component changes (e.g., incremental changes to components or "curative" measures), subsystem changes (e.g., efficiency improvements or process changes) and system changes (e.g., radical or "eco-effective" measures). The first two (incremental) categories focus on reducing environmental impacts whereas the third (radical) category focuses on biocompatible system redesign. That is, systems are created with components and subsystems that turn wastes into inputs. An example of system change is a passenger car fuelled by renewable energy with recyclable materials (from the perspective of an automaker).

User dimensions encourage environmentally sound use of products and services or behavioral change and may draw on user preferences to develop EIs. The term "user acceptance" refers to the way consumers use products and EI in this dimension encourages behavioral change among users that benefits the environment (e.g., recycling). "User development" refers to instances where EIs are initiated by users and can occur in tandem with manufacturers.

In a narrow sense, product-service dimensions include the development of services to reduce the environmental impact of a particular product (e.g., energy efficiency services) and measures that stimulate EI in supply chains. Carbon labels that display the amount of carbon dioxide embodied in a product can both encourage sustainable consumption and stimulate EI in the supply chain. In a broader sense, product-service dimensions encompass "a marketable set of products and services capable of jointly fulfilling a user's needs" (Goedkoop et al. 1999). Here the focus is on the delivery of a function (e.g., electric car leasing). In some instances, product-service combinations harness supporting networks (e.g., networks rather than chains of suppliers) and infrastructure (e.g., electric car-charging stations) to deliver this functionality (Mont 2002).

Governance refers to institutional or organizational measures that "resolve conflicts over environmental resources in both the public and private sectors" (Könnölä et al. 2008). Governance at the institutional level refers typically to policy and its role in stimulating innovation and overcoming technological lock-ins. At the business level governance typically includes relationships with key stakeholders such as governments that can assist in overcoming barriers to EI. Firms may also seek to create new organizational structures that facilitate EI. Joint ventures between automakers and utility companies that aim to develop infrastructure for charging electric vehicles are an example of such measures.

9.3 Methodology

This study utilizes mixed methods to examine the utility of the typology outlined above, which we used to analyze EI in large Swedish companies.

9.3.1 Sample and Data Collection

Our sample consists of large Swedish companies from a range of industries. We defined company size in terms of employees and focused on companies that operate in Sweden. We focused on companies that perform research and development activities in Sweden and export products. We compiled a sample of 92 publicly traded companies using employment data from Statistics Sweden (SCB).

We examined EI using corporate annual reports and sustainability reports. We used the most recently published reports in July 2012. We chose this method because it provides comprehensive and widely accessible data sources. Other methods such as interviews and surveys would not have provided the same level of access and reliability. However, our analysis reflects the level of detail companies provide in reports. Corporate reports are publicly available and target investor audiences and other stakeholders. Since reports are an instrument for external communication, they are designed to bolster corporate images, which may challenge the validity of our results. However, it is likely the case that all methods suffer this problem.

Initially we set out to organize companies into industry sectors using the Swedish Standard Industrial Classification (SNI), which is based on the Statistical Classification of Economic Activities in the European Community (NACE). This approach led to a large number of sector categories each with few entries and was thus not suitable for our analysis. Instead we organized the companies in our sample based on their main areas of business and product offerings as described in reports. We divided the companies into 11 reasonably homogenous groups: chemical-producing companies (8); consultancy and service companies (7); retail companies (3); food companies (4); construction companies (4); electrics and electronics companies (10); companies producing pulp, paper, and wood products (10); mining-, metal-, and material-producing companies (14); automotive companies (9); logistics and transport companies (6); and machinery and equipment companies (17).

9.3.2 Analysis

We identified every eco-innovative measure reported by each company during the year that preceded the publication of the report in question. We copied the provided information regarding each EI into a template for each company. We then coded this information using the typology described in Table 9.1. Our results thus reflect the number of eco-innovative measures under each dimension as the companies in our sample reported them.

At this stage it became apparent that some of the categories in the original framework (Carrillo-Hermosilla et al. 2009) understated certain EIs and overstated others. We thus expanded these categories to encapsulate the complexities of different eco-innovative measures in more detail (Table 9.2). We expanded the design dimension to include measures that focus exclusively on product and process developments. The main reason for this is that processes can also encompass complex technological systems that can be adapted at the level of individual components, subsystems, and systems.

We extended these distinctions to organizational processes. Here changes at the system level refer to changes to an entire company (e.g., energy management across all operations). The subsystem level refers to changes to operations (e.g., logistics or manufacturing) and the component level refers to changes to supporting processes

Table 9.1 Dimensions and categories of eco-innovative activities

Dimensions of eco-innovation	Categories
Product	Component change/addition
	Subsystem changes
	System changes
Production process	Component change/addition
	Subsystem changes
	System changes
Organizational process	Supporting procedures and processes
	Operational measures
	General policy and management
User	User acceptance
	User development
Value chain	Product services
	Other value chain (e.g., suppliers)
Governance	Partnership with other private company
	Partnership with university or similar
	Partnership with government/third sector

Derived from Carrillo-Hermosilla et al. (2009)

Table 9.2 Eco-innovative measures among large Swedish companies

Dimensions of eco-innovation	Categories		
Product	Component change/addition	119	5 %
	Subsystem changes	206	8 %
	System changes	108	4 %
Production process	Component change/addition	177	7 %
	Subsystem changes	155	6 %
	System changes	65	3 %
Organizational process	Supporting procedures and processes	328	13 %
	Operational measures	350	14 %
	General policy and management	252	10 %
User	User acceptance	34	1 %
	User development	58	2 %
Value chain	Product services	89	4 %
	Other value chain (e.g., suppliers)	187	7 %
Governance	Partnership with other private company	176	7 %
	Partnership with university or similar	112	4 %
	Partnership with government/third sector	78	3 %
	Total	*2,494*	

and procedures (e.g., environmental training for employees). We included both planned and implemented changes. Companies can implement changes (e.g., introducing environmental management systems), but they can also plan to implement changes (e.g., introducing targets for emission reductions). We included planned changes as they can potentially influence core values and corporate culture.

We also expanded on the governance dimension to include details of companies' collaborative activities. We categorized collaborations according to the type of partner involved. Here we established separate categories for collaborations with private companies, universities/research institutes, and government/third sector organizations.

9.4 Results

In this section we apply our typology to a sample of 92 large Swedish companies. Results are summarized in Table 9.2. Product and process changes amount to around 70 % of the total measures, with organizational processes the most populated dimension. Companies reported fewer EIs based on collaborations with users, suppliers, and other external actors.

In what follows, we examine the efforts of each group of companies vis-à-vis dimensions of EI. For each dimension we provide examples from the group of automotive companies in our sample. These examples are purely to exemplify the way in which we categorized EIs using the typology. The examples also illustrate the challenges of adopting such a framework. We chose automotive companies because automobiles are well-known products. This, we felt, would make our results more accessible to uninitiated readers in comparison to machinery and equipment companies, for instance, whose products and processes are less familiar.

9.4.1 Product and Process Dimensions

With the exception of organizational processes, large Swedish companies are most active in terms of product and process changes. As regards automotive companies, we categorized measures focused on the introduction of new materials within automobiles as product component changes. Examples include the development of wood fiber doors to reduce fossil fuel dependency and the introduction of nano-structured compounds to reduce weight. Here we had to impose system boundaries for each company. Companies such as Volvo are automakers that design and assemble vehicles. We treated vehicles as the "system" to which product changes apply. Wood fiber doors are components in larger systems (vehicles) and were thus considered as component changes.

We categorized measures as subsystem product changes where they focused on changes to subsystems within vehicles, such as the hybridization of the drivetrain or the use of renewable fuels. The Volvo Group launched trucks based on hybrid and methane-diesel technologies. We reasoned that the use of alternative fuels represents a subsystem change because it only involves changes to vehicles' drivetrains, not the entire vehicle.

We treated Volvo Cars' launch of an all-electric car as a systemic change since Volvo has traditionally focused on vehicles with internal combustion engines. In addition to the battery/motor, electric cars include a range of measures to reduce weight and increase the use of renewable materials and can thus be considered systemic changes. In contrast, Haldex introduced an electromechanical braking system to improve vehicle efficiency and safety. Whilst the braking system encompasses a subsystem change for an automaker such as Volvo Cars, it represents a system change for Haldex, whose main products are braking systems.

As regards process EIs, we categorized energy-saving measures such as automatic lighting systems and wastewater treatment systems as component-level technological process changes. An automatic lighting system is only a component change if one considers the entire production system. We categorized measures that variously focused on energy, emissions, recycling, and the use of chemicals as subcomponents of manufacturing processes as subsystem-level changes. SAAB launched an information technology (IT) tool to monitor chemical use in all production processes. We treated it as a subsystem change, but not a system change in that it does not fundamentally change the entire production system. Technological system-level changes to processes were much fewer in number than component- and subsystem-level changes.

Overall, the balance between product and process EIs shifts between company groups (Table 9.3). On average, the companies in our sample reported on 4.7 and 4.3 product and process EIs, respectively. Consultancy and service, automotive, construction, and retail companies reported on more product EIs than the other groups. In contrast, logistics and transport, food, and retail companies report higher numbers of process EIs and fewer numbers of product EIs. This is probably because some companies in our sample deliver mainly goods whereas others provide services. The companies in the logistics and transport group move items such as post and freight, and their main environmental impacts arise from transportation via land, sea, or air. It makes sense that such companies focus on process EIs since they are the main cause of environmental degradation.

Overall, companies noted few system-level changes. This is perhaps to be expected as such changes are more radical and risky. However, consultancy and service companies focused on system changes more than any other group. This may be a reflection of their role as a provider of expertise to other companies. It may also be that consultants do not bear the risks of the changes they advocate.

9.4.2 Organizational Process Dimensions

The companies in our sample engage mostly with organizational process changes (Table 9.2). We categorized the following types of measures among automotive companies as belonging to this dimension. Several automotive companies noted that they increasingly use renewable energy within production and manufacturing. We categorized the use of renewable energy as a change to supporting procedures

Table 9.3 Eco-innovative measures: average per category and per company

	Chemicals	Consultancy/ services	Retail	Food	Construction	Automotives	Logistics and transports	Machinery and equipment	Electric and electronics	Pulp, paper and wood	Mining, metals and other materials	Mean average
Product	4.9	10.3	6.3	4.5	6.5	7.8	0.2	4.2	3.7	3.3	3.3	*4.7*
Production process	5.6	2.4	9.7	6.8	2.5	2.9	6.2	2.9	3.0	5.9	4.9	*4.3*
Organizational process	7.1	7.6	16.3	6.0	9.3	7.4	13.8	11.1	13.9	11.4	8.5	*10.1*
User	1.0	2.1	1.3	4.3	2.5	1.0	0.0	0.6	1.2	0.5	0.1	*1.0*
Value chain	2.8	5.9	4.7	6.5	3.3	3.6	2.3	2.4	3.0	2.6	1.2	*3.0*
Governance	3.8	3.9	5.3	2.8	3.3	3.0	2.0	1.8	9.1	4.8	4.4	*4.0*

and processes because energy is only one resource used in the production process and because automotive companies purchase energy from electricity suppliers.

We characterized nontechnical measures that focus on a company's operational procedures as operational measures. The most common measures in this category focus on environmental management systems and ISO14001 certification. Companies also mentioned changes to corporate social responsibility routines. We treated these measures as operational when applied across a company's operations. In contrast, we characterized company-wide nontechnical changes as changes to general policy and management. Two companies introduced a code of conduct, and Toyota introduced "climate accounting." We categorized these measures as general policies and management because they influence the entire company.

A large number of organizational process changes are related to targets and statements of intent. We felt that this type of measure is somewhat overstated in company reports, which to some extent are intended to bolster corporate image. We included statements of intent in our analysis because they can potentially shape and influence organizational cultures, but with the risk that our findings overstate the importance of such measures. Their inclusion may explain why the retail, logistics/transport, and electric/electronic companies are more active in terms of organizational process changes than the other companies in our sample (Table 9.3). It may be the case that these groups of companies experience more stakeholder pressure than the others and thus do more to maintain a positive brand or image.

9.4.3 User Dimensions

We characterized measures that draw on user inputs as a source of EI under the user development dimension. The Volvo Group, for instance, introduced a database system for user feedback into product development. We characterized measures that seek to change user behavior under the user acceptance dimension. These measures include the provision of environmental information via labelling schemes and eco-driving training.

Companies reported on fewer measures that focus on collaborations with users than products and process changes (Table 9.3). Companies mentioned relatively few measures that involved users as the source of EI, even though user-driven sectors like food and consultancy/services have more focus on consumers than chemical companies. Our findings suggest that companies may not be aware of the potential to involve lead users as a source of EI. It may also be the case that end users are incapable of making contributions to developments in electronics or automobiles because of product complexity. User inputs may be limited to explicating demands regarding product functionality. However, several companies in our sample develop products for other businesses. This applies to automotive companies, where automotive suppliers comprise a large portion of our sample. Hence we suspect that full details of innovative procedures are not included in company reports.

9.4.4 Value Chain Dimensions

We categorized measures as product services when companies provide a service that aims to reduce products' environmental impacts. For example, the Volvo Group provides an IT service that allows users to calculate their carbon dioxide emissions. The distinction between product service and user acceptance is not always that clear, as both focus on user behavior.

Companies reported on interactions with the value chain in different ways. Some groups were more active in delivering product services to users whereas others focused more on activities with suppliers. We categorized measures that involve nonusers in the other value chain dimension. Automotive companies assess suppliers based on environmental criteria and in some instances select suppliers with ISO14001 certification. Toyota launched a "sustainable retailer program" to reduce energy use and emissions.

Overall, companies were moderately active as regards value chain measures, although food and consultancy/service companies were much more active than the rest (Table 9.3). In comparison to the user dimension, companies were more active as regards collaborations with suppliers. When compared to previous studies (e.g., Zaring and Hellsmark 2001), this study suggests that Swedish companies have expanded their efforts to collaborate with suppliers and pursue a more systemic approach. However, the majority of value chain measures focus on procurement policies, assessing and auditing suppliers' environmental credentials, or providing suppliers with a code of conduct. Examples of close and innovative collaborations in the value chain were harder to find.

9.4.5 Governance Dimensions

We categorized measures that involve a partnership with an external organization beyond the value chain under the governance dimension. Aside from memberships in numerous industry associations, automotive companies have established notable partnerships that focus on electrified vehicles. Kongsberg launched a joint venture with QRTECH that focuses on hybrid and electric drivelines, and Volvo Cars launched a strategic cooperation with Siemens that has a similar focus.

Overall, electric and electronic companies were the only group that reported on notably higher-than-average levels of governance measures (Table 9.3). Generally, companies collaborate more frequently with industrial partners than universities. However, the pulp and paper and the automotive groups partner more with universities than other groups. In contrast, consultancy, retail, construction, electric, and electronics companies collaborate more with industrial partners. It is unlikely that companies fully describe governance activities in corporate reports given the many forms of network collaboration. These include participation in trade associations and industry networks, informal collaborations with science partners, memberships in advocacy coalitions, contracts with consulting companies, agreements with key suppliers and dealers, and so on.

Notwithstanding, Swedish companies appear to have expanded collaborations with actors beyond the value chain (cf. Zaring and Hellsmark 2001). However, a large proportion of the governance measures described in this study encompassed partnerships that are not necessarily focused on actual innovations. These include memberships in industry associations and other coalitions such as the United Nations (UN) Global Compact. Companies also noted various partnerships with NGOs and governments or government agencies. We interpret these measures to be of importance as they can help build societal and legislative legitimacy for EIs and because these sorts of partnerships may reduce the risks and uncertainties related to radical measures.

However, companies mentioned relatively few governance measures that comprise collaborative partnerships which aim to develop new technologies. Those instances where companies did describe such measures offer great promise. As noted above, automotive companies have established ambitious partnerships with companies outside their traditional value chains related to the electrification of vehicles. Companies producing electric and electronic goods are involved in collaborative activities that use smart grids as an infrastructural development that can support other EIs. Several companies reported that their contributions to the Stockholm Royal Seaport smart grid project are supported by partnerships with universities and research institutes. They also reported on opportunities to provide venture capital to explore new technological developments and new business models. However, the general lack of these types of measures may be due to the fact that companies have not identified the potential to make environmental improvements via innovation networks or because of the various barriers to innovating via networks. Networking with other private companies may be risky in the context of competitive markets, for instance, implying a need for policy support in establishing networks.

9.4.6 The Greening of Swedish Industry?

Our results can be compared to previous studies that have examined environmental measures in Swedish industry, such as the Swedish Business Environmental Barometer (Terrvik and Wolff 1993; Wolff and Strannegård 1995; Adolfsson 1997; Belz and Strannegård 1997; Zaring and Hellsmark 2001). These studies show that Swedish companies focused mainly on end-of-pipe measures (component changes) and measures aimed at efficiency improvements in the form of reduced emissions to air, reduced waste, and reduced energy consumption during 1990–2000. Many companies also adopted environmental management systems and environmental reporting but seldom focused on green product development. These studies also show that internal actors were the most critical to environmental change and collaborations with external actors were fragmented along the value chain. Companies typically selected suppliers according to their environmental credentials and sought to impose environmental requirements on the supply chain.

Our findings suggest that the Swedish companies in our sample have expanded on these efforts. They have a strong focus on the development of efficient products and processes. They also work with suppliers in a systematic and standardized way via procurement policies, codes of conduct, and supplier auditing. The companies in our sample increasingly collaborate with partner organizations that are not part of their supply chain. However, our findings suggest that there is a lack of radical EI in Swedish industry. Our findings also suggest that Swedish companies have not yet overcome barriers to truly innovative network collaborations with suppliers, users, and other private sector partners. In other words, large companies appear to utilize internal knowledge and competences but have not yet established systemic approaches that may be more conducive to radical EI (Hart and Milstein 2003; Hart and Dowell 2011).

9.5 Concluding Remarks

This chapter modified an existing typology to analyze how large Swedish companies engage with eco-innovation. Our typology examines the complexities of the different dimensions of EI, which allows for a consideration of the systemic elements of EI from a company perspective. By examining collaborative partnerships that support EI, our modified typology creates linkages between the micro- and meso-levels in a manner that befits other studies on innovation systems. By including these additional dimensions, our modified typology can have various benefits for practitioners and policymakers. Practitioners working within companies can use this typology as an element of their business intelligence activities. That is, practitioners can examine EI in rival companies and other industries for benchmarking purposes as they seek to adopt best practices. This is particularly true for organizational dimensions of EI, which can support developments in products and technological processes.

Policymakers and governmental agencies can also use our typology to identify industrial strengths and weaknesses. The typology can be used to examine the relationships between different dimensions of EI, which can form the basis for efforts to address barriers to radical and systemic changes. Our findings suggest a lack of system-level (radical) changes and a lack of collaborative efforts with external partners in Swedish industry. Given that some radical EIs require concerted efforts from a range of actors, we argue that a networked approach to EI might be beneficial. Radical, systemic EIs require institutional support in the form of public policies (e.g., Markard and Truffer 2008; Dewald and Truffer 2011). In our opinion policies should support the creation of eco-innovation networks.

Whilst this study has produced relatively broad and comprehensive results, we feel that there are several aspects of EI that warrant further research. Whilst corporate reports are accessible and provide comprehensive databases, supplementary methods would provide more in-depth analyses. Mixed-method approaches using patent analysis, questionnaires, and interviews can (1) identify EIs across all dimensions and in more detail and (2) assist in identifying drivers and barriers to EIs. Mixed methods could also adopt longitudinal and industry-specific perspectives.

Acknowledgments This chapter is based on a wider study (Brunklaus et al. 2013) funded by the Swedish Governmental Agency for Innovation Systems (VINNOVA). We would like to thank Anna Sandström, Göran Andersson, and Jonas Brändström from Vinnova for their useful inputs. We would also like to thank Anika Regett and Gibran Vita for their assistance in collecting and analyzing data.

References

Adolfsson P (1997) Swedish business environmental barometer. In: Belz F, Strannegård L (eds) International business environmental barometer 1997. Cappelen Akademisk Forlag, Oslo

Andersen MM (2005) EI indicators, Background paper for the workshop on EI indicators. EEA Copenhagen, 29 Sept 2005

Andersen MM (2008) EI: towards a taxonomy and a theory. Paper presented at the 25th celebration conference 2008 on entrepreneurship and innovation – organizations, institutions, systems and regions. Copenhagen, CBS, Denmark, 17–20 June 2008

Belz F, Strannegård L (eds) (1997) International business environmental barometer 1997. Cappelen akademisk Forlag, Oslo

Brunklaus B, Hildenbrand J, Sarasini S (2013) Eco-innovative measures in large Swedish companies: an inventory based on company reports. Vinnova, VA 2013:03

Carrillo-Hermosilla J, del Rio GP, Könnölä T (2009) EI: when sustainability and competitiveness shake hands. Palgrave Macmillan, Hampshire

Dewald U, Truffer B (2011) Market formation in technological innovation systems – diffusion of photovoltaic applications in Germany. Ind Innov 18(3):285–300

Edquist C (2001) The systems of innovation approach and innovation policy: an account of the state of the art. Lead paper presented at the DRUID conference, Aalborg, 12–15 June 2001, under theme F: 'National Systems of Innovation, Institutions and Public Policies'

Faucheux S, Nicolai I (2011) IT for green and green IT: a proposed typology of EI. Ecol Econ 70:2020–2027

Goedkoop MJ, van Halen CJG, te Riele HRM, Rommens PJM (1999) Product service systems. Ecological and economic basics. The Hague, Den Bosch and Amersfoort: Pi!MC, Stoorm C.S. and PRé Consultants. Gothenburg Research Institute, Gothenburg

Hajer MA (1995) The politics of environmental discourse: ecological modernization and the policy process. Oxford University Press, Oxford

Hart SL, Dowell G (2011) Invited editorial: a natural-resource-based view of the firm fifteen years after. J Manage 37(5):1464–1479

Hart SL, Milstein MB (2003) Creating sustainable value. Acad Manage Exec 17(2):56–67

Hellström T (2007) Dimensions of environmentally sustainable innovation: the structure of EI concepts. Sustain Dev 15:148–159

Huber J (2008) Pioneer countries and the global diffusion of environmental innovations: theses from the viewpoint of ecological modernisation theory. Global Environ Change 18(3):360–367

Huesemann MH (2003) The limits of technological solutions to sustainable development. Clean Technol Environ 5:21–34

James P (1997) The sustainability circle: a new tool for product development and design. J Sustain Prod Des 2:52–57

Kemp R, Foxon, T (2007) Typology of EI: http://www.merit.unu.edu/MEI/deliverables/MEI%20D2%20Typology%20of%20EI.pdf, UM-MERIT: Measuring EI. Accessed 1 Apr 2013

Könnölä T, Carrillo-Hermosilla J, del Río Gonzalez P (2008) Dashboard of EI. Paper presented at DIME international conference "innovation, sustainability and policy", GREThA, University Montesquieu Bordeaux IV, France, 11–13 Sept 2008

Kronsell A (1997) Sweden: setting a good example. In: Skou Andersen M, Liefferink D (eds) European environmental policy: the pioneers. Manchester University Press, Manchester

Lundqvist LJ (2000) Capacity-building or social construction? Explaining Sweden's shift towards ecological modernization. Geoforum 31:21–32

Markard J, Truffer B (2008) Actor-oriented analysis of innovation systems: exploring micro–meso level linkages in the case of stationary fuel cells. Technol Anal Strateg Manage 20(4):443–464

Mol APJ, Sonnenfeld DA (2000) Ecological modernisation around the world: perspectives and critical debates. Frank Cass, London

Mont O (2002) Clarifying the concept of product-service system. J Clean Prod 10(3):237–245

OECD (2011) Better policies to support EI, OECD studies on environmental innovation. OECD Publishing. http://dx.doi.org/10.1787/9789264096684-en

Porter ME, van der Linde C (1995) Toward a new conception of the environment-competitiveness relationship. J Econ Perspect 9(4):97–118

Reid A, Miedzinski M (2008) EI–final report for sectoral innovation watch, for Europe Innova, Technopolis group

Rennings K (2000) Redefining innovation – EI research and the contribution from ecological economics. Ecol Econ 32:319–332

Sarasini S (2009) Constituting leadership via policy: Sweden as a pioneer of climate change mitigation. Mitig Adapt Strat Glob Chang 14(7):635–653

Schumpeter JA (1934) The theory of economic development. Harvard University Press, Cambridge, MA

Terrvik E, Wolff R (1993) Svenska miljöbarometern företag 1993. Gothenburg Research Institute, Gothenburg

Uba K (2010) Who formulates renewable-energy policy? A Swedish example. Energy Policy 38:6674–6683

Wolff R, Strannegård L (eds) (1995) The nordic environmental business barometer. Bedriftøkonomens Førlag, Oslo

Zaring O, Hellsmark H (2001) The Swedish business environmental barometer 2001. Gothenburg Research Institute, Gothenburg

Chapter 10
Integrated Environmental Management Tools for Product and Organizations in Clusters

Greening the Supply Chain by Applying LCAs and Environmental Management Systems with a Cluster Approach

Tiberio Daddi, Marco Frey, Fabio Iraldo, Francesco Rizzi, and Francesco Testa

Abstract In industrial clusters, cooperation among different actors operating in the same sector provides theoretical arguments in favor of the coordination of life cycle thinking initiatives, such as the sharing of green supply chain practices and the implementation of product-oriented environmental management systems, as well as other evolved environmental management practices. Grounding on the Eco-innovation "Imagine" project, this chapter describes a case study on the practical applicability of a cluster-based approach to life cycle management and, in particular, to life cycle assessment, one of the most innovative product-related environmental management tools. The project focused on four industrial clusters located in the Tuscany region (Italy), which have been involved in the implementation of local innovative environmental policies and tools, based on a life cycle approach, to improve their competitiveness. In-field evidences are discussed and a roadmap is shown for supporting the coordination of life cycle management in industrial clusters.

Keywords Environmental management • Life cycle assessment • Industrial clusters • Eco-innovation • Supply chain management

---

T. Daddi • F. Rizzi • F. Testa (✉)
Institute of Management – Sant'Anna School of Advanced Studies, Piazza Martiri della Libertà 33, 56127 Pisa, Italy
e-mail: tiberio.daddi@sssup.it; francesco.rizzi@sssup.it; f.testa@sssup.it

M. Frey • F. Iraldo
Institute of Management – Sant'Anna School of Advanced Studies, Piazza Martiri della Libertà 33, 56127 Pisa, Italy

IEFE – Institute for Environmental and Energy Policy and Economics, Bocconi University, Milan, Italy
e-mail: frey@sssup.it; f.iraldo@sssup.it

S. Azevedo et al. (eds.), *Eco-Innovation and the Development of Business Models*, Greening of Industry Networks Studies 2, DOI 10.1007/978-3-319-05077-5_10,

10.1 Introduction

Industrial clusters, i.e., geographical concentrations of firms operating in the same supply chain (SC), still represent a promising research field with regard to the connections between interorganizational approaches, environmental management, and SC management.

In industrial clusters, the coordination and cooperation among different actors operating in the same sector (and not even necessarily linked to the territory) tend to favor coordination of life cycle thinking initiatives (OECD 1999), sharing of green SC practices (Srivastava 2007), implementation of product-oriented environmental management systems (Klinkers et al. 1999), as well as other evolved environmental management practices (Iraldo et al. 2013).

This phenomenon cannot be explained merely by a greater sense of awareness and responsibility towards the environmental impacts that the industrial firms operating in clusters have, "outside the boundaries" of their production sites as well (Ross and Evans 2002). Rather, it seems to be the consequence of an intentional cooperative dynamic in relation to environmental issues, which actively involves the cluster companies. In fact, there is increasing evidence that cluster-based interorganizational green management practices are boosting the development of new and more effective methodologies and approaches in environmental management.

Several studies have shown that a wide range of internal and external factors may push companies to "extend" the scope of their environmental management to upstream and downstream SC activities. Such factors include the need to respond to increasing pressures from external stakeholders (e.g., consumers or institutions), the need to ensure compliance with more stringent environmental requirements, the willingness to align corporate strategies to societal needs, or simply a strategy to opportunistically gain a competitive advantage (Sharfman et al. 2009; Darnall et al. 2008; Nawrocka 2008). Corbett and Decroix (2001) emphasize that the need to extend environmental management practices to the SC is recurrently felt by larger and smaller companies when pursuing the aim of improving their environmental performances.

Unfortunately, despite several theoretical insights, there is still much to be done in order to support companies in considering interorganizational dimensions alongside intraorganizational ones when setting their competitive strategies and operational routines in the environmental and sustainability "arena."

Demonstration projects play a fundamental role here in bringing theoretical contributions into daily managerial practices.

The case study presented in this chapter describes the experience of the EU-funded "Imagine" project – Innovations for a Made Green in Europe – aimed at applying some innovative environmental policy tools in industrial clusters, creating an effective opportunity for small and medium enterprises (SMEs) and local business communities to pursue sustainability objectives. The project focuses on four industrial clusters located in the Tuscany region (Italy), supporting them in implementing local innovative environmental policies and tools, based on a life cycle approach, to improve their competitiveness. Evidences of the applicability of a cluster approach to the implementation of life cycle assessment tools are discussed and the policy implications are examined.

10.2 Theoretical Background

According to the scientific literature of the late 1990s, there is no doubt that interorganizational stewardships have a relevant role in addressing sustainable development pressures (Angell and Klassen 1999; Bansal and Roth 2000; Hart and Milstein 1999; Porter and Van der Linde 1995; Srivastava 1995; Sharma and Vredenburg 1998).

In the operations management area, systemic perspectives are recognized as particularly important when investigating SC approaches and management logic from an environmental point of view. In particular, interorganizational strategies are reported as solutions for coping with complexities associated with product, remanufacturing, testing, evaluation, returns volume, timing, and quality in closed-loop SCs (Guide et al. 2003; Rizzi et al. 2012). Thus, not surprisingly, integration and "contamination" of environmental and SC management studies are widely recognized as important when confronting the area of the management of sustainability (Srivastava 2007).

Unfortunately, such issues have been studied in depth so far only in relation to certain phases or activities of the SC viewed individually (e.g., green design, green procurement, reverse logistics, etc.) (Srivastava 2007; Sharfman et al. 2009). While a comprehensive theoretical framework in the field of green SC management is under development, a number of management studies in the field of interorganizational management argue that high interdependence and diversification of functional teams generally lead to greater exploration of possibilities of actions, at the cost of increased difficulties in coordination. Examples were provided by Patrick and Echols (2004) for product evolution, Rivkin (2000) for firm development, Rivkin and Siggelkow (2003) and Levinthal and Warglien (1999) for organizational design, Gavetti et al. (2005) for strategic analysis, Choi et al. (2001) and Choi and Krause (2006) for SC management, and Lorenzen (2002) for industrial cluster coordination. In this framework, coordinating actions between corporate environmental management systems (EMS) is likely to increase effectiveness and efficiency of interorganizational exchanges. Darnall et al. (2008) argue that the skills and capabilities needed to manage a green SC are synergistic and complementary with those that characterize the implementation of an EMS.

The process of a gradual "opening" of an EMS is recognized as a significant innovation in environmental management practices that requires the gradual assimilation by enterprises and their management of a logic inspired by the so-called life cycle thinking. Product-oriented EMSs (POEMSs) (Klinkers et al. 1999) are a typical expression of the gradual extension of the objectives and scope of EMSs in a life cycle perspective.

Non-collaborative relationships between the company itself and the actors that influence environmental impacts of production processes or products usually make ineffective the implementation of traditional corporate environmental management. By this logic, the goal of the extended EMS is not only to manage relations with the outside world but also to foster and promote actions and interactions through which many actors manage the impacts related to the different stages of the product's life cycle (Sharfman et al. 1997).

Of course, POEMSs present both benefits and limitations. On the one hand, taking into account the entire life cycle, POEMSs can provide numerous opportunities to reduce the environmental impact associated with the products, either through unilateral ad hoc actions or through joint efforts involving different stakeholders along the chain (Sharfman et al. 2009) that have the capability to help achieving tangible improvements in performance (van Berkel et al. 1999; Charter and Belmane 1999; Brezet and Rocha 2001). On the other hand, the process of "opening up" the EMSs is not immediate or "painless" for businesses. In fact, life cycle approaches are often characterized by a strictly "engineering" perspective based on the technical reconstruction of all the steps needed to make a product. This does not reflect the economic and commercial complexity of the value chain, which consists of the spectrum of relationships between SC actors that operate in the company's various target markets (Heiskanen 2000). Studies show that the main difficulty that companies face when implementing a product-oriented approach is precisely that the networks involved in the analysis and the companies' network of relationships do not coincide (Fuller 1999).

The environmental impacts of a product often affect actors and stages of the life cycle with which the producing company seldom has direct contacts, making it difficult to manage these relationships as a system goal. It is worthwhile to note that commercial intermediaries, such as industrial distributors of raw material and intermediate products, wholesalers, major retailers, and suppliers of secondary materials, introduce a "brokering" stage that contributes to loosening the links between the manufacturer and the other actors who play a key role in product management. This results in a reduced ability to influence (or simply interact with) the value chain (Ammenberg and Sudin 2005; Fuller 1999). But these are not the only obstacles in building cooperative, or at least "coordinated," relationships throughout the SC. First, there are obstacles to the flows of communication and information necessary for the development of life cycle logic in many large companies also because of difficulties related to the compatibility between business information systems, confidentiality and control over information, different languages and routines, etc. Second, transaction costs that the company bears in order to implement an eco-friendly product-oriented collaboration (e.g., attributable to the need to negotiate and reach agreements, define common measures for environmental improvement, etc.) play a role that should not be underestimated (Sinding 2000). In fact, while intraorganizational eco-friendly initiatives can be developed according to a transparent economic analysis, interorganizational ones are often associated to search, information, bargaining, and enforcement costs, i.e., those costs that are necessary to design and develop a useful collaboration but that can, sometimes, prevent it.

The experience of these difficulties has contributed significantly to the development of methodological and management tools aimed not only at accounting for the impacts associated with the life cycle of the product but also at involving the suppliers in the assessment and quantification dynamics of these impacts. Among these tools, the methodologies of life cycle assessment (LCA) are predominant. LCA has been extensively used to optimize closed-loop SCs as well as to improve product design and stewardship (e.g., Krikke et al. 2004; Sarkis 2001; Sroufe et al. 2000). LCA combines the possibility to use standardized datasets with the possibility to create along the entire SC the information and communicational channels necessary to collect and process specific data and information that allow the quantification of the interactions between the "product system" analyzed and the environment (Pesonen 2001; Krikke et al. 2004; Sarkis 2001; Sroufe et al. 2000).

There are a number of factors that could make LCA an effective tool for improving the environmental performances throughout its SC (Hagelaar and van der Vorst 2001). Among these, first, LCA must be considered a context-dependent tool, i.e., its results are closely tied to the definition of the objectives and scope at the start-up of the evaluation process. Thus, each application requires a fine-tuning of the agreements among the actors involved (i.e., nature and extent of relations, the reasons for the agreements, property of the outcomes, etc.). Second, strict requirements have to be set, to preserve not only trust and openness but also transparency in the data sharing and consistency in their production and management. Third, LCA requires motivations deriving from shared and long-lasting environmental goals. The design of an LCA guides the involved actors in a careful analysis of goals deriving from both external factors (e.g., competition within the sector or between sectors, regulatory conditions, pressure from stakeholders, etc.) and internal factors (e.g., interest in developing new knowledge, need for reliable accounting performance, etc.).

Unfortunately, setting these goals is particularly difficult when sustainability is the target (Lehtonen 2004; Elkington 1998). Sustainable development is an intrinsically complex dimension that requires coordination of social, environmental, and economic dimensions. An LCA (or any other environmental tool) that focuses on environmental parameters disconnected from social concerns may sometimes even be counterproductive to overall sustainability goals. As a consequence, since it is very difficult to understand interdependencies when social factors are involved, most studies do not attempt to do so. For example, Handfield et al. (1997), Melnyk et al. (2003), and Zhu and Sarkis (2004) have shown that environmental practices positively affect the firm's operational performance, but they do not analyze social factors. Therefore, the implications of LCA for strategy have to be carefully analyzed. Considering that risk management is widely recognized as a core business function, the inability to identify key stakeholders and potential social outcomes is becoming an increasingly important challenge, yet one insufficiently addressed by traditional management approaches (Stone and Brush 1996).

From an analytical perspective, there are two complications: it may be difficult to identify not only interdependencies between parameters but also the key parameters. Conversely, beyond the reasons more closely tied to proactive and "value-driven" environmental management, the key drivers for the adoption of life cycle approaches in cooperation with suppliers usually refer also to the uncertainty of information that governs the nature and extent of environment-related impacts (Vermeulen and Ras 2006; Sharfman et al. 2009). In other words, the complexity and difficulty of the decision-making processes frequently created by this uncertainty are both drivers and constraints for the adoption of LCA in eco-management practices.

From a theoretical perspective, interorganizational initiatives in industrial clusters could help to manage costs and benefits of comprehensive life cycle approaches. Territorial approaches could serve the objective of supporting analytical LCA, i.e., so as to account for immediate physical flows like resources, material, energy, and emissions, etc., using average data for each unit process. Even more, they could support the so-called consequential LCA, i.e., the analysis of how physical flows can change as a consequence of an increase or decrease in demand for the product system under analysis. In fact, shared resources and goals could help in overcoming barriers to the inclusion of unit processes inside and outside of the product's immediate system boundaries and to the collection of economic data to measure physical flows of indirectly affected processes (Earles and Halog 2011). To this end, the so-called life cycle costing provides useful guidance on how to integrate "conventional" accounting with an approach that allows for the identification of longer-term strategic opportunities and efficiency margins.

Unfortunately, interorganizational collaborations are not a spontaneous phenomenon. According to Levinthal and Warglien (1999), the synchronization of behavior across organizations is crucial to facilitate cooperation. In fact, when actors only consider the payoff implications of their local actions and ignore the entire topography of the landscape, they "myopically adapt" and only see illusory improvements. In other words, environmental management programs at the territorial level that are highly specialized but disconnected may not be as effective as management for sustainable development, which is based on understanding, seeking out, and exploiting broad interdependencies. This recalls the theoretical contribution of Gavetti et al. (2005), who recommend analogical reasoning based on accumulated (multifaceted) experience and alertness, which are valuable resources to be shared in far-seeing collaborations.

Finally, it is possible to account for some additional evidences related to the implementation of tools linked to life cycle thinking in the logic of SC management. When the LCA is really integrated into the EMS within a logic of SC dynamic management, it is to be assumed that those companies that use and promote this approach are able to affect the environmental impacts or influence the behavior of actors that are external to the "boundaries" of the companies' organization. The use of LCA within SC relationships could provide important contact points, synergies, and complementarities with an "extended" EMS and its accounting for "indirect environmental aspects."

The generation of added value through elaborations of available data is made possible, thanks to the cooperative relationship among the companies in the SC. Product-oriented logic can effectively engage the environmental management system of a company that operates as a producer or as a customer in any type of SC. This logic provides crucial support for the customer relationship management and bridges the gaps of the system when it comes to identifying the customers' and stakeholders' needs, defining modes of interaction, handling complaints and returns (for environmental reasons), reviewing the contract (which can include requirements concerning the product's impact), and measuring their satisfaction. These are essential elements in order to assess the environmental competitiveness and "green" marketing strategies implemented by the company itself.

In this regard, it is finally worthwhile to mention a specific opportunity in terms of marketing and environmental communication introduced by the possibility to certify the environmental impact of a product of a "local" chain or a group of producers on the basis of an international scheme based on the ISO 14025 standard. In detail, the EPD (Environmental Product Declaration) international system, currently managed by a body comprising representatives of some EU countries, offers the possibility to certify, alongside the environmental performances of a product or service of a single firm, also an "average" or "typical" territorial product. Once the EPD is validated and registered, it can be used effectively as a tool for communication and marketing, by linking it to the local product, the chain, or the sector of which you want to enhance the environmental benefits.

In order to shed some light on the potential solutions to the contrast between barriers and opportunities for the adoption of a cluster approach to life cycle thinking, this study addresses the following research question: is it possible to develop an LCA at cluster level so as to provide firms with a shared tool that is suitable for driving life cycle thinking initiatives?

The section that follows presents a demonstration of how the aforementioned theoretical background can be translated into practice at the territorial level.

10.3 Methodology

10.3.1 Context of the Study: The Project Imagine

The application of an LCA assessment of typical products of a cluster of SMEs was developed within the Imagine project – Innovations for a Made Green in Europe – a project funded by the EU Eco-innovation program. The project, started in October 2009 and completed in October 2012, aimed to develop and apply an innovative method to support the implementation of an LCA-based cluster environmental management approach, oriented to identify, improve, and communicate the environmental performance of a typical product of a cluster of SMEs. The aim of this approach was twofold: (1) support SMEs, usually affected by lack of human and financial resources, in performing an LCA to identify the environmental hotspots of their production processes and plan and realize actions in order to improve their efficiency in the use of resources, and (2) define communication tools to be used at the cluster level that summarize the environmental performance of the typical products in their life cycle and can be used to clearly inform consumers and clients on the environmental quality of the local products.

Data for answering the research question were gathered through the following steps:

- Identification of suitable methodologies for life cycle assessment
- Identification of the standards to be used in the validation process
- Desk analysis of the sector, based on available literature
- Definition of the protocols for the development of the life cycle assessment at cluster level
- Stakeholder engagement
- Collection of data
- Elaboration of data
- Validation of the results
- Discussion on their relevance for environmental management practices at cluster level

Within the project, this cluster approach to LCA was applied in four industrial clusters located in the region of Tuscany (Italy), which represent the whole fashion SC:

- Industrial cluster of Prato (textile sector)
- Industrial cluster of Santa Croce sull'Arno (tannery sector)
- Industrial cluster of Empoli (clothing sector)
- Industrial cluster of Lucca (shoe production sector)

The replication of the study in four clusters serves the scope to test both the construct validity (i.e., whether the information gathered in each step reflects dynamics that are not sector-specific) and the internal validity (i.e., whether the adoption of LCA tools is influenced by external and spurious relationships).

These sectors have been involved in testing the approach through the creation of SME-oriented LCA tools and models which, in the future, will be imitable and transferable in other similar SCs and in other European clusters. The approach relies basically on a cooperative and "modular" life cycle management of common environmental problems among four clusters that are linked with each other as part of the same SC. The products of Prato (textile fabrics) are used in Empoli for the clothing sector, while the leather produced in the cluster of Santa Croce is used both in Lucca for shoes and in Empoli for leather clothing. This approach aimed to enhance competitiveness in a high-quality industrial sector (fashion) as well as in the other "typical" traditional sectors that characterize the whole European productive system, to make it more integrated and to provide an advantage over external competitors.

To test the reliability of the research design, findings have been discussed with local stakeholders. From an operational perspective, the first step of the approach was the implementation, in each involved cluster, of the requirements of the EMAS Regulation (Reg. n. 1221/2009/CE) at the cluster level (Daddi et al. 2010), as a preliminary step towards an LCA-based approach. The actions foreseen by EMAS and implemented in a synergistic way at the cluster level by the Imagine project are the following:

- Setup of a Promotion Committee composed of public and private actors representing the collective interests of a cluster and involved in the local environmental policies and strategies such as municipalities, provinces, control authorities, trade associations, companies which provide local environmental services (i.e., waste collection and treatments, purification of industrial wastewater), and local environmental associations

- Carrying out a Cluster Environmental Review, identifying and evaluating the most significant environmental aspects in the cluster (as a basis for the primary data needed as an input for the subsequent LCA)
- Definition, by the Promotion Committee of an Environmental Policy for the whole industrial cluster, identifying the commitment of all the main local actors towards the continual improvement of the environmental performance of products and processes
- Elaboration and drafting of a Cluster Environmental Programme containing the detailed actions and measurable targets for operationally pursuing continuous improvements
- Promotion and carrying out of specific initiatives addressed to local actors (SMEs, suppliers in the local chain, service providers, local authorities, etc.) aimed at satisfying the commitments undertaken with the shared program

These steps allowed us to gather evidences on both the feasibility of the implementation of the cluster approach to LCA (i.e., through achieving the drafting of an LCA report in conformity to the reference standards) and its practical relevance as a driver for life cycle thinking initiatives (i.e., through observing its use during the elaboration of the Cluster Environmental Programme) (Fig. 10.1).

10.3.2 Coordination of Life Cycle Assessment (LCA) at Cluster Level

Grounding on this first step, the following level of activities was directly related to the product environmental performance in the SC, which has been the main aim of the project. The idea underpinning the Imagine approach has been that of developing a "cooperative" LCA on which to build the policies and actions to be implemented in the clusters for improving the performances throughout the life cycle of the targeted products. This group of activities mostly regarded the dissemination of the life cycle thinking approach in the companies of the clusters involved, through some supporting actions carried out by the Promotion Committee in order to perform four "modular" LCAs of the average products, representing the characteristics and impacts of the local SC.

The term LCA methodology was first coined during a SETAC (Society of Environmental Toxicology and Chemistry) conference in 1990 in Vermont (USA) and is defined as "an objective process of evaluation of environmental burdens associated with a product (...) through identifying and quantifying energy and materials used and waste released into the environment, to assess the impact of these uses of energy and materials and releases into the environment and to evaluate and implement environmental improvement opportunities. The assessment includes the entire

	Stage 1: Definition of the research question	Stage 2: in strument development	Stage 3: Data gathering	Stage 4: Analyze data	Stage 5: Disseminate
Methodology	Analysis of current literature available in English from ISI and Scopus	Review of available LCA methodologies; Action-research oriented collaboration with the Promotion Committee of each cluster	Technical analysis of environmental performances at cluster level; questionnaires at company level; discusions with relevant stakeholders	Commercial LCA software; discussion of LCA report (action-research)	Discussion on the pros-cons and cost-effectiveness of the cluster approach to LCA
Outputs	Literature revies; Research question	LCA protocol; Database on environmental performances at cluster level; in-field observation	Life cycle inventory; evidences on stakeholders' reactions to LCA report	LCA report; adoption of LCA outcomes during the elaboration of the Cluster Environmental Programme	Identification of the weaknesses and the needs for further research

Fig. 10.1 Methodological approaches employed in the five stages of the case study research (Stuart et al. 2002; Seuring 2008)

lifecycle of the product (…), including extraction and processing of raw materials, manufacture, transport, distribution, use, reuse, recycling and final disposal" (SETAC 1991).

The first LCA studies were undertaken in the late 1960s and covered some aspects of the life cycle of materials and products to highlight issues such as energy efficiency, consumption of raw materials, and waste disposal.

The development of LCA methodology culminated in the codification of a family of standards ISO 14040 (Environmental Management – Life Cycle Assessment), published in 1997.

In our study, we performed four LCAs, one for each typical product of the four involved clusters, as well as the drafting of the corresponding EPDs. To elaborate the mentioned four LCAs, the data have been collected from a total of 34 companies located in the four clusters. The companies were identified according to the representativeness of their productive characteristics. In particular for each cluster the data gathering phase involved 6 companies of industrial cluster of Prato (textile sector), 6 companies of industrial cluster of Empoli (clothing sector), 19 companies of industrial cluster of Santa Croce sull'Arno (tannery sector), and 3 companies of industrial cluster of Lucca (shoe production sector). The different number should be linked to the characteristics of the productive process of the involved companies. So, for example, the companies of Lucca had an integrated process that covers all phases and also the sum productive capacity has been considered representative of the whole production of the cluster. On the contrary, the involved companies located in the tannery clusters were mainly small enterprises, and each of them represents only a specific phase of the production process. For this reason, we needed to involve a higher number of companies to assure the representativeness of the productive process.

The results of LCA studies allowed for the identification of some key performance indicators measuring the average impact of the production of the characterizing products of the four involved clusters. The LCA study was developed in compliance with ISO 14040 standards (ISO 2006a, b).

According to ISO, LCA is a technique for assessing the environmental aspects and potential impacts throughout the life cycle of a product or process or service, which is divided into four phases:

1. Setting the goals and boundaries of the system (goal and scope definition – ISO 14041)
2. Data collection (inventory analysis – ISO 14041)
3. Environmental impact assessment (impact assessment – ISO 14042)
4. Interpretation of results and improvement (improvement analysis – ISO 14043)

10.3.3 Goal and Scope Definition

The goal of the LCAs was to assess the potential environmental impacts of the four average products and of their production chain. The results of the study are used for a twofold objective: on the one hand, to define an environmental declaration including the average environmental performance of one of the characterizing products in each involved cluster and, on the other hand, to identify the environmental requirements of a local product ranking scheme, which aims to stimulate the environmental improvements in firms operating in each cluster.

This tool was also conceived to support local policies for greening the market, as suggested by several background documents published by the European Commission (2013).

The analysis was performed to evaluate the average production processes of the following products characterizing the involved cluster – leather, recycled wool, leather and wool coats, and leather shoes – so as to estimate the environmental impacts on the basis of a representative sample of companies and products and build on these results a consistent and appropriate requirements, tailored to the specificities of the local production.

10.3.4 Life Cycle Inventory

The most time-consuming step in the implementation of an LCA is the collection of data for the life cycle inventory. However, this stage is of utmost importance since, as often reported in the literature (Iraldo et al. 2013), there is a lack of reliable primary data. This step includes the data collection and the quantification of the interactions between the investigated system and the environment. These interactions comprise the use of natural resources, emission into atmosphere, and discharges into water or on ground (Rubik et al. 2008). This is an iterative analysis since the data collection could determine new data needs or some changes in the study's design (i.e., objective, scope, boundaries of the system).

For this study, companies from each cluster were involved in compiling a questionnaire developed for the purpose of collecting, for each unit process, quantitative data on inputs and outputs (raw materials, energy and water consumption, waste production and transportation). Companies were involved through the local trade associations who helped organize public meetings to explain the general objectives of the Imagine project and the aim of the LCA study and to distribute the questionnaire, describing it in detail. Questionnaires were then collected by e-mail a few weeks later.

10.4 Results and Implications at Cluster Level

The LCAs were performed on four typical products of the clusters: leather, recycled wool (so-called cardato), leather and wool coats, and leather shoes. As a first, relevant, result, no major differences emerged from the replication of the study in the four different contexts.

As mentioned above, the LCA did not focus on a single organization, but aimed to represent the average environmental impact of the typical product of each cluster. The application of life cycle assessment allowed to estimate the average environmental impact of the five selected products produced in the involved clusters by calculating the value of each of the following key performance indicators:

- Global warming potential measures the level of warming of the atmosphere caused by human activities and it is accounted in kg CO_2 equivalents (Christensen 2009).
- Acidification potential measures the level of acids and compounds which can be converted into acids released by the investigated production processes. That contributes to death of fish and forests, etc. It is presented in kg SO_2 equivalents (ISO 14042).
- Ground-level ozone potential measures the effect of human activities on the formation of photochemical smog caused by the release in the atmosphere of volatile organic compounds (VOCs). It is measured by kg ethane equivalents (ISO 14042).

Table 10.1 KPI (impact category) of leather and wool production from LCA

KPI	Unit	III°	0	I°	II°
Leather (1 square meter)					
Global warming pot	kg CO_2 eq.	29.44	26.77	24.09	20.1
Acidification pot	kg SO_2 eq.	0.308	0.28	0.252	0.2
Ground-level ozone pot	kg ethene eq.	0.121	0.11	0.099	0.1
Eutrophication	kg PO_4 eq.	0.011	0.01	0.009	0.0
GER	MJ	40.85	37.14	33.42	27.9
Water footprint	l	349	317	285	238
Waste production	kg	3.19	2.9	2.61	2.2
Recycled wool (1 kg)					
Global warming pot	kg CO_2 eq.	30.74	27.95	25.15	21.0
Acidification pot	kg SO_2 eq.	0.11	0.10	0.09	0.08
Ground-level ozone pot	kg ethene eq.	0.011	0.01	0.009	0.0
Eutrophication	kg PO_4 eq.	0.231	0.21	0.189	0.2
GER	MJ	497	452	407	339
Water footprint	l	11,283	10,257	9,232	7,693
Waste production	kg	36.48	33.17	29.85	24.9

- Eutrophication is cause by the enrichment of nutrients (i.e., noxigen) in the water that determine algal bloom and, as a consequence, oxygen depletion and death of fishes. It is measured by kg NO_3 equivalents (Beccalli et al. 2010).
- Gross energy requirement (GER) measures total energy consumption, both from renewable and conventional energy sources.
- Water footprint measures the total water consumption needed for the production of a product, including direct and indirect uses.
- Waste production is the total waste generated for the production of a product including both hazardous and nonhazardous waste.

The column 0 in Tables 10.1 and 10.2 shows the results of LCA studies of the five selected products.

The results of LCA studies were not used to define "environmental quality standards" but as a starting point for the definition and implementation of innovation processes to achieve environmental improvements. According to in-field observations, it seems reasonable that each firm can compare its environmental performance with the "average value" measured in the cluster production and SC and accordingly rank its impact. This information can support the planning of strategic and operative actions in order to reduce the firm's "environmental footprint" (e.g., by improving its resource efficiency or by saving water or energy). In other words, a manager of a company operating in the cluster can both understand the environmental impact of the firm's product and also identify opportunities to increase efficiency and trigger a cycle of continuous improvement.

In each cluster, starting from the results of the LCA, four performance levels were defined for leather, recycled wool, coat, and shoe production in order to allow each firm of the clusters to identify their own position compared to the average level of the cluster. This resulted in being a very useful tool for starting to coordinate

Table 10.2 KPI (impact category) of final products from LCA

KPI	Unit	III°	0	I°	II°
Leather coat (1 unit)					
Global warming pot	kg CO_2 eq.	177	161	145	121
Acidification pot	kg SO_2 eq.	1.463	1.33	1.197	1.00
Ground-level ozone pot	kg ethene eq.	0.517	0.47	0.423	0.4
Eutrophication	kg PO_4 eq.	0.077	0.07	0.063	0.1
GER	MJ	3,023	2,748	2,473	2,061
Water footprint	1	1,561	1,419	1,277	1,064
Waste production	kg	132	120	108	90
Recycled wool coat (1 unit)					
Global warming pot	kg CO_2 eq.	52	48	43	36
Acidification pot	kg SO_2 eq.	0.70	0.64	0.576	0.48
Ground-level ozone pot	kg ethene eq.	0.022	0.02	0.018	0.0
Eutrophication	kg PO_4 eq.	0.077	0.07	0.063	0.1
GER	MJ	330	300	270	225
Water footprint	1	1,423	1.294	1,164	970
Waste production	kg	2.79	2.54	2.28	1.9
Leather shoes (1 pair)					
Global warming pot	kg CO_2 eq.	7.18	6.53	5.87	4.89
Acidification pot	kg SO_2 eq.	0.07	0.07	0.06	0.05
Ground-level ozone pot	kg ethene eq.	0.022	0.02	0.018	0.02
Eutrophication	kg PO_4 eq.	0.0022	0.002	0.0018	0.0015
GER	MJ	148	135	121	101
Water footprint	1	145	132	119	99
Waste production	kg	3.92	3.57	3.21	2.67

Table 10.3 Environmental performance ranking

Performance level		LCA results	Impact category	Suggested time for improvements
I°	Excellent performance	KPIs are 25 % lower than the value of the average performance of the cluster	Limited impact	3–5 years
II°	Good performance	KPIs are 10 % lower than the value of the average performance of the cluster	Mow impact but improvable	2–3 years
0	Average performance	KPIs have a similar value of the average performance of the cluster	Average impact	1–2 years
III°	Low performance	KPIs are 10 % higher than the value of the average performance of the cluster	Significant impact	<1 year

environmental improvement programs at cluster and corporate levels. Specifically, the standard level (0) indicates the value emerging from the LCA analysis of the identified typical products of the clusters. Each organization can thus compare the key performance indicators resulting from own LCA with the standard level and verify in which position it is. Three performance levels were defined by the members of the Promotion Committee in each involved cluster. The 3rd performance level means that a firm performance is lower than the average value in the cluster, the 2nd performance level means that a firm performance is slightly better than the average value, and the 1st level means that a firm performance is significantly better than the average value (see Table 10.3 for details).

10.5 Discussion

The methodology proposed in the case study fosters a cooperative and integrated approach to environmental management at the cluster level, by involving all relevant local actors and stakeholders in the actions for the improvement of the cluster (and the related SC) environmental performance. This approach relies on the cooperative nature of the business relations and on the potential synergies among companies, on the similarity of the processes under the technical point of view, on the good social relations, and, finally, on the tight interactions between stakeholders that exist within the "industrial cluster."

Since the firms located in the same industrial cluster are similar (as to process technologies, organizational structures, and managerial approach) and must confront the same environmental problems, it is possible for them to strongly rely on synergies that already exist at the cluster level to manage these problems effectively and, even more, to improve their competitiveness based on this.

Moreover, organizations that belong to the same cluster have to face the same regulations and interact with the same stakeholders and with the same SC. This is particularly important when the cluster becomes the dimension in which "green supply chain management" tools are applied, such as the LCA.

The life cycle approach applied in the Imagine project took into consideration the whole range of resource flows and environmental pressures associated with the products of the involved clusters, from an SC perspective. It included all stages from the acquisition of raw material to processing, distribution, use, and end-of-life processes and all relevant environmental impacts, health effects, resource-related threats, and burdens to society.

Availability of primary data (mostly collected by questionnaires and direct interviews) and secondary (but specific) data resulted in being sufficient for the development of a good quality inventory. What is more, the proposed approach resulted in being effective in exposing any potential trade-offs between different types of environmental impacts associated with specific characteristics of the products, and the policy decisions and management strategies that can influence these (e.g., design choices or SC management criteria).

In the described framework, the methodology of the project aimed to enable SMEs operating in the EU fashion sector to use the same cluster-based approach, and connected guidelines, to address their innovation strategies towards sustainability goals, towards environmental excellence.

Final markets of fashion products are increasingly showing that they appreciate product environmental quality. There is a growing share of consumers that is reward sustainability-oriented producers and, especially, the products they offer on the market. This is why an LCA-based approach has been successful in the Imagine project: it helped companies in the cluster to communicate their environmental performance to those consumers and clients that are more sensitive and eager to valorize it in their purchasing decisions.

In addition to prompting the environmental quality of the cluster products, the proposed approach strongly relied on a second strategic "pillar": guarantee and control. The tools adopted in the Imagine project – EMAS and EPD – both have a strong focus on the guarantees they can provide. They both rely on third-party certification mechanisms that are perceived by the stakeholders as a safeguard for the credibility of the environmental excellence achieved by the organizations or by the products that are able to obtain them. This explains why the companies of the four clusters involved in the Imagine project trusted EMAS (that was already known and diffused among the most advanced companies of the clusters) and the EPD. The latter, even if less diffused than EMAS, was seen as a trustable guarantee on the environmental performance of the product and, especially, an effective way to valorize them in the competitive arena.

Apart from the results connected with the actions oriented towards green products, the Imagine cluster experiences allowed the territorial areas involved to achieve important outcomes both at collective and at firm level.

At the cluster level, the project supported the acquisition of the EMAS Recognition of the Italian EMAS Competent Body for the four clusters involved. This special kind of recognition is issued by the national EMAS Competent Body only in Italy. It represents a further national development of article n. 37 of EMAS Regulation n. 1221/2009 (European Parliament and Council 2009) "Cluster and step-by-step approach" and states the following: "Member States shall encourage local authorities to provide, in participation with industrial associations, chambers of commerce and other concerned parties, specific assistance to clusters of organizations to meet the requirements for registration as referred to in Articles 4, 5 and 6. Each organization from the cluster shall be registered separately (…)."

In September 2011, the four involved clusters obtained this official recognition as an award for the environmental policies carried out in the clusters. Thanks to the obtainment of this recognition, the industrial clusters will be able to improve their competitiveness by setting up new green marketing actions, based on what they consider a real market opportunity, and could be even able to attract external investments, lured by the social and environmental "reputation" of the cluster. For example, this recognition could be used by the cluster as a benefit and a competitive advantage of the localization towards those companies that want to be located in an area with a high environmental management capacity (Daddi et al. 2012).

In addition to this, the LCAs on the main products of the four clusters helped local and regional policymakers in the identification of the improvement opportunities for local industrial processes and technologies, as well as for the entire supply chain. On one hand, the local policymakers, as members of the Promotion Committee described in the previous section, understood the relevance of the information emerging from the LCAs, and they fully exploited and valorized this information in developing ambitious Cluster Environmental Programmes that, contrary to what could be expected, were very much focused on product supply chain, rather than limited to innovation in local processes. On the other hand, the regional policymakers benefitted from the use of LCA as a source of information concerning the links and environmental interdependencies and mutual influences between four industrial clusters all located in Tuscany, and therefore, the results were very much suited to be used as a basis for regional environmental improvement plans.

At the firm level, the Imagine experience has enabled single companies to achieve considerable environmental benefits. Thanks to the application of the EMAS cluster approach, nine companies obtained the individual EMAS registration before the end of the project. The support activities carried out in the four clusters let the involved SMEs to overcome some barriers that they usually have to tackle in order to obtain EMAS registration. For instance, the identification of a common cluster audit team has surely been an effective support for the registered SMEs. The audit team carried out environmental compliance audits and trained the internal auditors in the involved SMEs. Common management and technological solutions were identified at the cluster level and disseminated in the local SMEs to stimulate and sustain innovation and environmental improvement, a prerequisite of EMAS.

Furthermore, for each involved cluster, a panel of common performance indicators were identified in order to support the SMEs in measuring their environmental impacts and to facilitate benchmarking between the companies of the same cluster. The indicators took into account the "core performance indicators" suggested by the EMAS Regulation and included additional more sector-specific indicators as well. Finally a common structure of an EMAS Environmental Statement was elaborated in order to provide the whole cluster with a common tool that any company can easily adapt to its needs and specificities and use it as a green communication tool.

10.6 Conclusion

The added value of the Imagine project experience, described in the previous sections, could be considered significant from several points of view, both for practitioners and for researchers.

First, at the methodological level, the Imagine project strengthened and helped spreading the cluster approach for environmental management, especially by positively testing the application of LCA for typical "average products" of a cluster. It has to be emphasized that, for many reasons, this approach has proven to be much more effective than supporting single companies in the development of LCAs for their own specific products. Our work clearly shows that, by applying a cluster approach based on both environmental management systems and LCA, the companies located in the cluster are not only able to obtain and use the results of this "integrated" process to improve their environmental competitiveness, but they can also benefit from many support tools deriving from this cooperative approach. Each company that belongs to the clusters participating in the Imagine project has been able to acquire a reasonably precise information on the "average" impact of the typical product of the corresponding cluster, as well as a range of operational tools enabling them to use the LCA for:

- Improving the design of their products and the performances of their SC
- Benchmarking their product-related and life cycle-based performances with the other co-opetitors operating in the same cluster
- Using a "basket" of common indicators on key environmental performances for the sector and, particularly, for the cluster
- Relying on a managerial and organizational common structure in the cluster (thanks to the twined EMAS cluster approach) that enables single companies to cooperate and create synergies in many innovation processes, especially relating to the product, its SC, and its life cycle

Moreover, in addition to the shared resources grounded on the "cluster LCA," a set of tools to support SMEs in performing an "individual" LCA of their own product and in implementing EMAS to their organization was also created. The experience of the Imagine project demonstrated "in-field" that for SMEs a feasible way to carry out an LCA at reasonable costs is to strongly rely on a cluster LCA (i.e., a study carried out on an "average" typical product of the cluster) "adopting and adapting" the model to their own characteristics. In other words, an SME can save time and resources by using an LCA model created at the cluster level on the basis of the "average" product and by customizing it, i.e., by simply adding or better specifying the input data to the model, concerning the specific product and its life cycle.

The collective or "shared" resources, on one hand, and the support tools for the single companies, on the other, yielded interesting effects on market and competitiveness-related variables as well.

A first relevant effect of the described approach is that working at the cluster level has been particularly effective in supporting the achievement of environmental improvements and innovation in products and processes. This was done mostly by successfully coupling LCA with an EMAS cluster approach, which is explicitly aimed at continuous improvement. By mixing LCA and EMAS at the cluster level, the outcome was to push the companies in the cluster towards the goal of continuously improving the environmental performance in their SC and for all the relevant life cycle phases. In doing so, the LCA of the "average" typical product of the cluster has been used as a guideline to orient and drive the improvement decision-making process in the right directions along the SC. This resulted in the activation of some fundamental networks (pillars of the cluster approach) with suppliers and other key actors in the SC (e.g., the specialized technology providers at the local level).

A second consequence of promoting the cluster approach has been to considerably improve the environmental communication strategies by the SMEs of the cluster. Prior to the Imagine project, few companies from the involved clusters provided reliable claims and environmental statements to the market and the stakeholders at large, whereas many others did not, which also created an unlevel playing field, particularly for SMEs, which do not have the same possibility to marketing their products as large firms. The aim of making a "cluster LCA" available for all the companies in the cluster was primarily to improve the capability of smaller companies to use credible data and indicators in their external communication and marketing actions.

The Imagine project experience demonstrated how that cluster approach can be applied to valorize typical Italian products (such as apparel and shoes), allowing them to leverage their market appeal not only towards their traditional characteristics of quality and design but also promoting their "green" peculiarities. In fact, this experience showed that the application of a clear and transparent methodology to evaluate the environmental impact of the products, such as the LCA, can be easily applied by SMEs as well and can generate several benefits and advantages at the economic and social level.

First, communicating environmental information transparently to the market and consumers enhances fair competition and fosters a more regulated internal market, reducing "greenwashing" practices.

Second, the adoption of LCA can generate some benefits even for very small companies, such as the reduction of costs by revealing inefficient uses of energy and identifying excess waste; increase in revenues by identifying new business opportunities by gaining access to markets that require or give preference to businesses that adopt LCA; and the use of science-based standards, providing standard compliant information to suppliers and consumers in order to establish a competitive advantage.

Third, the approach also had an effect on the operating costs and business conduct. Since all the supporting actions of the Imagine project were grounded on a territorial basis and guided by a "local governance," which intentionally stressed and valorized the strong links that exist in the cluster or in the related SC, the SMEs located in the cluster were really much more able to experience significant economic savings compared to SMEs that operate outside the cluster dynamics.

Finally, the enhancement of LCA by adopting a cluster approach can also have valuable social effects. The spreading of the life cycle approach to evaluate the environmental impact of products and to better communicate their performance to the market may increase the demand for certain skills. LCA experts, green marketing professionals, and third-party verifiers are the professions that could be positively affected by these actions. This was only partially observed during the Imagine, due to the limited scope of the project, but some preliminary signals on the market were emerging, especially in the segment of services and consultancy, where the know-how and the expertise locally available definitely improved. Moreover, many employees were trained during the project and a general higher environmental awareness was achieved in the involved clusters.

Despite the achievement of some of these positive economic, social, and environmental effects during the development of the Imagine, one of the limitations of this study is that the project ended recently and, consequently, the authors were not able to monitor in depth its outcomes. Therefore, a future research opportunity can be to further assess the effects of the "shared resources" and the supporting tools developed in the four involved clusters.

References

Angell L, Klassen R (1999) Integrating environmental issues into the mainstream: an Agenda for research in operations management. J Oper Manage 17:579–598

Bansal P, Roth K (2000) Why companies go green: a model of ecological responsiveness. Acad Manage J 43:717–736

Beccali M, Cellura M, Iudicello M, Mistretta M (2010) Life cycle assessment of Italian citrus-based products. Sensitivity analysis and improvement scenarios. J Environ Manage 91:1415–1428

Brezet H, Rocha C (2001) Towards a model for product-oriented environmental management systems. In: Charter M, Tischner U (eds) Sustainable solutions. Greenleaf, Sheffield

Charter M, Belmane I (1999) Integrated Product Policy (IPP) and Eco-Product Development (EPD). J Sustain Prod Des 10:17–29

Choi TY, Krause DR (2006) The supply base and its complexity: implications for transaction costs, risks, responsiveness, and innovation. J Oper Manage 24:637–652

Choi TY, Dooley KJ, Rungtusanatham M (2001) Supply networks and complex adaptive systems: control versus emergence. J Oper Manage 19:351–366

Christensen TH (2009) C balance, carbon dioxide emissions and global warming potentials in LCA-modelling of waste management systems. Waste Manage Res 27:707–715

Corbett CJ, De Croix GA (2001) Shared-savings contracts for indirect materials in supply chains: channel profits and environmental impacts. Manage Sci 47:881–893

Daddi T, Testa F, Iraldo F (2010) A cluster-based approach as an effective way to implement the environmental compliance assistance programme: evidence from some good practices. Local Environ 15:73–82

Daddi T, Tessitore S, Frey M (2012) Eco-innovation and competitiveness in industrial clusters. Int J Technol Manage 58(1–2):49–63

Darnall N, Jolley GJ, Handfield R (2008) Environmental management systems and green supply chain management: complements for sustainability? Bus Strategy Environ 18:30–45

Earles JM, Halog A (2011) Consequential life cycle assessment: a review. Int J Life Cycle Assess 16:445–453

Elkington J (1998) Cannibals with forks: the triple bottom line of 21st century. New Society Publishers, Gabriola Island

European Parliament and Council (2009) Regulation (EC) No 1221/2009 of the European Parliament and of the Council of 25 November 2009 on the voluntary participation by organisations in a Community eco-management and audit scheme (EMAS), repealing Regulation (EC) No 761/2001 and Commission Decisions 2001/681/EC and 2006/193/EC. Available from: http://eur-lex.europa.eu/legal-content/EN/TXT/PDF/?uri=CELEX:32009R1221&from=EN

European Commission (2013) Communication from the commission to the European parliament and the council: building the single market for green products- facilitating better information on the environmental performance of products and organisations. (COM(2013) 196 final) European Commission, Brussels. Available from: http://eur-lex.europa.eu/LexUriServ/LexUriServ.do?uri=COM:2013:0196:FIN:EN:PDF

Fuller DA (1999) New decision boundaries: the product system life cycle, in Sustainable Marketing. SAGE, Thousand Oaks/London/New Delhi

Gavetti G, Levinthal D, Rivkin J (2005) Strategy making in novel and complex worlds: the power of analogy. Strateg Manage J 26:691–712

Guide VDR, Jayaraman V, Linton JD (2003) Building contingency planning for closed-loop supply chains with product recovery. J Oper Manage 21:259–279

Hagelaar GJ, van der Vorst JG (2001) Environmental supply chain management: using life cycle assessment to structure supply chains. Int Food Agribus Manage Rev 4(4):399–412

Handfield RB, Walton SV, Seegers LK, Melnyk SA (1997) Green value chain practices in the furniture industry. J Oper Manage 15:293–315

Hart S, Milstein M (1999) Global sustainability and the creative destruction of industries. Sloan Manage Rev 41:23–33

Heiskanen E (2000) Managers' interpretation of LCA: enlightenment and responsibility or confusion and denial? Bus Strategy Environ 9:239–254
Iraldo F, Testa F, Bartolozzi I (2013) An application of Life Cycle Assessment (LCA) as a green marketing tool for agricultural products: the case of extra-virgin olive oil in Val di Cornia, Italy. J Environ Plann Manage (ahead-of-print), 1–26. doi:10.1080/09640568.2012.735991
ISO (2006a) ISO 14040:2006, environmental management – life cycle assessment – principles and framework. International organisation for standardisation. ISO Publishing, Geneva
ISO (2006b) ISO 14044:2006, environmental management – life cycle assessment – requirements and guidelines. International organisation for standardisation. ISO Publishing, Geneva
Klinkers L, van der Kooy W, Wijnes H (1999) Product-oriented environmental management provides new opportunities and directions for speeding up environmental performance. Greener Manage Int 26:91–108
Krikke H, Le Blanc H, Van de Velde S (2004) Product modularity and the design of closed-loop supply chains. Calif Manage Rev 46:23–39
Lehtonen M (2004) The environmental social interface of sustainable development: capabilities, social capital, institutions. Ecol Econ 49:199–214
Levinthal DA, Warglien M (1999) Landscape design: designing for local action in complex worlds. Organ Sci 10:342–358
Lorenzen M (2002) Ties, trust, and trade: elements of a theory of coordination in industrial clusters. Int Stud Manage Organ 31:14–34
Melnyk SA, Sroufe RP, Calantone R (2003) Assessing the impact of environmental management systems on corporate and environmental performance. J Oper Manage 21:329–351
Nawrocka D (2008) Inter-organizational use of EMSs in supply chain management: some experiences from Poland and Sweden. Corp Soc Responsib Environ Manage 15:260–269
OECD (1999) Boosting innovation: the cluster approach. OECD Publishing, Paris
Pesonen HL (2001) Environmental management of value chains. Greener Manage Int 2001(33): 45–58
Petrick IJ, Echols AE (2004) Technology road-mapping in review: a tool for making sustainable new product development decisions. Technol Forecast Soc Change 71:81–100
Porter M, Van der Linde C (1995) Green and competitive: ending the stalemate. Harv Bus Rev (September-October):120–134
Rivkin J (2000) Imitation of complex strategies. Manage Sci 46:824–844
Rivkin J, Siggelkow N (2003) Balancing search and stability: interdependencies among elements of organizational design. Manage Sci 49:290–311
Rizzi F, Bartolozzi I, Borghini A, Frey M (2012) Environmental management of end-of-life products: nine factors of sustainability in collaborative networks. Bus Strategy Environ. doi:10.1002/bse.1766
Ross S, Evans D (2002) Use of life cycle assessment in environmental management. Environ Manage 29:132–142
Rubik F, Scheer D, Iraldo F (2008) Eco-labelling and product development: potentials and experiences. Int J Prod Dev 6:393–419
Sarkis J (2001) Manufacturing's role in corporate environmental sustainability. Int J Oper Prod Manage 21:666–686
Seuring SA (2008) Assessing the rigor of case study research in supply chain management. Supply Chain Manage Int J 13:128–137
Sharfman MP, Ellington RT, Meo M (1997) The next step in becoming "green": life-cycle oriented environmental management. Bus Horiz 40:13–22
Sharfman MP, Shaft TM, Anex RP Jr (2009) The road to cooperative supply-chain environmental management: trust and uncertainty among pro-active firms. Bus Strategy Environ 18:1–13
Sharma S, Vredenburg H (1998) Proactive corporate environmental strategy and the development of competitively valuable organizational capabilities. Strateg Manage J 19:729–753
Sinding K (2000) Environmental management beyond the boundaries of the firm: definitions and constraints. Bus Strat Environ 9(2):79–91

Society of Environmental Toxicology and Chemistry (SETAC) and SETAC Foundation for Environmental Education Inc. (1991) A technical framework for life – cycle assessment. Society of Environmental Toxicology and Chemistry and SETAC Foundation for Environmental Education Inc., Washington, DC (Workshop held in Smugglers Notch, Vermont, August 18–83, 1990)

Srivastava P (1995) Ecocentric management for a risk society. Acad Manage Rev 20:118–137

Srivastava SK (2007) Green supply-chain management: a state-of-the-art literature review. Int J Manage Rev 9:53–80

Sroufe R, Curkovic S, Montabon F, Melnyk SA (2000) The new product design process and design for environment. Int J Oper Prod Manage 20:267–291

Stone M, Brush C (1996) Planning in ambiguous contexts: the dilemma of meeting needs for commitment and demands for legitimacy. Strateg Manage J 17:633–652

Stuart I, McCutcheon D, Handfield R, McLachlin R, Samson D (2002) Effective case research in operations management: a process perspective. J Oper Manage 20:419–433

Van Berkel R, van Kampen M, Kortman J (1999) Opportunities and constraints for product-oriented environmental management systems (P-EMS). J Clean Prod 7:447–455

Vermeulen WJ, Ras PJ (2006) The challenge of greening global product chains: meeting both ends. Sustain Dev 14(4):245–256

Zhu Q, Sarkis J (2004) Relationships between operational practices and performance among early adopters of green supply chain management practices in Chinese manufacturing enterprises. J Oper Manage 22:265–289

Chapter 11
Toward Joint Product–Service Business Models: The Case of Your Energy Solution

Andrea Bikfalvi, Rodolfo de Castro Vila, and Xavier Muñoz

Abstract The revolution generated by the emergence and spread of information and communication technology (ICT) has spawned a series of new technologies, novel applications, and ultimately innovative business models. This chapter presents a case study of a new technology business venture focusing on energy management systems and its evolution over the period of a decade. Transforming technology into functional products, which unite science and the marketplace, is the first challenge to any new technology-intensive business venture. The firm that is the subject of this study has a trajectory that passes through different stages, from pure product orientation to pure service orientation, to developing product–service joint modes of operating and finally to continuously evolving product-related services. Recommendations for practitioners refer to aspects related to business model analysis and reconsideration of value generation in the form of holistic solutions, including technology-based products complemented with complex, value-added services that emerge in response to ever-changing sophisticated customer demands. The case is also rich in the different types of innovation discussed: ecology, product, service, and business model innovations being the key types.

Keywords Servitization • Business model • Innovation • Energy • Spin-off

---

A. Bikfalvi (✉) • R. de Castro Vila
Department of Business Administration and Product Design, University of Girona, Escola Politècnica Superior, C/Mª Aurèlia Capmany, 61 Campus de Montilivi, 17071 Girona, Spain
e-mail: andrea.bikfalvi@udg.edu; rudi.castro@udg.edu

X. Muñoz
General Manager of Business DSET, Office A.2.16, C/Pic de la Peguera, 15, 17003 Girona, Spain
e-mail: xmunoz@dset-solutions.com

S. Azevedo et al. (eds.), *Eco-Innovation and the Development of Business Models*, Greening of Industry Networks Studies 2, DOI 10.1007/978-3-319-05077-5_11,

11.1 Introduction

Europe's growth strategy for the next decade is formulated in the Europe 2020 initiative (EC 2013) according to which, "Europe 2020 strategy is about delivering growth that is: smart, through more effective investments in education, research and innovation; sustainable, thanks to a decisive move toward a low-carbon economy; and inclusive, with a strong emphasis on job creation and poverty reduction. The strategy is focused on five ambitious goals in the areas of employment, innovation, education, poverty reduction and climate/energy."

While innovation has early and continuously been acknowledged as an important contributor to socioeconomic welfare and development, energy has been moving up the priorities recently, passing from an initial ecological concern toward general policy-making for industry and business. As a consequence, the combination of innovation and energy has become a powerful binary, setting the context for ambitious political and business targets. Eco-innovation is the emerging term for innovations generating increased sustainability and resource efficiency.

Environmental sustainability, understood broadly as development that "meets the needs of the present without compromising the ability of future generations to meet their own needs," and resource efficiency, best described as "improving the efficiency and effectiveness of how we use resources, i.e., using less to do more, and causing less impact from those resources we do use" (EC 2011), are key issues in the new European journey in which all Quadruple-Helix participants – academia, business, policy-makers, and users/citizens – have a proactive role.

It is in this scenario of positive political and institutional disposition, and overall societal and public consciousness, complemented with the proliferation of Information and Communication Technologies (ICT), that business opportunities emerge and are transformed into useful solutions. However, there have been few cases of large-scale eco-innovation to date. Dissemination of good practice, through case studies, can be an initial step toward enhancing awareness and knowledge of sustainable practices and consumption.

The objective of the present chapter is to present the case of a company called Your Energy System (YES) which develops technology-based products, focusing on improving energy efficiency, through an intelligent energy management system (EMS) with the overall aim of optimizing energy consumption.

The contribution of this case study is mainly related to the identification of key success factors in the field of eco-innovation, with the emergence of a new business model that intelligently combines value-added services and technologically intensive and complex products. This model ultimately translates into product–service systems (PSS).

Companies continually strive to increase production, while in recent years it has become evident that providing products alone is insufficient to remain competitive (Yu et al. 2008). According to various authors (Baines et al. 2009; Morelli 2003), the way to approach this new production system that combines products and services has followed one of two routes that have now finally converged: the first route is "servitization" which is the evolution of product identity based on material content to a position where the material component is inseparable from the service system, and the other route is "productization," when the service component evolves to include a product or a new service component marketed as a product. Thus, the PSS can be seen as the convergence of these two trends, which end in the consideration of a product and a service as a single offering.

A dynamic approach is appropriate and a chronological logic is applied to demonstrate the evolution of the company from its birth in a university research group to the present day, in order to highlight the appropriateness of the strategy of bundling products and services as a way of responding quickly to changing customer needs. The case highlights four crucial steps through which the business passed, analyzing facts, barriers, and enablers that are used to demonstrate the power of the final configuration.

We have called the different phases (1) the knowledge-orientation phase, (2) the product-orientation phase, (3) the user-orientation phase, and (4) the solution-orientation phase. Critical junctions represent intermediate steps or transition stages. They are defined in the spin-off literature as those difficulties that businesses have to overcome in order to pass from one development stage to another (Vohora et al. 2004). Before describing the phases, a note on the methodological approach is pertinent.

11.2 Methodology

In the general field of innovation studies, the application of a range of methodological initiatives and a variety of approaches has produced an increased maturity in the field. A recent survey of innovation studies (Hong et al. 2012) attempts to systematize and describe the existing initiatives. Despite the strong technological basis, data collection about eco-innovation data is still in its infancy. Due to the high cost of surveys, varying definitions and understanding of the concepts, and frequent use of umbrella or aggregate terms, it is not possible to rely on systematic measurements using standard instruments that are valid for all industries, sizes of enterprises, or countries.

It is in this scenario that the qualitative approach is selected, as appropriate for answering "how"- and "why"-type questions, because knowledge in the field of eco-innovation is distinct from other more traditional products and processes which may be considered innovative.

Moreover, case research enables researchers to immerse themselves in rich data and reflect on the longitudinal or dynamic progress of an establishment or phenomenon. Cases are descriptions of particular instances of a phenomenon that are typically based on a variety of data sources (Yin 1994), and cases can range from historical accounts to contemporary descriptions of recent events (Eisenhardt and Graebner 2007).

Secondary sources of information, from the Internet, press releases, and the media, were collected by examining the corporate website, annual reports, and various texts (Bikfalvi and Serarols 2008) of the participating company. This data collection was especially useful in helping us become familiar with the firm and its mission, vision, and activities.

Since our primary aim was to answer the what, how, why, who, where, and when of the phenomenon under investigation, an interview guideline was prepared and three interviews were conducted between February and April 2013 to clarify all the details. The primary methodological instrument was an open-ended interview guide that included a limited number of concrete topics and which was used during the face-to-face interviews in the respondents' daily workplace.

The interviews lasted between half an hour and one and a half hours, and the CEO and founder of the company as well as two key employees were invited to participate in the working sessions. Their previous academic background made them sensitive to our request and open to requests for cooperation. All interviews were recorded and fully transcribed. This procedure was followed by a within-case interpretation, meaning that two of the authors independently analyzed each interview and looked for patterns and themes previously agreed upon. A table template (see later in Fig. 11.2) was used, and an additional document links the topics with relevant literal citations. This chapter includes a considerable number of quotes from the original data to preserve transparency in the descriptions and to provide valuable illustration.

The explanatory power of a single case remains limited; in general terms, its main usefulness lies in its practicality and closeness to reality, as Siggelkow – a defender of case studies – states "A paper [based on case research] should allow a reader to see the world, and not just the literature, in a new way" (Siggelkow 2007: 23).

11.3 The Case

In this section, we describe the case from a process perspective, describing the different phases of development. This is followed by a detailed review of critical junctures. At the end we list, and briefly comment on, the different innovations in the case study and how they coexist. The outline of this section is depicted in Fig. 11.1.

11.3.1 Phases

Over the period of more than a decade, YES evolved through a variety of phases, shown below under four main headings, representing key periods in its evolution: (1) from knowledge to product, (2) from product orientation, (3) from service orientation, and (4) from service to integral solutions.

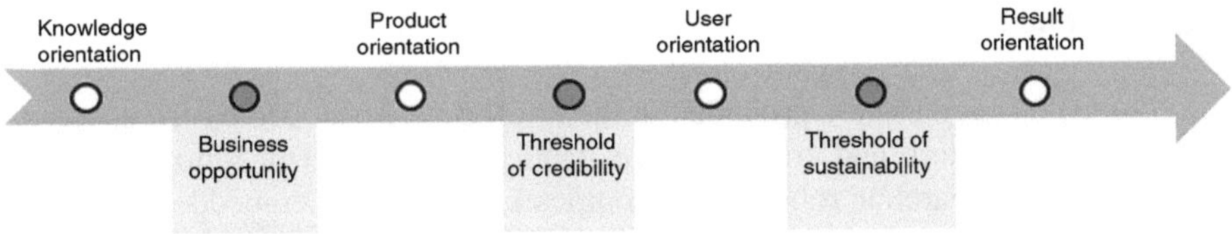

Fig. 11.1 Case outline – phases and critical junctures (Note: The upper concepts represent phases of evolution; the lower concepts represent critical junctures)

11.3.1.1 The Knowledge-Orientation Phase

The former head of the Department of Computer Engineering at the University of Girona, Spain, is a very ambitious, visionary, and active man, who wanted to promote technology transfer between the university and industry. After obtaining the results of their research, he told colleagues, "By the end of the year each of you could have your own company." The results of this statement did not take long to surface. Within a few months, two people from research group dealing with underwater robotics and artificial intelligence created a company called YES. The origin of the company, a university spin-off at that time, had the typical characteristics, limitations, and advantages of being born in a public entity and starting the journey of commercializing its research.

The majority of universities consider that the rules used for promotion on the professorial scale conflict with the entrepreneurial culture. This cultural problem reflects the sensitive issue of the ultimate ends of university research. This issue has long been avoided because of the strong influence of the "scientific" paradigm on the academic culture. According to that paradigm, the sole purpose of academic research is to increase and enhance human knowledge, regardless of any practical application. This paradigm recognizes only two ways of exploiting knowledge: (1) publications (i.e., books, articles, or conferences) that contribute to the collective and cumulative process of knowledge production and (2) education that provides students with opportunities to learn the latest scientific findings and discoveries. According to this conception, academic research is clearly a public good.

This paradigm has contributed to a system of values that is deeply rooted in the academic culture and that opposes the commercialization of research through spin-offs. In this system of values, three features seem particularly difficult to change: (1) the "publish or perish" drive, (2) the ambiguous relationship of researchers to money and (3) the "disinterested" nature of academic research.

Transforming a remote-controlled underwater robot with ecological functionalities into an unmet market need was the real challenge at this point. The entrepreneurial team decided to start to explore applying remote control and embedded solutions, principles, and technologies and transformed them into a product targeting ecological and energy-related functionalities. At that stage the robot, as a product, was abandoned.

11.3.1.2 The Product-Orientation Phase

The core technology used is grounded in the know-how developed by the group in embedding computers and free software applications (Linux-uClinux) obtained during years of research in robotics and computer research. An embedded system is a special-purpose system in which the computer is completely encapsulated in the device it controls. Unlike a general-purpose computer, such as a personal computer, an embedded system performs one or a few predefined tasks, usually with very specific requirements. Since the system is dedicated to specific tasks, design engineers can optimize it, reducing the size and cost of the product. Some also have real-time performance constraints that must be met, for reasons such as safety and usability; others may have low or no performance requirements, allowing the system hardware to be simplified to reduce costs.

Embedded systems are often mass produced, so the cost savings may be multiplied by millions of items. The rise of the World Wide Web has given embedded designers another quite different option, by providing a web page interface over a network connection. This is useful for remote, permanently installed equipment and avoids the cost of a sophisticated display, providing complex input and display capabilities when needed, on another computer.

Specifically, the team adapted the know-how they had developed in the robotics research group in embedded solutions to a commercially feasible application: remote Internet-based supervision systems for installed equipment like communal boilers (central heating), security, meters, and other applications.

The real innovation is the application of embedded technology to a field that did not employ this technology previously. The remote Internet-based supervision application is now a very promising market, exhibiting exponential sales growth. The benefits of controlling and supervising any application remotely, such as cost reduction and efficiency, are obvious, according to the founders: "We are very innovative in our target market by adapting embedded technology to work with other applications."

The functioning of technology can be summarized in six different steps:

- Gather information through meters (flow, temperature, electricity, etc.).
- Detect any malfunction of the device (there are certain anomalies that can be solved without human intervention; thus, the system can correct it, e.g., a low operating temperature of the boiler).
- Solve the problem automatically by the system.
- Store all this information in the central memory.
- Send alerts and other information through an Internet connection.
- Human intervention in case it is needed.

At the time the company was created, they were strongly focused on developing the product, and no attention was paid to possible related value-enhancing applications. The team assumed that the customer would be the one to use the tool to generate all the data. These data were converted into information, and this information would be further used in a decision-making process.

At this stage all attention was focused on transforming the technology into a functional product. They generated two differentiated products. In the beginning, YES focused on offering embedded solutions in custom energy applications to control communal boilers to provide hot water and central heating. Their first product, called DSE1 (Device to Save Energy – Home), was applied to a block of flats in the university's neighborhood. DSE1 was designed for blocks of flats that had a communal boiler which met the heating and hot water requirements of each apartment. The system permits remote control of the supply of hot water and heating, thereby achieving a significant energy saving. Moreover, DSE1 allows the measurement of each apartment's consumption. The final objective of such supervision is to improve the system's efficiency and to facilitate its maintenance. In addition, all the equipment was remotely supervised through online Internet-based technology.

After their experience with this product, the founding team thought that these embedded solutions could also be applied to other fields, such as hotels. Therefore, they developed a product to manage all the services available in a hotel room. These services include the control of the air conditioning, minibar, emergencies, fire alarms, opening doors, blinds, etc. This product was named DSE2 (Device to Save Energy – Hotel). DSE2 is a system that controls and monitors hotel rooms. It offers energy savings and better comfort and quality of services in the room. The core of the system is managed by an electronic board named DSE2. This is an intelligent device capable of simultaneously controlling different rooms. Located in different areas of the hotel, a communication bus board allows each module to manage the extensions placed in each room. Because it uses a modular hardware design, the system is easily adaptable to the needs and requirements of any hotel. The following facilities offered by DSE2 can be highlighted:

- Air-conditioning control of every room: an interval of temperature is established in order to avoid excessive energy consumption. The system can distinguish whether the rooms are occupied or not.
- Devices to save energy: for example, there are mechanisms that switch off the air conditioning when they detect that doors are open.
- The minibar can be monitored.
- Security warnings: intrusion, emergency, fire, etc.

A computer server monitors, collects, and stores the data provided by the electronic board (DSE2) about the current state of each room. It also allows storing and analysis of archival data in real time. In addition, the current state of each room can be accessed and consulted from any computer with Internet access. This software application is flexible and adaptable to any user's needs and helps the generation of strategies for saving energy.

At this time they realized they were certainly very focused on the product, and they left the management of the information generated by the product in the hands of the customer. The correct application of the information generated by DSE2 covered many requirements related to improving energy management, but the customer could not or did not know how to handle this information correctly.

Based on the definition of the business idea, the entrepreneurs investigated the main needs this product should meet. According to the managers: "We wanted to provide a product that served the following needs: energy saving, above all, real-time control of the building, the possibility to exactly determine the individual consumption of each apartment, record keeping that permits the analysis of historical data, analysis of global profitability of the installation/equipment, real-time alerts and warnings … And all these, through embedded systems which are cheaper than using microprocessor technology."

Once the first product was successfully installed, the team directed its efforts to two different activities: (1) the development of DSE2, due to a purchasing order that came from a local hotel via the original university research group, and (2) marketing actions to launch DSE1, the first product commercialized by the company.

The products' potential was evident, but the team was facing problems in the sense that reality did not meet expectations, since the product did not have the expected success. This made them refocus their business, so that it was clearly aimed at improving energy management. The key question the founders asked themselves and which finally changed their vision was: "How is it that a product that is so good and so useful for managing and saving energy does not have spectacular success?"

11.3.1.3 The User-Orientation Phase

At this stage new competitors from abroad were appearing on the scene. The environment was changing for remote supervision companies. This change in the environment significantly affected the business: their competitors caught up with EMS-embedded technology by using better microprocessors and embedded systems capable of offering the same benefits at a lower cost.

Nevertheless, the main problem was probably the long maturation time before sale conversion. In the building industry, it takes at least 1 or 2 years from the planning stage until the building is constructed. Despite this problem of cash conversion, it was clear that the problem was not in the marketing strategy of the product, nor in the people who did it: "The problem was that we couldn't sell our product any more."

Given the previous arguments, the team realized that the business opportunity they were planning to exploit would not provide enough revenue to survive in the short run. At this point, YES was in a very delicate situation: costs were greatly exceeding revenues, sales forecasts had been too optimistic, marketing actions had not succeeded and new competitors were appearing on the scene. Therefore, a different strategy was needed.

Furthermore, the team decided to increase its customized software development activities until they would reach the break-even point or they would find more funding to conduct a new sales strategy. In this "new" strategy, focused only on the service, they even put aside their initial product. It was a clear application of the pendulum law.

Gradually, client companies having DSE1 and DSE2 products started asking for an extension of the service they were provided.

11.3.1.4 The Result-Orientation Phase

According to the CEO, "If one thing was clear for us is it was that we had to stay in the field of energy efficiency. We always saw a huge potential in it. The question was how if not with the product/s we have been developing over years." Initially, and in response to client companies' requests, YES offered primary services of information management and exploitation (of the information collected and stored by the product). It was at that point that an important step toward the product–service system (PSS) was made. The ultimate PSS objective is to increase a company's competitiveness and profitability (Geng et al. 2010), while another PSS objective is to reduce the consumption of products through alternative scenarios of product use.

PSS represents an opportunity to shift from selling products to selling solutions through the use of these products. Thus, it becomes an opportunity to create value for the provider and to differentiate from competitors and customers by delivering personalized products and services, as already identified in the literature. They convert their production to "servitization," as the evolution of product identity based on material content to a position where the material component is inseparable from the service system.

The future at this moment was clear for YES. They offered a service through the installation of their product in the buildings and facilities and, as a new business line, in existing facilities. Thereby they could improve the energy cost by managing and monitoring buildings better. In addition, they opened a new business related to the maintenance of the installation, the update in terms of control and the adequacy of policy actions in order to improve the efficiency of installations. Initial product marketing was transformed into service marketing.

11.3.1.5 Summary

Figure 11.2 summarizes the main phases and relevant characteristics of each.

11.3.2 Critical Junctures

In this section, we describe the different critical phases the company went through to reach its present status.

11.3.2.1 Opportunity Recognition and Entrepreneurial Orientation

Opportunity recognition is the ability to synthetize scientific knowledge with an understanding of markets, and it is a very important step. Identifying components of the concept "remote-controlled underwater robot with environmental functionalities" that would work as a solution for business made the team decide to leave their core field of robotics and automation behind and focus on the extremes: remote control on the one hand and the environment on the other. Thinking back the CEO

	Phase 1 Knowledge-orientation	Critical Juncture · Opportunity recognition and entrepreneurial orientation	Phase 2 Product-orientation	Critical Juncture · Credibility threshold	Phase 3 User-orientation	Critical Juncture · Sustainability threshold	Phase 4 Results-orientation
Business							
Self-definition Characterisation	Research and teaching staff		ICT business University spin-off		Service provider business University spin-off		Integral solution provider; ICT business; New technology business venture
Relationship with university Research Group	High		Medium-high		Low		Origin of the business
Target sector	Underwater robotics with environmental functionalities		Construction Hotel		Information management Data mining		Energy efficiency
Strategy and competitive advantage							
Cost/Price	Low importance		Medium importance		Medium importance		High importance
Others: Innovation, flexibility, quality, customisation, service	Innovation		Product		Service		Product, Service, Customization
Services							
Offered			Basic training ;Technical maintenance		Basic training; Technical maintenance; Installing; Energy manager		Basic training; Technical maintenance; Financing; 24 hours attention; Energy manager; Technology infrastructure assessment; Renting; Report; Installing; Energy audits; Optimisation solution proposal
Clients			Hotels Construction companies		University SME		Public administration Big energy companies
Products offered	Electronic device		DSE1 (Home), DSE2 (Hotel)		Software ERPs		
Relation to ECO	Underwater robot with ecological functionalities		Monitor energy consumption		Exploit energy consumption monitoring data		Decision-making, solution proposal and solution implementation

	Phase 1 Knowledge-orientation	Phase 2 Product-orientation	Phase 3 User-orientation	Phase 4 Results-orientation
		Critical Juncture · Opportunity recognition and entrepreneurial orientation	Critical Juncture · Credibility threshold	Critical Juncture · Sustainability threshold
Revenues				
Euros (€)		400.000 €	600.000 €	750.000 €
Product (%)	na	100%	80%	20%
Service (%)	na	0%	20%	80%
Distribution by client				
Public (%)	na	0%	0%	20%
Private (%)	na	100%	100%	80%
Geographical distribution				
Region	na	100%	0%	0%
Spain	na	0%	100%	90%
International	na	0%	0%	10%
Research and Development				
Function	Research	Research	Research and Development	Research, Development and Innovation
Personnel	3	3	2	1
Personal				
Full time	na	2	14	11
PhD	na	2	1	1
Degree obtained at	na	University of Girona	University of Girona	University of Girona
Vocational training	na	0	0	1

Fig. 11.2 From knowledge to PSS • a phased approach

explains: "People always say that having an entrepreneurial background or history in the family has a certain influence and can be helpful when detecting an opportunity. My father had his own business...Meanwhile, my department was saturated with academic staff and they communicated to me that my chances of following a university career were really low. At this point, I started liking the idea of becoming an entrepreneur both by necessity and by opportunity." A research internship spent in a foreign country and the support received from the closest colleagues turned into factors favoring the decision. The prestige of the research group in terms of publications, patents, awards of excellence, contract research, and the potential of the technology developed during the PhD years complemented the previous reasons.

Entrepreneurial orientation and commitment are the contrast between the venture champion as developer of the business and academic individuals who lack entrepreneurial capabilities. The abyss between those in the "Ivory Tower" and those willing to find "immediate and dirty solutions" is huge. It is common, up to a certain point, to all companies emerging from higher education institutions (HEIs). The collision, as it was defined by the CEO, between the academic world and enterprise reality made university researchers grow up fast: "It was like making three MBAs in 6 months. Even attending a new venture creation management post-graduate course, my daily attributions and work tasks were going faster than the teaching/learning process."

It is often recognized that the support received from certain key persons at this stage is fundamental. The group leader and co-founder of the company as well as an ex-PhD student, friend of the founder and co-founder of YES, initiated the entrepreneurial journey. The positive reading of the circumstance was that they were part of the same situation and apparently had the same aspirations. As time passed, discrepancies in interest and other job opportunities meant that they grew apart and became more distant from the business. The CEO remembers, "One day I was informed that he did not want to be an entrepreneur and I fully accepted and respected his choice. Regarding my own situation, exiting the university was a gradual process, during a time I combined teaching tasks with the daily business activity ... however I realised that you are in or you are out of the business. At that stage I decided that I'm in!"

11.3.2.2 Credibility Threshold

This stage refers to the entrepreneur's inability to gain access and acquire an initial stock of resources necessary for the business to function.

Before the crisis hit Spain, public funding was an important source of finance for new university ventures. Seed capital provided by the regional development agency in cofinanced conditions, the Torres Quevedo program funding for incorporating PhDs in business, technological centers, entrepreneurial associations, and science and technology parks with the primary aim of industrial research or technological development were only some of the subsidies received by YES that helped the company to struggle through a series of initial obstacles.

In the CEO's vision, financial resources are fundamental: "The crisis affects us relatively little because since our initial existence we have been facing continuous crisis. They say when things don't work change them. Change implies doing things differently or doing new things … that ultimately translates into expenditure related to these new things or new ways of doing things. Subsidies were available some time ago, the good news is that we had them then and they are not vital for us now."

11.3.2.3 Sustainability Threshold

This phase refers to generating revenues from customers and milestone payments from collaborative agreements for investments from existing or new investors.

Each time more clients were pressuring toward help in their decision-making processes, meaning exploiting energy consumption-related information is a useful and user-friendly format. At this point the team started raising the question to its clients: "Here you have the hardware, what do you want from it? What information and in which form it would be useful in decision-making based on energy audits?"

11.4 Orchestrating Innovation

The OSLO Manual (OECD 2005) defines product innovation as the introduction of a good or service that is new or significantly improved with respect to its characteristics or intended uses. This includes significant improvements in technical specifications, components and materials, incorporated software, user friendliness or other functional characteristics.

11.4.1 Product Innovation

The term innovation is generally associated with product innovation which most often refers to a physical good. In people's common understanding, this is the most clearly identified type of innovation. This assumption is complemented by the findings of a recent and comprehensive review on the state of research in the field of innovation management. Keupp et al. (2012), after reviewing 342 articles, show that product innovation received major attention in the field of research.

For the particular case of YES, their initial objective was to create the new device DSE1 first, followed by DSE2. The new product was new to the firm and also new to the market the firm was operating in. It is a combination of existing technology in the field of underwater robotics and artificial intelligence applied in a totally new field, mainly buildings. The founder's PhD years represent a strong research and development phase that, after one and a half years, turned the products of the doctoral degree into functional products.

At present DSE1 is an energy management system for buildings using collective boilers, and it enables the measurement, regulation, and remote control of energy consumption. The core of the system is the electronic circuit DSE1. The board is installed in each apartment/house, and it permits the individual monitoring of energy consumption.

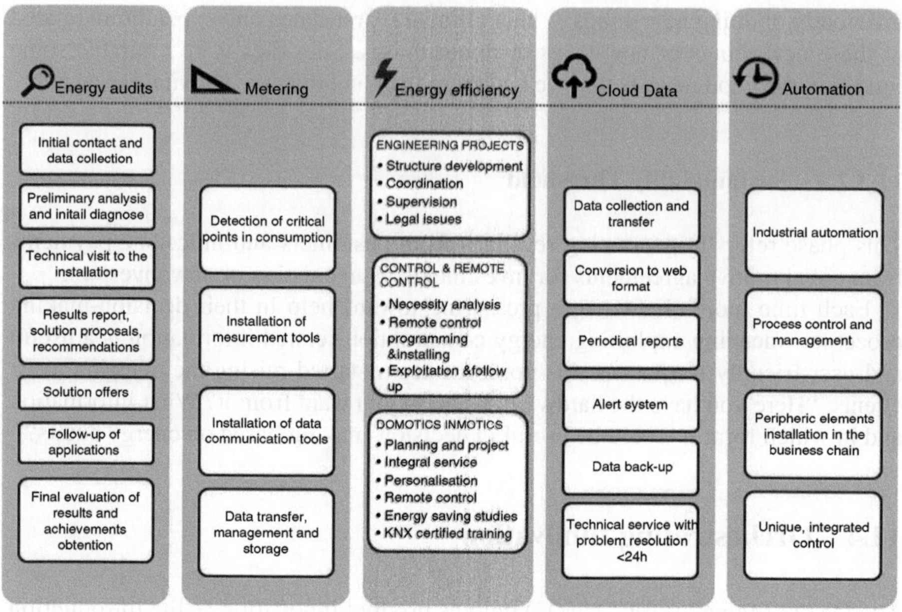

Fig. 11.3 Services offered

11.4.2 Service Innovation

The same manual states that innovations in services can include significant improvements in how they are provided (e.g., in terms of their efficiency or speed), the addition of new functions or characteristics to existing services or the introduction of entirely new services (OECD 2005: 48). In 2013, the full range of services provided by the energy management business are shown in Fig. 11.3. Most of them were offered after 2012, and they account for 80 % of the turnover of the company.

11.4.3 Eco-innovation

According to the EU, "Industry remains one of the core driving forces behind the technical and technological innovation required for greater sustainability and resource efficiency. Eco-innovation is the emerging term for innovations specifically directed at this area." The term is comprehensive in the sense that it can refer to goods, services, and processes, without differentiating between them.

The Eco-innovation Action Plan is the formal policy instrument to promote initiatives in the field of sustainability and resource efficiency. According to this document, "Eco-innovation in companies leads to reduced costs, improves capacity to capture new growth opportunities and enhances their reputation among customers.

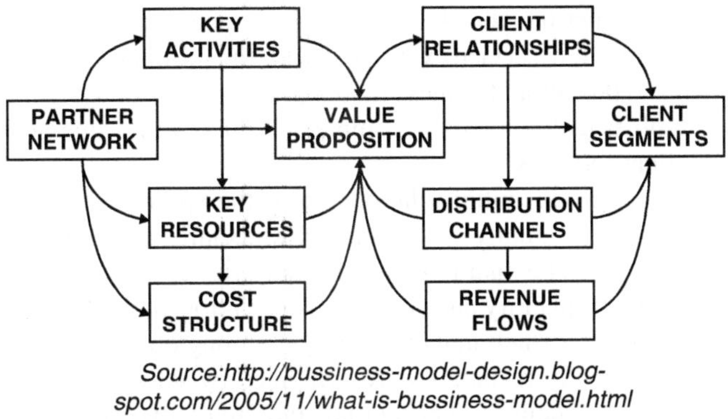

Fig. 11.4 Osterwalder's 9-point decomposition of a business model (Source: Chesbrough 2010)

It is therefore a powerful instrument, combining reduced negative impact on the environment with a positive impact on the economy and society." Eco-innovation supports innovative products, services, and technologies that make better use of resources, while reducing Europe's ecological footprint.

Overall, YES appears as a double player in the issue of eco-innovation:

- As an eco-innovator creating and exploiting a business based on solutions aimed toward energy consumption optimization
- As an eco-innovation booster since the integration of its solutions into client sites creates eco-innovation in all implementing organizations

Impact assessment shows that Europe is in a situation best described by the phrase, "A lot has been done, but there is still lot to do," and the absolute value impact of individual business, either as solution providers or eco-innovative solution integrators, is only beginning. Proactive awareness is best summarized in the phrase, "Everybody counts."

11.4.4 Business Model Innovation

Chesbrough (2010), citing Osterwalder's business model approach, writes about business model innovation, a renewal, or change in any or all areas of the business model (see Fig. 11.4).

In order to clarify the innovative business model associated to the energy management systems, we center our attention on the description of its three major areas most affected by innovation which are:

Value proposition and key activities – YES initially offered the DSE products with two basic associated services: basic training and technical maintenance. While marketing actions were intensive, the desired results were elusive. A deviation from the core product happened in a subsequent phase and the team developed software and some new services were added to the products. This service phase shifted to what is the actual business model product–service system in which the company combines technologically complex products with value-added services, based upon client requests.

Partner network – a strong initial relationship to the spinning-out university and research group has gradually lost intensity, especially after the first sales and business customers. The main focus of YES was to target big corporations through a variety (in quantity and typology) of relationships: customer, preferred distributor, provider, partner, and collaborator, among others.

Revenue flows – originally revenues were generated by selling products and in the next step software licenses. The actual and innovative form is generating revenues as a part of savings generated by energy consumption and reduction due to YES products.

11.5 Findings and Discussions

11.5.1 Lessons Learned

Lesson 1 – The entrepreneurial adventure. Driven by a series of strong motivations and circumstances – entrepreneurial background, willingness to have something of one's own, insecure future at the university, strong technology, and knowledge base – YES was born in 2004 as a university spin-off commercially using knowledge generated in a research group. At present it has 11 full-time employees and offers integral solutions in the field of energy management, software development, and ICT-related services. Its core business consists of exploiting a product–service system. The transition from product to an integrated bundling of product and services is described in the present case study paying attention to the main phases and critical stages catalyzing them.

Lesson 2 – Energy as an emergent field for business. In the relatively recent past, energy was often characterized by environmentalists who were vehement in their discourse and action. ICT, policy-making, and general societal interest make the subject popular in almost all fields. Business did not balk in identifying opportunities either as solution providers or, increasingly as time went by, as solution integrators/users going beyond the limits of legal minima and combining efforts to achieve a better future.

Lesson 3 – Eco-innovation is a must for all. Energy solutions in the business field either aim to do things better, to do the same or more with less, or to change the way to do things. These terms used in popular contexts correspond to energy efficiency and environmental sustainability, necessarily through innovation. Eco-innovation appears as a transversal possibility for any type and size of business, whatever their location. Although implementation in the short term has an inherent cost, the benefits in the mid-long term are exponential.

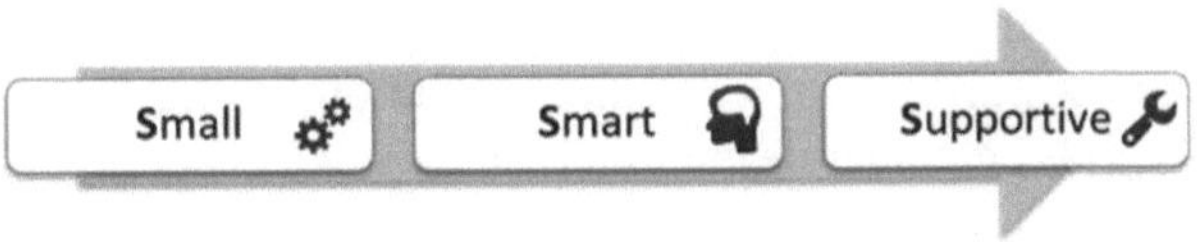

Fig. 11.5 3S model of ICT for eco-innovation

Lesson 4 – Small businesses offer big solutions. Traditionally businesses in the field of energy have been big corporations, usually multinational conglomerates operating on a monopolistic basis. Now reality is changing. Even civil society has contributed in some countries, like Germany, the Netherlands, and Spain, and cooperative-style organizations are reaping the early fruits of their actions. The business sector is diverse, and small- and medium-sized companies complement big multinationals. Our story of a small young business shows the advantages of flexibility, adaptability, continuous learning, ability to respond, low structural costs, and ultimately success.

Lesson 5 – Key success factors: network, position, and reputation. At regional level, as the initial operating area of a new business venture, it is crucial to be part of relevant industry platforms, clusters, networks, and associations. YES is part of the Catalan Energy Efficiency Cluster. Being small, having big brothers is an advantage. One key strategy followed by YES is the entrance and consolidation of its position through a close relationship with organizations with a good reputation and strong market position in the field of energy (Gas Natural, Endesa, Kamstrup, Barcelona Energy Agency). The ability of YES to win different prizes and subventions in competitive public calls from the regional development agency (ACC1Ó) has positively contributed in generating a reputation and increasing user confidence. This aspect is especially important for a new business since users are the ultimate barometer for an innovation to be successful.

11.5.2 ICT for Eco-innovation

For the specific field under analysis, evidence from the case study points to three vital aspects of how ICT can be used for generating eco-innovation (see Fig. 11.5).

In the following we describe the pillars of a 3S model that stands for small, smart, and supportive.

- Small: the reduction in size of PCs and electronic devices has led to the possibility of the creation of embedded systems. These are special-purpose systems in which the computer is completely encapsulated by the device it controls. Unlike a general-purpose computer, such as a personal computer, an embedded system performs one or a few predefined tasks, usually with very specific requirements. Since the system is dedicated to specific tasks, engineers can optimize it, reducing the size and cost of the product they are integrated in.

- Smart: adding the adjective smart to concepts (e.g., smartphones, smart cities) necessarily involves computer and digital aspects. In the field of energy, smart grids are under discussion. "Smart grid" generally refers to a class of technology people are using to bring utility electricity delivery systems into the twenty-first century, using computer-based remote control and automation. Ultimately smart grid means "computerizing" the electric utility grid (OE 2013). The number of applications that can be used on the smart grid once the data communication technology is deployed is growing rapidly as inventive companies create and produce them. Benefits include enhanced cybersecurity, handling sources of electricity like wind and solar power and even integrating electric vehicles onto the grid. The companies making smart grid technology or offering such services include technology giants, established communication firms, and even brand new technology firms, such as the one we describe.
- Supportive: ICTs are a powerful tool to generate, capture, store, manage, exploit, analyze, read, share, and transmit data and information which is gradually increasing in quantity in all kinds of organizations. In all these areas, solutions are required in order to avoid "infoxication," a term introduced by Spanish expert Alfons Cornellà and describing a situation where the received/available information exceeds the ability to manage it. ICT is supportive for modernization, optimization, and automation either of traditional processes or of entirely new ones. Being a means and not an objective, its implementation has to generate clear advantages in terms of new solutions compared to previous versions. It is at this stage that ICT remains supportive and not decisive. Whatever the technology behind the solution, users are key actors, playing a decisive role in the solution uptake. Technology, in general, and ICT, in particular, has to be functional and has a key role in an initial phase of developing a solution. Once this aspect is achieved, human and organizational aspects of style – such as resistance to change, mentality, and consumption culture – become fundamental.

The combination of small, smart, and supportive represents the ICT backbone of the technology needed for the system to perform optimally.

11.6 Conclusions

The case study described in the present book chapter is the story and not the history of Your Energy System and is intended to generate value in the form of holistic integral solutions, including technology-based products complemented with complex services. The example is rich in innovation types evolving and coexisting at present.

The implications for management concern continuous business model analysis and value reconsideration. We attempt to show the evolution of a business model by studying how a firm has been able to respond to a changing environment, trying to find a fit between the sophisticated demands of customers and a high-technology-based product. The ultimate result is that the company adopts complex business solutions that complement the product with high value-added services. Through the case study, the authors aim to demonstrate that combined solutions, rather than single options, lead to improved result.

Going beyond traditional innovation typologies – product, process, organizational, and marketing – we also illustrate that there are at least nine key areas in which companies can use innovation to further maintain or improve their competitiveness. Both eco-innovation and business model innovation are transversal typologies and can offer areas of improvement that may be worthwhile to consider.

Servitization, briefly consisting in complementing the product offer with services, is a trend that can bring a variety of advantages for business. Service offering contributes to additional revenues, differentiation from competitors, long-term client relationships and, last but no less important, difficulties in imitating (as compared to products). By offering integrated product–service solutions, businesses transform into one-stop shops where customers find concrete solutions that are adapted to their problems. This further contributes to loyalty and reputation.

Finally, for those having a dream of entrepreneurial opportunities, we show how the spin-off modality of research commercialization is one viable possibility, especially for research developed in a unit that is part of a public higher education institution. Even on a secondary level, the case is a success story of a founding team able to translate their PhDs into an innovative solution targeting traditional fields.

References

Baines TS, Lightfoot HW, Kay JM (2009) Servitized manufacture: practical challenges of delivering integrated products and services. Proc Inst Mech Eng B J Eng Manuf 223(9):1207–1215

Bikfalvi A, Serarols C (2008) Remote internet-based supervision systems, S.L. In: Medina-Garrido JA, Martínez-Fierro S, Ruiz-Navarro J (eds) Cases on information technology entrepreneurship. IGI Publishing, Hershey, pp 36–62

Chesbrough H (2010) Business model innovation: opportunities and barriers. Long Range Plan 43:354–363

EC – European Commission (2011) Sustainable industry: going for growth & resource efficiency. Rotterdam, The Netherlands. Retrieved 10th of June 2013 from http://ec.europa.eu/enterprise/policies/sustainable-business/files/brochure_sustainable_industry_150711_en.pdf

EC – European Commission (2013) Europe 2020 in a nutshell – priorities. Retrieved 10th of June 2013 from http://ec.europa.eu/europe2020/europe-2020-in-a-nutshell/priorities/index_en.htm

Eisenhardt KM, Graebner ME (2007) Theory building from cases: opportunities and challenges. Acad Manag J 50(1):25–32

Geng X, Chu X, Xue D, Zhang Z (2010) An integrated approach for rating engineering characteristic's final importance in product-service system development. J Comput Ind Eng 59(4):585–594

Hong S, Oxley L, McCann P (2012) A survey of the innovation surveys. J Econ Surv 26(3):420–444

Keupp MM, Palmié M, Gassmann O (2012) The strategic management of innovation: a systematic review and paths for future research. Int J Manag Rev 14:367–390

Morelli N (2003) Product-service systems, a perspective shift for designers: a case study the design of a telecentre. Des Stud 24(1):73–99

OE – Office of Electricity Delivery & Energy Reliability U.S. Department of Energy (2013) Smart grid. Retrieved 10th of June 2013 from http://energy.gov/oe/technology-development/smart-grid

OECD (2005) The measurement of scientific and technological activities Oslo manual. Guidelines for collecting and interpreting innovation data. OECD, Paris
Siggelkow N (2007) Persuasion with case studies. Acad Manag J 50(1):20–24
Vohora A, Wright M, Lockett A (2004) Critical junctures in the development of university high-tech spinout companies. Res Policy 33(1):147–175
Yin R (1994) Case study research: design and methods, 2nd edn. Sage, Thousand Oaks
Yu M, Zhang W, Meier H (2008) Modularization based design for innovative product-related industrial service. In: IEEE international conference on industrial engineering and engineering management, Singapore, pp 48–53

Chapter 12
Business Model Innovation for Eco-innovation: Developing a Boundary-Spanning Business Model of an Ecosystem Integrator

Anastasia Tsvetkova, Magnus Gustafsson, and Kim Wikström

Abstract Biogas production and its use as traffic fuel are discussed in this chapter as an example of a system eco-innovation that is struggling to become implemented in a focal municipality. The biogas producer and distributor as the owner of the "core technology" have the potential to become the integrator of a functioning ecosystem required for the innovation to succeed. The company's business model, however, should be transformed to incorporate the radical and system nature of the eco-innovation as well as create a business solution that would make the technology profitable. The aim of this chapter is to demonstrate how an ecosystem integrator can develop a boundary-spanning business model that is capable of integrating the multitude of stakeholders into a working biogas-for-traffic solution, thereby achieving a system change. The chapter is based on the results of two research projects during which a sustainable local biogas-for-traffic solution and the business model of the ecosystem integrator were developed together with the major ecosystem stakeholders. The main principle of developing such a business model lies in considering business models of the relevant stakeholders and managing uncertainties pertinent to their integration into the biogas-for-traffic ecosystem.

Keywords Eco-innovation • Ecosystem integrator • Boundary-spanning business model • Biogas

A. Tsvetkova (✉) • M. Gustafsson • K. Wikström
Industrial Management, Åbo Akademi University, Piispankatu 8, Turku, FI-20500, Finland

PBI Research Institute, Aurakatu 1 B, Turku, FI-20100, Finland
e-mail: anastasia.tsvetkova@abo.fi; magnus.gustafsson@abo.fi; kim.wikstrom@abo.fi

S. Azevedo et al. (eds.), *Eco-Innovation and the Development of Business Models*, Greening of Industry Networks Studies 2, DOI 10.1007/978-3-319-05077-5_12,

12.1 Introduction

Replacement of fossil fuels by renewable fuels in transportation is seen today as one of the major means to reach sustainability goals. Biogas is already commonly used as green traffic fuel, bringing additional environmental benefits such as reduced levels of CO_2 emissions, nitrogen emissions, noise, and particle emissions (Gustafsson et al. 2011). Taking a more holistic perspective, if biogas is produced from waste materials and is consumed locally, then the lifecycle emissions are significantly lower due to the efficient nutrient cycling and short transportation distances (Tsvetkova and Gustafsson 2012). Moreover, since the low-energy content of biomass limits its transportation radius, the localized character of biogas production is also more practical in an economic sense.

In order to develop a biogas-for-traffic solution that is sustainable, it is important to reconsider the fuel production value chain as the linear and centralized way of producing fossil fuels is often blueprinted for biofuel production. This leads to the environmental and social problems normally associated with fossil fuels: competition for resources, a shift of the environmental burden from one country to another, and an increase in waste generation, to name only a few (Mirata et al. 2005). Most importantly, however, the biofuel industry struggles to be economically sustainable, feasible, and competitive when organized in a centralized way. If, however, it is structured in a distributed and cyclic manner, the benefits of industrial symbiosis can be realized by utilizing waste for biofuel production and implementing well-planned cycling of nutrients. This, in turn, requires a shift in the overall system of how biofuel is produced and consumed, that is to say, a radical system innovation.

The technical basis for sustainable biogas production is already well developed and technical integration is achieved by optimizing material and energy flows between the stakeholders. In contrast, and as is often the case, the business integration presents the greatest challenge for any system innovation (Liinamaa and Wikström 2009). This is especially true for a biogas-for-traffic production business, because there is a need to integrate a significant number of stakeholders that are currently unrelated to the energy and fuel industry. One example illustrating this problem can be seen in the fact that the suppliers of biomass generally belong to such industries as food production, farming, or waste treatment. For these companies the supply of biomass is outside their main business activities. These activities are normally seen as the disposal of waste, the smart utilization of by-flows, or, in rare cases, the secondary business of producing biomass for biogas production. While integration of traditionally unconnected industries is the main reason why industrial symbiosis is beneficial (Chertow 2000), it leads to complications in integrating the suppliers into the biogas production ecosystem.

Another challenge is related to the technology shift that potential biogas consumers need to make. Utilization of biogas as a traffic fuel is severely limited by the need for different forms of system investment, i.e., investment into many other elements of the whole value chain rather than only the biogas production facilities. Such changes would first of all include the acquisition of gas-driven vehicles by potential customers and the need for investment in a biogas distribution infrastructure. These requirements tend to lead to a “chicken and egg” or “cartel of fear” situation (Geels 2005), where business actors are afraid to make the first step due the investment risks.

The biogas producer, as the owner of core technology, can become the integrator of the ecosystem required for a functioning biogas-for-traffic solution. However, in order to address the outlined challenges, it is necessary to rethink the traditional fuel production business model. The core capability of an ecosystem integrator lies beyond technical integration; it requires business and social integration skills so as to create a resilient and feasible solution. These goals can be achieved by developing a boundary-spanning business model (Wikström et al. 2010), which builds on a deep understanding of the business models of the other ecosystem stakeholders.

The aim of this chapter is to discuss what is critical for the business model of an ecosystem integrator in order to facilitate the biogas-for-traffic eco-innovation. A boundary-spanning business model is proposed as a means to ensure the feasibility of this solution. Furthermore, we describe how such a business model can be developed through incentivizing various actors in the ecosystem, redistributing system benefits, and sharing the necessary investments and burden of risk.

12.2 Theoretical Background

12.2.1 Radical and System Eco-innovation

Eco-innovation can be defined as an innovation that improves ecological, economic, and social sustainability (Carrillo-Hermosilla et al. 2010). However, since sustainability is hard to define or propose a form for, there are no ideal criteria for what can be termed eco-innovation (Boons and Leudeke-Freund 2013). Bioenergy, as such, can be a solution to the energy challenges of the present and the future. However, the major obstacle is to make the relevant solutions economically sustainable and to incorporate them into the current markets and infrastructures. Thus, it can be concluded that until a bioenergy solution is feasible, it cannot be rightfully ascribed as being sustainable even if it solves environmental and social challenges.

Innovations that require a shift within larger socio-technical systems, rather than simply the introduction of a new product onto the market, can be called “system innovations.” This phenomenon is widely discussed in the literature on general innovation (Geels 2005) and eco-innovation in particular (Loorbach 2010). The interest in systems around innovations is based on the understanding that a new product, a service, or a combination of these needs to be embedded in a complex “landscape” comprised of social, regulative, economic, and infrastructural elements. These elements are often established in connection with other technologies, creating a “lock-in” in the currently prevailing technological regimes that are difficult to break (Geels 2002, 2005; Nelson and Winter 1982; Rip and Kemp 1998; Sartorius 2006).

Moreover, environmental sustainability of products and services can only be defined by considering the complex socio-technical system within which they are produced and utilized (Gaziulusoy et al. 2013).

It is noted that many “eco-innovations” remain at an incremental level (Larson 2000; Wagner and Llerena 2011) and are the result of isolated companies striving to optimize their production and operations. This leads to “suboptimizing” and to completely ignoring the systemic perspective (Boons 2009). In order to fulfill sustainability criteria, incremental innovation is not enough: current socio-technical systems are incompatible with the goals of sustainability and therefore they require redesigning (Ceschin 2013). Radical innovation is a better way to bring new, greener technologies into operation and to restructure unsustainable modes of production. This, however, requires a larger system-wide effort (Boons et al. 2013) and an extensive reconnection of traditionally defined industries (Moore 1996).

The necessity to build or reconfigure the socio-technical system when radical innovations are introduced (Geels 2002, 2005) goes hand in hand with the need for creating functioning *industrial* ecosystems when sustainable biofuels are produced (Tsvetkova and Gustafsson 2012). It is, however, important to note that the original notion of industrial ecosystems stems from the field of industrial ecology, which is inspired by natural ecosystems and the material and energy roundputs within them (Benyus 1997; Chertow 2000; Ehrenfeld and Gertler 1997). While the smart exchange of material and energy flows is the way to reach environmental sustainability, the business dimension of such ecosystems is much more extensive than the system of the enterprises physically exchanging the material flows. One example of a stakeholder that might be outside of the obvious industrial ecosystem is a gas vehicle dealer; the absence of direct material or energy exchange does not eliminate the crucial role of this actor in bringing the biogas-for-traffic innovation to a successful implementation. It can therefore be concluded that industrial ecosystems for biogas production need to be embedded in a larger business ecosystems. In this chapter, we refer to an ecosystem as the term that encompasses both the industrial ecosystem, i.e., the means of organizing material and energy roundput, and the larger business or socio-technical ecosystem that defines the economic and social sustainability of the solution.

Geels (2005) argued that companies are not able to affect the larger landscape, but rather that they can establish new socio-technical regimes when “windows of opportunity” appear. We believe that while certain opportunities, such as the rising price of fossil fuels, are able to foster the use of biogas as traffic fuel, the biogas company can try to shape the business landscape (Ceschin 2013) through innovating its own business model (Amit and Zott 2010; Chesbrough 2010). Further, we discuss how new business models help in making this shift to new industrial structures, thereby promoting radical and system eco-innovations.

12.2.2 Business Model as a Vehicle for Innovation

According to Boons et al. (2013), the current industry structures and business models employed by companies are not sustainable, and, at the same time, the move towards more sustainable production modes is limited due to the lack of practical and theoretical knowledge. While it has been acknowledged that companies are the drivers of radical eco-innovation, the research focus has largely remained on a policymaker's level (Ceschin 2013). This has left such topics as how companies actually implement this type of innovation and what it means for their business not being sufficiently well researched (Boons et al. 2013).

Recently, there have been attempts to connect the theory on business models with eco-innovation (Boons and Wagner 2009; Boons and Leudeke-Freund 2013). It has been noted that companies striving to make radical system innovations need to shift the innovation effort from the products or processes they control to the larger systems they are part of. Moreover, they need to actively construct the appropriate business model and engage the relevant stakeholders in such a development (Boons et al. 2013).

In academic literature, the discussion concerning "sustainable" business models has paid much attention to dematerialization (Halme et al. 2004, 2007; Mont 2002, 2004) and the switch from owning to delivering functionality (Ceschin 2013). Indeed, new business models based on, e.g., car sharing and energy use optimization are able to decrease energy and fuel consumption significantly. The focus of this type of business model innovation is mainly related to the social dimension, and technological artifacts are not considered to be crucial to the innovation, even if they are included (Ceschin 2013). The decrease in fuel usage can solve only part of the energy problem because complete dematerialization is not possible. Fuel still needs to be produced in a manner that is sustainable, and success in this aim is largely dependent on business integration.

Even though the need is regularly articulated for any innovation to be embedded into large socio-technical systems (Geels 2002, 2005; Boons et al. 2013), the business side of the innovation still requires more attention. The organizational level of the business model appears to be the most suitable level at which to analyze how radical and system eco-innovations can be realized. In order to make a transition to sustainability, companies need to change their whole "business mindset" or "business design" and acquire a focus on the future, rather than make decisions based only on the current situation (Gaziulusoy et al. 2013). We propose that the idea of boundary-spanning business models can be a step in this direction, helping companies to embed sustainability and systems thinking into the way they create value and interact with the stakeholders (Loorbach and Wijsman 2013). Companies would thereby help shape the biogas-for-traffic industry into being sustainable both environmentally and economically.

Boons and Leudeke-Freund (2013, p. 15) defined how business model innovation could contribute to eco-innovation:

> Business model change on the organizational level is about the implementation of alternative paradigms other than the neoclassical economic worldview that shape the culture, structure and routines of organizations and thus change the way of doing business towards sustainable development; a sustainable business model is the aggregate of these diverse organizational aspects.

Business models are seen in this chapter as a vehicle for eco-innovation. Since business models are concerned with the outcome, the implementation, and the tactics of strategies (Casadesus-Masanell and Ricart 2010), they can serve as a tool for implementing radical system innovations such as the focal biogas-for-traffic solution.

There are various views as to which elements constitute a business model. The core elements that appear in various frameworks depict the value creation process or offering and the value capturing mechanism or revenue model (Afuah and Tucci 2001; Linder and Cantrell 2000; Teece 2010). Another element agreed upon by many researchers on business models is the articulation of a target market, including customers, groups, or segments (Chesbrough and Rosenbloom 2002; Magretta 2002; Osterwalder et al. 2005; Teece 2010). Capabilities are required in order to deliver the promised value to the customer, and a cost structure largely affects the value capturing and profitability of the business (Afuah and Tucci 2001; Halme et al. 2007; Osterwalder et al. 2005).

In this chapter, the business model concept is chosen in order to analyze the way in which the biogas business can be organized. The reason for this choice is because business models comprise a solid building block within the ecosystem, allowing a structural observation to be made as regards the value creation. The following elements of the business model appeared to be the most relevant when analyzing the business of the ecosystem integrator: offering, revenue model, customer, capabilities, and cost structure.

The business of an ecosystem integrator might be the most challenging to describe since multiple stakeholders, revenue streams, and business logics need to be interconnected within one industrial ecosystem. Gulati et al. (2012, p. 573) refer to the business model of something larger than a company – a meta-organization – and discuss the main characteristics of a meta-organization as follows:

> …each agent has its own motivations, incentives, and cognitions, but unlike in a traditional business firm, they are not linked via a framework of formal authority associated with employment contracts.

In order to embrace the complex value creation within the ecosystem and to make them resilient, there is a need for the ecosystem integrator to employ a business model that is flexible and inclusive. Wikström et al. (2010) describe boundary-spanning business models that are designed to manage a large number of complex and demanding relationships. These business models imply that companies simultaneously operate in various environments through the use of multiple business models. Flexibility is thus achieved since the business models can be interconnected in different ways. Boundary-spanning business models draw on the different actors' strengths in order to achieve value through tighter cooperation and commitment to the common goal in one or another way. Similarly, focusing solely on the biogas business will not help in implementing the biogas-for-traffic solution, since other crucial ecosystem participants have various business logics and interests. However, the ability to understand their business models and incentivize them based on that knowledge (Raven 2005; Turner and Simister 2001) can serve as the basis for the integrator's business model which will "lock-in" the stakeholders in a sustainable manner.

In this chapter, we demonstrate how an ecosystem integrator can build a boundary-spanning business model through acknowledging the complexity of the business environment required for its implementation and, most crucially, shaping it in order to succeed. Such a business model is expected to incorporate the system character of the biogas-for-traffic eco-innovation, which is the focus of the study.

12.3 Method and Research Design

This chapter presents the findings of two research projects aimed at developing a solution for a Finnish municipality that would allow the utilization of locally produced biogas as traffic fuel. The research process was largely based on a combination of clinical research and design science approaches. Clinical research originates from the research tradition of action research and implies engaging in solving problems that are relevant to the industry (Coget 2009; Coghlan 2000, 2009; Schein 1993, 1995, 2008; Schön 1995). In this mode of research, the researchers help companies to diagnose and solve problems. Thus, the main aims of a clinical inquiry include solving a clinical problem and triggering organizational change (Schein 1995). The main feature of such an approach is that tight cooperation with business actors occurs throughout the process and is iterative. In the pursuit of developing a feasible and sustainable biogas-for-traffic solution, the involvement of relevant stakeholders in its design was crucial for ensuring the applicability of research results and the commitment of the companies to make the change.

During the projects the researchers attempted to reveal challenges and opportunities in implementing a biogas-for-traffic ecosystem and, together with the major stakeholders, developed a design proposition (Romme 2003), i.e., a sustainable business solution. The focus on design required by the design science approach (Romme 2003; van Aken and Romme 2009) was reflected in the way the design proposition was developed further; the final solution changed as the discussions with the practitioners helped to validate, reject, or furnish ideas.

One of the key elements for creating the target ecosystem was the business model of the company that would integrate it. The analysis of the major challenges and the means by which they can be triggered led us to the idea of boundary-spanning business models (Wikström et al. 2010) that can foster ecosystem integration in biogas business. Thus, the development of the ecosystem integrator's business model was based on considering the business models of the involved actors: their earning logics, value propositions, and benefit and risk sharing.

Collaboration with the relevant stakeholders was established and developed during the research process by conducting several rounds of interviews, joint workshops, and meetings. During the projects around 40 interviews were conducted with the following business actors: municipal authorities, biogas producers, gas distribution companies, truck and bus operating companies, financers, vehicle dealers, and many others. The choice of the stakeholders was based on the need for their involvement in the planned ecosystem. The underlying idea was that in order to establish beneficial and resilient industrial symbiosis, it was important to understand the responsibilities, gains, risks, and commitments of every ecosystem participant. The stakeholders that were interviewed initially included the companies directly involved into the major material flows, i.e., biogas, biomass, and digestate flows. Later, certain actors that are not directly involved in material exchanges within the ecosystem appeared to be crucial for establishing the business solution required for the biogas-for-traffic ecosystem. Such actors included vehicle dealers, financing institutions, and municipalities. The choice of which company representatives were to be interviewed was driven by their ability to make the decisions and changes required in order to become a part of the ecosystem. For example, in the case of the biogas producing company, the top management was interviewed, while in the case of vehicle dealers, cooperation with sales managers of the relevant types of vehicles was sufficient.

As the design proposition for the target industrial ecosystem evolved, more discussions with the same stakeholders were required in order to validate, test, and improve the design. Communication with these actors included short meetings, telephone calls, electronic mailing, and workshops. Thus, constant validation of the research ideas and co-creation with the stakeholders were at the core of the research. The main reason for this was an endeavor to develop a scientific knowledge capable of being prescriptive and actionable (Schön 1995). As a result of this study, discussion was started among the key stakeholders as regards creating a company that would become the integrator of a biogas-for-traffic ecosystem.

12.4 Results

12.4.1 The Case of "Biogas-for-Traffic"

The biogas-for-traffic solution, which is the focus of this chapter, was designed for a Finnish municipality that has around 200,000 inhabitants. A biogas production plant treating sewage sludge already existed in the area. The produced fuel, however, was only utilized for heat and power production, providing low-energy output and low value to the producer. The utilization of biogas as traffic fuel was seen as more beneficial not only for the biogas producer but also for the local municipality. It was deemed to be beneficial if the operators of heavy vehicles could be persuaded to switch to biogas as this would reduce the environmental impact of the traffic in the municipality and potentially solve the current feasibility problems of the public transportation system.

The annual quantities of sludge, already supplied by the wastewater treatment plant for biogas production, were considered sufficient to produce the amount of biogas at the initial stages of the ecosystem's development. However, the need for other sources of biomass was predicted at later stages when consumption of biogas increases. Apart from other types of organic waste, these additional sources of biomass would include green biomass provided by farms.

A gas distribution infrastructure does not yet exist in the focal area. This limits the possibilities of natural gas backup in case there is not enough biogas in the area. However, to a certain extent, this is a benefit, since the absence of competition means that the price for biogas does not need to be dictated by the market price of natural gas.

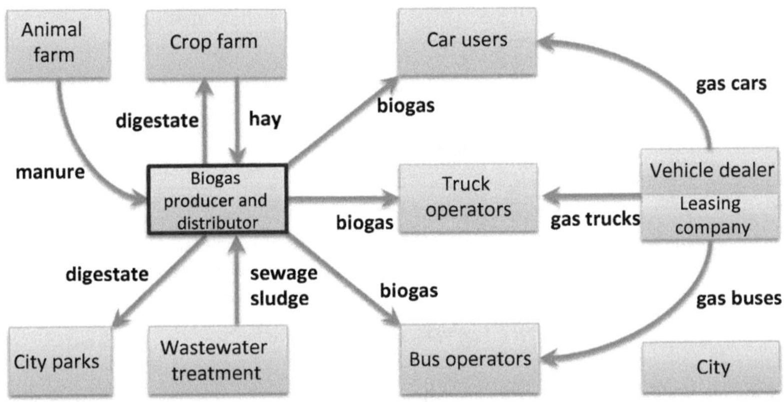

Fig. 12.1 Target "biogas-for-traffic" ecosystem and the key stakeholders (the ecosystem integrator is marked with a *bold line*)

The major stakeholders and material flows in the target biogas-for-traffic ecosystem are visualized in Fig. 12.1. The company producing and distributing biogas is seen here as the ecosystem integrator. Although the biogas production company already exists in the area, the integrating company can be a newly established actor that would manage the biogas distribution and, more importantly, manage the ecosystem integration. To ensure the integration of biogas production and distribution functions, the biogas producer can become a shareholder in the integrating company.

12.4.2 Establishing the Ecosystem

One of the major prerequisites for establishing the target biogas-for-traffic ecosystem is the construction of a biogas upgrading and distribution infrastructure. This requires large investments to be made by the ecosystem integrator. The profitability of this investment, however, is significantly dependent on the investment of potential biogas consumers into biogas-driven vehicles. Thus, in order to secure the consumption of biogas as soon as the upgrading and distribution infrastructure is in operation, the integrating company needs to ensure that potential consumers acquire gas-driven vehicles well in advance.

The potential consumers include businesses that operate fleets of vehicles, such as transportation companies, waste management companies, delivery companies, and also individuals. For the ecosystem integrator, it is reasonable to put diverse efforts in establishing collaboration with these consumers for a number of reasons. First of all, the consumption level of one heavy vehicle – a truck or a bus – is significantly greater when compared to a passenger car. This means that ensuring the switch of one bus to biogas is able to generate much more fuel sales compared to one car. Second, companies such as the bus operators, waste management, or delivery

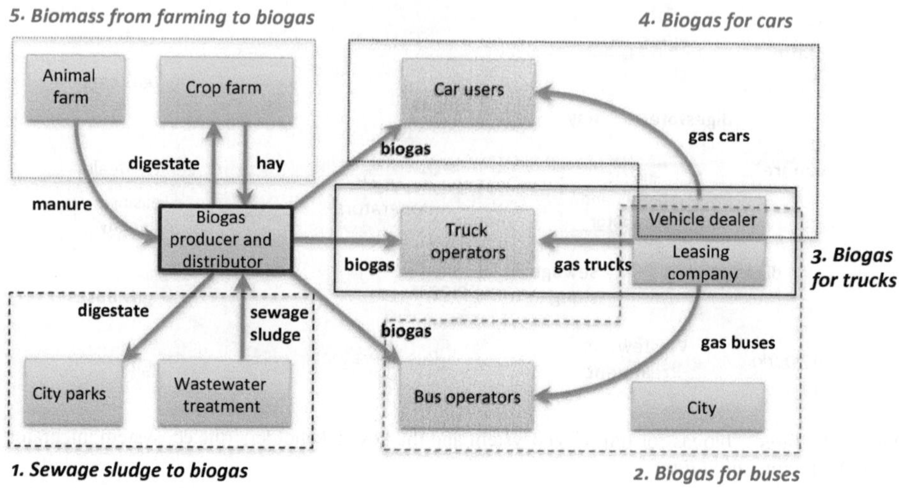

Fig. 12.2 Cooperation models in the target ecosystem

companies can generally take "fleet decisions" where a larger number of vehicles are purchased at once. Therefore, it is reasonable for the ecosystem integrator to focus on companies operating heavy vehicles at the first stage of developing the biogas-for-traffic ecosystem.

However, the potential consumers employ different business models and therefore need to be motivated to switch to biogas-driven vehicles in various ways. Similarly, potential biomass suppliers, which in the focal area include the wastewater treatment company and farms, have rather different business logics and cannot be integrated into the ecosystem in the same way. These stakeholders primarily operate within other industries and do not perceive the supply of biomass as their core business.

The role of the municipality is also crucial not only because its influence on public transportation through tendering but also because municipal authorities play a significant role in issuing permits, parking, and other areas that are important in order to establish the targeted ecosystem. Therefore, the commitment and involvement of the local authorities was another prerequisite for a successful ecosystem implementation.

Based on the differences in the business models of the stakeholders, the ecosystem integrator needs to implement different means to ensure their commitment to becoming part of the ecosystem. As a result, the overall business model of the ecosystem integrator can be divided into several distinct sub-models as envisioned in Fig. 12.2. Models 1 and 5 represent the biomass supply, while models 2, 3, and 4 represent the collaboration required for ensuring biogas consumption. We further describe the specificity of each model and the way the system integrator can build collaboration with the relevant stakeholders through employing a boundary-spanning business model.

12.4.2.1 Securing Biogas Consumption

The business model of public transportation companies, i.e., bus operators, is characterized by their direct dependency on municipal tendering (see model 2 in Fig. 12.2). If the city authorities make a decision to switch all public transportation to renewable fuel, the bus operators would have to renew their bus fleets with gas-driven vehicles in order to win the tender.

Another important feature is that buses operate locally and do not require a large fuel distribution infrastructure; a filling station that is located close to the bus depot or at the depot can serve their needs. Furthermore, if biogas is provided at a relatively low price, the bus operators will be able to be competitive from a cost perspective and become less dependent on fossil fuel price fluctuations.

In this type of model, the role of the city is crucial, because it is the ultimate decision-maker that can initiate investment into gas-driven buses. Long-term contracts for biogas at a fixed low price are the mechanism that can make the decision more attractive for the city. At the same time, this ensures certain minimum consumption of biogas during the first years of biogas distribution.

The biogas producer's main liability is to provide biogas at an agreed price, in agreed volumes, and of proper quality. The purity of biogas appeared to be a critical issue for the bus operators and the city, which is the ultimate financier of the public transportation services. Their concern was that malfunction of buses and refused guarantee claims caused by poor-quality biogas can significantly increase the operational costs of public transportation. In order to mitigate this risk, the biogas distributor and bus dealers need to coordinate the acceptable quality of biogas so that the vehicles can remain within their guarantee. As a result, a more reliable product, i.e., gas-driven buses, can be offered to the consumers.

It is possible that the risk of owning gas-driven buses is still too high for the bus operators to bear. The solution might be in establishing cooperation with the bus dealers and leasing companies in order to offer leasing packages to the operators. While the bus dealers would be able to increase the sales of gas-driven vehicles, the bus operators would be spared from the ownership risk at the initial stages.

Tight collaboration with vehicle dealers is required in order to make the overall investment into gas-driven buses more attractive to the bus operators. First, it allows the customers to be assured as regards the technical reliability of the biogas technology, which is new to them. Second, the sales effort can be united if the dealers are able to offer not only the vehicles but also long-term fixed-price fuel contracts on behalf of the biogas distributor. Finally, the bus dealers can offer a better maintenance infrastructure and service agreements when they are able to sell larger volumes of buses at once.

Companies that operate trucks, such as delivery and waste management companies, constitute another large group of potential biogas consumers (model 3 in Fig. 12.2). Since these business actors are not dependent on tendering, the decision to switch to biogas technology is purely business driven for them. Fuel costs have a large share in their operating costs and are also a risk factor. The global trend is that fossil fuel prices are continuously increasing, and therefore, biogas business can make a unique offering to transportation companies in this respect. Similar to the previous model, the biogas distributor can offer long-term and fixed-priced contracts for truck operating companies. This is in line with both parties' business models and provides the basis for dividing the burden of system investment between many stakeholders. The extra volumes of produced biogas can be sold at a higher price to customers that are not strategic, for example, to individual consumers. Such an arrangement is possible due to the specific nature of a biofuel business organized in a distributed manner. Localized production of biogas is commensurable with the local demand for fuel and is rather independent from the fluctuations on the fossil fuel market. Thus, a new type of customer offering can be developed. This is an example of how the biofuel industry can gain a competitive advantage by deviating from the traditional fuel production value chain.

Furthermore, the utilization of biofuel can help truck operators gain a "green" image for their business. Concern about their environmental impact is becoming increasingly more relevant to waste management and delivery companies. Thus, if they switch to biogas technology, they would be able to change their business model to become more sustainable. The strength of such motivation to use renewable biofuel, however, depends on the current priorities of a company.

The third type of biogas consumers includes car users, such as taxi owners, delivery companies, and individuals (see model 4 in Fig. 12.2). They can benefit from direct cost savings when replacing gasoline or diesel with biogas if a competitively priced biogas is offered. However, since cars consume much less than heavy vehicles, volume discounts are not reasonable in attracting these consumers. In addition, the price difference between a gas-driven passenger car and a gasoline or diesel car is relatively small. Therefore, investment in the vehicle does not constitute such a large challenge as it does for other groups of consumers. However, car users generally require a more developed distribution infrastructure, which clearly requires a greater investment. The system integrator can make such an investment at a later stage of the business development and recoup the investment costs with a higher but still competitive price for biogas.

Various benefits offered by the municipality, such as free parking for clean vehicle users and vehicle tax deductions, can serve as another motivation for switching to biogas-driven cars. The city would benefit from promoting sustainable fuel use in the city, since the increase in the overall biogas consumption would reduce biogas production and distribution costs. As a result, the costs of biogas-based public transportation would decrease.

12.4.2.2 Biomass Supply

Sewage sludge is already used for biogas production in the focal case (model 1 in Fig. 12.2). Therefore, there is no effort required to establish this cooperation. The business model of the biogas producer is built on the receiving of gate fees for treating sludge, which is a waste material, by anaerobic digestion. In this case, biogas can be seen as a by-product. When biogas is sold as traffic fuel, however, utilizing sludge means that the biogas producer has to acquire the main resource for the biofuel production at a negative price. The main challenge for the integrator is therefore to keep this model functioning and to secure the continuation of the sewage sludge supply. The level of the gate fee needs to be balanced so that it does not become more attractive for the wastewater treatment company to dispose of sludge in some other ways, for example, at a landfill. Alternatively, the city authorities are able to influence this problem by introducing a strategy for waste management where sewage sludge is dedicated to renewable fuel production.

The quality of biomass used for biogas production defines the properties of another product – the digestate. Since it constitutes a significant material flow, digestate disposal requires proper management to ensure the feasibility and sustainability of the overall biogas-for-traffic business. Although technically and legally digestate produced from sewage sludge can be used for fertilization, the main barrier is usually of social origin: the farmers perceive it as dangerous. Therefore, there is a need either to change perceptions about this product or to find another innovative way to utilize it. For example, the option to use the digestate for landscaping faces smaller social resistance. However, it is also less feasible for the biogas producer.

At the moment when the volume of sewage sludge does not satisfy biogas production needs, there is an opportunity to establish cooperation with crop and animal farms in order to utilize hay and manure (see model 5 in Fig. 12.2). This is far more challenging compared to the previous model, because it also requires changes in the farms' business models and additional investments. For example, crop farmers can dedicate a number of their fields to growing biomass that would be used for biogas production instead of growing grain. This means the biogas producer needs to buy the biomass at a price that would cover the costs of a "lost opportunity" for the farmer. A more beneficial way is to motivate the farmers to change to a new mode of crop production: the rotation of food crops and grasses over a number of years combined with replacing synthetic fertilizers by the digestate. This way the farmer can achieve the same overall level of food crop yields while reducing the need for synthetic fertilizers.

Additional costs can be caused by the need to harvest the grasses, which would otherwise be left in the fields. The biogas producer can bear the costs of this harvesting, thereby decreasing or eliminating the investment needs for the farmer. These additional costs will be recouped, however, since the extra volume of biogas produced from green biomass is intended for car users paying higher price for biogas. Furthermore, the biogas producer can sell digestate to the farmers, which would decrease the overall costs of acquiring green biomass.

Animal farms that can provide manure have motivations similar to sludge providers, because manure is perceived as waste if there are not enough fields on which to spread it. However, those farms which are able to sell their manure as a fertilizer can be incentivized by offering a certain price for the manure or, instead, by receiving back digestate that has similar or even better fertilization properties.

12.5 Discussion

To assist the transformation into sustainable societies, there is a need to improve the efficiency of material and energy cycles. This gives a significant potential for renewable energy produced from biomass. However, to bring these technologies into practice, the cooperation between different groups of actors, often having various goals and visions, needs to be well established. The attempt of this research was to demonstrate how such cooperation could be triggered by employing new business models.

A crucial factor is how commitment is built inside these complex systems, since the stakeholders generally belong to various industries and thus are not traditionally used to cooperating (Gustafsson et al. 2011). For example, it is more challenging to create a strong connection between farming and the biogas business than between biogas production and fuel distribution businesses. The mechanisms proposed in the chapter focused mostly on the business models of the stakeholders and particularly on their earning logic. We argue that the final technical planning of the system needs to be done only after the "business engineering" stage. Planning in reverse order potentially results in a solution that is not viable in the real world.

The greatest challenge for the establishment of a biogas-for-traffic business in the focal area was the absence of a gas distribution system and network. Gas distribution generates a need for significant capital investment to be made in order for the system to work. The cornerstone is not the cost of the investment, but rather the risk that the system will not succeed. Slow adoption of biogas onto the market is risky and unfeasible. Thus, the introduction of the product needs to be relatively rapid and has to start with a considerable volume of consumption. For this to be possible, however, the investments by different stakeholders need to be taken more or less simultaneously.

By understanding and interconnecting with the relevant stakeholders' business models, the integrating company is able to ensure that all the business models are connected and the cooperation will be sustainable. Thus, the different approaches to various consumer groups take the form of differentiated pricing and contractual models. Table 12.1 summarizes how the ecosystem integrator employs various business models in order to interconnect the relevant stakeholders.

From the supplies side, understanding that the difference in the types of biomass necessarily leads to various modes of cooperation and value streams is crucial for securing a stable biomass supply. In the case of biomass obtained from farming, the cost structure of the biogas production is completely different compared to sewage sludge utilization. Therefore, the business logic in cooperating with a wastewater treatment plant and a farmer creates various implications for the business model of the ecosystem integrator. While sewage sludge can be obtained at a negative price, in the case of green biomass, there is a need to establish a win-win situation that would decrease the biomass purchasing costs.

Many of the models described in Sect. 12.4.2 imply significant involvement by other actors able to influence the decision of potential biogas consumers to switch to biogas.

Table 12.1 The boundary-spanning business model of the biogas-for-traffic ecosystem integrator

Biomass supply	Elements in the business model of the ecosystem integrator		Biogas consumption
Connection to the business models of the stakeholders	Biogas production	Biogas distribution	Connection to the business models of the stakeholders
Sub-model 1. Sewage sludge to biogas		*Sub-model 2. Biogas for public transport*	
Wastewater treatment company supplies sewage for biogas production at a competitive gate fee level The city commits to utilizing digestate for landscaping in parks and construction in order to improve the stability of the local biogas-for-traffic solution	*Revenue model:* Gate fees for waste utilization Biogas as a valuable by-product *Cost structure:* The biomass has a negative price	*Offering and revenue model:* Long-term contracts for biogas supply at a fixed price Fuel contracts are sold together with vehicles	Bus operators adhere to tendering criteria and acquire gas buses. Savings on fuel costs recoup investment and maintenance costs The city organizes tendering for cleaner transportation based on biogas in order to improve local public transportation Bus dealers offer packages that include vehicles, maintenance, leasing, and fuel guarantee
Sub-model 5. Biomass from farming to biogas		*Sub-model 3. Biogas for truck operators*	
Crop farms switch to crop rotation, utilize digestate as a fertilizer, and supply biomass for biogas production. Cost structure and value of the end products improved Animal farms supply manure for biogas production at a certain price	*Revenue model:* Biogas sales as traffic fuel *Cost structure:* Need to purchase hay or organize the harvesting Delivery of digestate to crop farms	*Offering and revenue model:* Long-term contracts for biogas supply at a fixed price Fuel contracts are sold together with vehicles Green fuel	Truck operators invest in gas trucks. Savings on fuel costs recoup investment and maintenance costs Truck dealers offer packages that include vehicles, maintenance, leasing, and fuel guarantee
		Sub-model 4. Biogas for car users	
Overall implications on the business model of the ecosystem integrator: Differentiated pricing Joint offerings with vehicle dealers Different production costs depending on biomass utilized Higher involvement of certain actors, such as the city authorities Gradual investment into infrastructure enlargement		*Offering and revenue model:* A higher but still competitive price for fuel Wider availability of fuel Benefits granted by the city authorities	Car users acquire gas cars. Fuel savings recoup the relatively higher investment in gas cars City introduces free parking and tax reductions for "clean vehicle" users in order to improve the stability of the local biogas-for-traffic solution

These actors include local authorities. Vehicle dealers are another example of actors who are able to increase the feasibility of the overall investment into biogas technology. It is therefore crucial to identify and build collaboration with these actors in one way or another.

The ecosystem integrator employs different roles in the ecosystem depending on which type of collaboration is being discussed. Moreover, the role of an ecosystem integrator might not always be clearly articulated. What makes an integrator is the fact that the business model of this company is able to integrate stakeholders into a working ecosystem by establishing various interfaces among them.

As a result of the differentiation in the capabilities of the ecosystem integrator, the business model acquires the flexibility necessary in order to integrate actors that are traditionally outside of the renewable fuel production value chain. Thus, a key feature of the integrator's business model is in balancing the act of integration and continuous adaptation as the business ecosystem evolves. This is crucial when system innovations are implemented and new more sustainable industries emerge. The ecosystem integrator is able to achieve both the goals of integration and adaptation through employing a boundary-spanning business model. Namely, instead of following a rigid business model, the integrating company strives to "reflect" the business environment and the business models of relevant stakeholders in its own business model.

Depending on the business model of a respective ecosystem actor, the integrator finds a way to make collaboration beneficial for both parties. One common mechanism is the management of uncertainty regarding investments. This can be done, for example, by cooperating with leasing or vehicle sales companies which would then bear the risk of owning a vehicle. Another mechanism is joint offerings such as vehicles being sold together with a maintenance contract and guaranteed fuel contract. By making this value proposition, both the biogas distributor and the vehicle dealer can make their core products more valuable. This reflects a view on eco-innovation and system innovation that implies that products and services are valuable and sustainable only within a larger business or socio-technical system (Gaziulusoy et al. 2013). The ultimate customers are, in turn, spared from the uncertainty related to the new biogas technology, since they are guaranteed maintenance and the availability of fuel once they invest in gas-driven vehicles. The third type of mechanism is joint production planning, as in the case of green biomass farming. The ecosystem integrator cannot establish a beneficial collaboration with a farmer since the current business model of the farmer does not allow for it. The purchase of hay especially grown for biogas production would significantly increase the biogas price. However, there is significant potential for combining farming with biofuel production if the business model of the former is altered. The farmers can decrease their own costs by efficiently rotating fields and selling biomass to the biogas producer. Moreover, such food can be graded as organic in the future and thus become a more valuable product.

Geels (2005) proposes that linking with other developing technologies can promote innovations. In the case of biogas, for example, the gas-driven vehicles are presently neither a niche technology nor a dominant technology in the focal country.

This technology has been coevolving together with natural gas technology, which has not become sufficiently widespread in the area; the availability of natural gas has been physically limited to a few cities, making gas vehicles unattractive to potential users. The promotion of biogas use in vehicles is possible since the vehicles are already available on the market. In turn, biogas business can facilitate the development of the gas vehicle market as the biofuel becomes widely available.

As demonstrated in this chapter, integration is achieved by considering the stakeholders' business models: operational and capital cost structures, earning logics, and the industry structures they operate within. After this process, the connections between the stakeholders can be established by various commercial, social, or technical mechanisms. Certain mechanisms allow the interlocking of the stakeholders' business models while also recognizing that they belong to a different value chain. Other mechanisms, however, result in drastic changes in the stakeholders' business models in order for them to fit better into the target ecosystem. Certainly, these mechanisms will vary from location to location, but the strategy of *spanning the boundaries* of a company's own business model can be applied to business planning as regards other solutions based on industrial symbiosis. Certain contractual models may be derived from the empirical case analysis and applied in other locations with minor adjustments. This would allow the replicating of sustainable eco-innovations based on industrial symbiosis without the need for high adjustment costs. Thus, the economies of repetition (Davies and Brady 2000; Hellström 2005) and consequently cost saving can be achieved.

The process of building a boundary-spanning business model for the ecosystem integrator starts with analyzing and understanding the business models of the relevant stakeholders. After this, the collaboration mechanisms are developed based on this information. The aim of these mechanisms is to integrate the actors into the biogas-for-traffic ecosystem in a way that is beneficial for all parties. However, certain business ideas might only be the starting point for developing a really sustainable way of cooperating. The engagement of the ecosystem stakeholders into a discussion on benefits, risks, and responsibility sharing will shape the final business idea of the integrating company. Thus, the value of the process for the business model development proposed in this chapter is provided not only by the concrete examples of collaboration mechanisms but also in explicating the mindset and the processes required for their development.

We argue that boundary-spanning business models are able to incorporate the systems thinking into the way industrial activity is organized. It is demonstrated in this chapter how the ecosystem integrator is able to connect to other stakeholders' business models, not only through vision development and knowledge sharing (Baas 2011) but also through concrete business mechanisms. Since business models deal with the company level and its interfaces with the business environment, the adoption of boundary-spanning business models by companies is a practical way to increase the probability of success when implementing system eco-innovations. Moreover, addressing specifically the investment and ownership burden can mitigate the initial investment uncertainty that is relevant for any radical and system innovation.

12.6 Conclusions

The research presented in this chapter contributes to the theoretical field of system eco-innovation (Boons and Wagner 2009; Gaziulusoy et al. 2013) in two ways. First, it discusses how fundamental change, required for making industrial structures more sustainable, can be driven by industrial symbiosis. Using the example of the biogas industry, the chapter demonstrates how the cycling of materials along with increased collaboration among local stakeholders can be the basis for a feasible biogas-for-traffic business as opposed to the traditional centralized way of producing and distributing fuels. Second, the role of the business model of an integrating company is discussed. Admitting the potential of business model innovation to reinvent industry structures and drive system innovation (Boons et al. 2013; Gulati et al. 2012), we present the way a boundary-spanning business model of a biogas company can integrate the ecosystem required for a sustainable biogas-for-traffic solution.

Since renewable fuels have different production logic compared to fossil fuels, the overall industry logic needs to comply with the nature of the fuel. Moreover, the different business logic is able to generate benefits, such as fuel price stability, fuel supply security, and a positive effect on environmental, social, and economic sustainability. However, since this new logic requires a system change (Geels 2005), the old business models need to be replaced by more inclusive and flexible ones – boundary-spanning business models. The approach utilized in this research stresses the need to address the feasibility of eco-innovation by configuring the ecosystem around it. While environmental challenges and governmental policies will affect the adoption of greener technologies, the smart design of value and monetary flows along with material and energy cycling is the way to bring sustainable eco-innovations more rapidly into being.

The major recommendations to the companies that strive to implement system eco-innovations can be formulated as follows:

- It is crucial to realize that any innovation's success is dependent on the business and socio-technical ecosystem surrounding it. Moreover, an ecosystem usually spans the boundaries of traditionally defined industries (Moore 1996) and therefore is challenging to embrace using old business logics and models.
- To implement a system eco-innovation, a company needs to become an ecosystem integrator, if not overtly then in terms of its role within that ecosystem. This, in turn, requires a boundary-spanning business model that would facilitate the increased cooperation and positive interdependency with the relevant stakeholders.
- A boundary-spanning business model can be built by considering and addressing the most critical issues in the stakeholders' own business models when radical and system innovations are implemented: operation and capital costs, investment risk, and market uncertainty. Such a business model can be built based on the business mechanisms that reduce these uncertainties and which will create value and cash flows among the ecosystem actors.

The managerial implications, therefore, include the need to perceive a company striving to implement a system eco-innovation as an integrator of the ecosystem it would operate within. Upon understanding this, a company needs to actively shape its business environment. Considering the complexity of the business and socio-technical ecosystems around any technology (Geels 2002; Loorbach 2010), the effort can still come from a company level and is achieved by opening the boundaries of its business model and increasing cooperation with the stakeholders. In order to control and influence the ecosystem formation, companies need to redesign value flows by various mechanisms in order to make collaboration beneficial for all involved parties. Additionally, since a radical system innovation does require significant industry restructuring, it is not possible for a company to accomplish it on its own. Collaboration and open discussion are an important part of building the business model for an ecosystem integrator.

References

Afuah A, Tucci CL (2001) Internet business models. McGraw-Hill/Irwin, New York

Amit R, Zott C (2010) Business model innovation: creating value in times of change. IESE Business School of Navarra, Barcelona. IESE working paper, No. WP-870

Baas L (2011) Planning and uncovering industrial symbiosis: comparing the Rotterdam and Östergötland regions. Bus Strateg Environ 20:428–440

Benyus JM (1997) Biomimicry – innovation inspired by nature. William Morrow and Company, New York

Boons F (2009) Creating ecological value. An evolutionary approach to business strategies and the natural environment. Elgar, Cheltenham

Boons F, Leudeke-Freund F (2013) Business models for sustainable innovation: state of the art and steps towards a research agenda. J Clean Prod 45:9–19

Boons F, Wagner MA (2009) Assessing the relationship between economic and ecological performance: distinguishing system levels and the role of innovation. Ecol Econ 68(7):1908–1914

Boons F, Montalvo C, Quist J, Wagner MA (2013) Sustainable innovation, business models and economic performance: an overview. J Clean Prod 45:1–8

Carrillo-Hermosilla J, del Río P, Könnölä T (2010) Diversity of eco-innovations: reflections from selected case studies. J Clean Prod 18:1073–1083

Casadesus-Masanell R, Ricart JE (2010) From strategy to business models and onto tactics. Long Range Plan 43:195–215

Ceschin F (2013) Critical factors for implementing and diffusing sustainable product-service systems: insights from innovation studies and companies' experiences. J Clean Prod 45:74–88

Chertow MR (2000) Industrial symbiosis: literature and taxonomy. Annu Rev Energy Environ 25:313–337

Chesbrough H (2010) Business model innovation: opportunities and barriers. Long Range Plan 43:354–363

Chesbrough H, Rosenbloom RS (2002) The role of the business model in capturing value from innovation: evidence from Xerox Corporation's technology spin-off companies. Ind Corp Chang 11(3):529–555

Coget JF (2009) Dialogical inquiry: an extension of Schein's clinical inquiry. J Appl Behav Sci 45(1):90–105

Coghlan D (2000) Interlevel dynamics in clinical inquiry. J Organ Chang Manag 13(2):190–200

Coghlan D (2009) Toward a philosophy of clinical inquiry/research. J Appl Behav Sci 45:106–121

Davies A, Brady T (2000) Organizational capabilities and learning in complex product systems: towards repeatable solutions. Res Policy 29(7–8):931–953

Ehrenfeld J, Gertler N (1997) Industrial ecology in practice: the evolution of interdependence at Kalundborg. J Ind Ecol 1(1):67–79

Gaziulusoy AI, Boyle C, McDowall R (2013) System innovation for sustainability: a systemic double-flow scenario method for companies. J Clean Prod 45:104–116

Geels F (2002) Technological transitions as evolutionary reconfiguration processes: a multi-level perspective and a case-study. Res Policy 31:1257–1274

Geels F (2005) Technological transitions and system innovations: a co-evolutionary and socio-technical analysis. Elgar, Cheltenham

Gulati R, Puranam P, Tushman M (2012) Meta-organization design: rethinking design in interorganizational and community contexts. Strateg Manag J 33:571–586

Gustafsson M, Stoor R, Tsvetkova A (2011) Sustainable bio-economy: potential, challenges and opportunities in Finland. Sitra, Helsinki

Halme M, Jasch C, Scharp M (2004) Sustainable homeservices? Toward household services that enhance ecological, social and economic sustainability. Ecol Econ 51(1–2):125–138

Halme M, Anttonen M, Kuisma M, Kontoniemi N (2007) Business models for material efficiency services: conceptualization and application. Ecol Econ 63(1):126–137

Hellström M (2005) Business concepts based on modularity: a clinical inquiry into the business of delivering projects. Doctoral dissertation, Åbo Akademis förlag, Åbo

Larson AL (2000) Sustainable innovation through an entrepreneurship lens. Bus Strateg Environ 9:304–317

Liinamaa J, Wikström K (2009) Integration in project business: mechanisms for knowledge integration. Int J Knowl Manag Stud 3(3/4):331–350

Linder JC, Cantrell S (2000) Changing business models. Institute for Strategic Change, Accenture, Chicago

Loorbach D (2010) Transition management for sustainable development: a prescriptive, complexity-based governance framework. Governance 23(1):161–183

Loorbach D, Wijsman K (2013) Business transition management: exploring a new role for business in sustainability transitions. J Clean Prod 45:20–28

Magretta J (2002) Why business models matter. Harv Bus Rev 80:86–92

Mirata M, Nilsson H, Kuisma J (2005) Production systems aligned with distributed economies: examples from energy and biomass sectors. J Clean Prod 13:981–991

Mont O (2002) Clarifying the concept of product-service system. J Clean Prod 10(3):237–245

Mont O (2004) Product-service systems: panacea or myth? Doctoral dissertation, IIIEE Lund University

Moore JE (1996) The death of competition: leadership and strategy in the age of business ecosystems. Harper Business, New York

Nelson RR, Winter SG (1982) An evolutionary theory of economic change. Belknap Press, Cambridge, MA

Osterwalder A, Pigneur Y, Tucci C (2005) Clarifying business models: origins, present and future of the concept. Commun Assoc Inf Sci 16:1–25

Raven RPJM (2005) Strategic niche management for biomass. Eindhoven University of Technology, Eindhoven

Rip A, Kemp R (1998) Technological change. In: Rayner S, Malone EL (eds) Human choice and climate change, vol II, Resources and technology. Battelle Press, Columbus, pp 327–399

Romme AGL (2003) Making a difference: organization as design. Organ Sci 14(5):558–573

Sartorius C (2006) Second-order sustainability conditions for the development of sustainable innovations in a dynamic environment. Ecol Econ 58(2):268–286

Schein EH (1993) Legitimating clinical research in the study of organisational culture. J Couns Dev 71:703–708

Schein EH (1995) Process consultation, action research and clinical inquiry: are they the same? J Manag Psychol 10(6):14–19

Schein EH (2008) Clinical inquiry/research. In: Reason P, Bradbury H (eds) Handbook of action research. Sage, London, pp 226–279

Schön DA (1995) Knowing-in-action: the new scholarship requires a new epistemology. Change 27:27–34

Teece DJ (2010) Business models, business strategy and innovation. Long Range Plan 43(2):172–194

Tsvetkova A, Gustafsson M (2012) Business models for industrial ecosystems: a modular approach. J Clean Prod 29–30:246–254

Turner JR, Simister SJ (2001) Project contract management and a theory of organization. Int J Proj Manag 19(8):457–464

van Aken JE, Romme G (2009) Reinventing the future: adding design science to the repertoire of organization and management studies. Organ Manag J 6:5–12

Wagner M, Llerena P (2011) Eco-innovation through integration, regulation and cooperation: comparative insights from case studies in three manufacturing sectors. Ind Innov 8(18):747–764

Wikström K, Artto K, Kujala J, Söderlund J (2010) Business models in project business. Int J Proj Manag 28:832–841

Part III
Future Directions: Eco-innovation Initiatives

Chapter 13
A New Methodology for Eco-friendly Construction: Utilizing Quality Function Deployment to Meet LEED Requirements

William L. Gillis and Elizabeth A. Cudney

Abstract Leadership in Energy and Environmental Design and building commissioning are two eco-innovations that have contributed to the long-term efficiency and sustainability of new-building construction. The Leadership in Energy and Environmental Design system promotes green, efficient, and sustainable design and construction. Building commissioning is a quality process used to verify that the owner's project requirements are being met by the final design, construction, and operations and maintenance. Quality function deployment has been successfully used in product development to capture the voice of the customer, translating it into engineering characteristics and then carrying the parameters into production and service to ensure the voice of the customer is met with the final product. To provide the next level of eco-innovation for commissioning and Leadership in Energy and Environmental Design, the four-phase quality function deployment model will be adapted to, and integrated with, the commissioning and Leadership in Energy and Environmental Design processes. The adapted model can effectively link the project phases to ensure the owner's project requirements for Leadership in Energy and Environmental Design are met with the final building. The primary objective of this research is to develop an integrative and systematic methodology to adapt the four-phase quality function deployment model to the commissioning process of new-building construction, to provide practitioners the steps to take the adapted model through the commissioning process as a means to oversee the entire design and construction quality process, ensuring the owner's Leadership in Energy and Environmental Design goals are integrated into the design and carried through construction to the operations and maintenance phase for long-term efficiency and sustainability.

Keywords Quality function deployment • House of quality • Commissioning • Construction • Leadership in Energy and Environmental Design • Eco-innovation

---

W.L. Gillis • E.A. Cudney (✉)
Missouri University of Science and Technology, Rolla, MO 65409, USA
e-mail: wgillis@mst.edu; cudney@mst.edu

S. Azevedo et al. (eds.), *Eco-Innovation and the Development of Business Models*, Greening of Industry Networks Studies 2, DOI 10.1007/978-3-319-05077-5_13,

13.1 Introduction

The construction of a new nonresidential building can represent one of the largest investments any business will face. With most investments, it is desired to maximize value and minimize cost and this holds for most new-building construction. Trends are driving building designs and construction methods that are considered "sustainable" and/or "green"; however, the benefits of going sustainable/green vary greatly depending on the needs of each potential building owner. Along with these new eco-innovative design trends are more complex building systems, higher first cost of construction, and greater risk of not meeting the owner's expectations with the final product.

13.1.1 LEED

With energy costs rising and the need for energy efficiency and environmental protection, the US Green Building Council (USGBC) developed and implemented a program to define and measure green buildings. This program known as Leadership in Energy and Environmental Design (LEED) is a rating system that building owners must utilize to qualify their projects for registration as LEED certified with the Green Building Certification Institute. LEED 2009 has several areas of certification, which include New Construction, Existing Buildings, Commercial Interiors, Core and Shell, Schools, Retail, Health Care, Homes, and Neighborhood Development. The focus of this research is limited to LEED 2009 for New Construction (LEED-NC). All of the certification areas use a point system with 100 base points and 10 bonus points. Four levels of certification may be attempted (Table 13.1). The certification categories and the corresponding points are then divided among seven categories (Table 13.2).

Each of the categories contains prerequisites that must be met prior to any points being awarded, and each prerequisite must be met by any project applying for certification. Each LEED for New Construction-certified building will have the prerequisites in common, but how the certification points are accomplished is left to the owner, architect, and engineering teams to determine. Two additional consulting groups who are added to the team on many LEED projects are a LEED Accredited Professional (LEED AP) and a commissioning authority (CxA).

Table 13.1 LEED certification levels and required points (USGBC 2009)

Certification level	Required points
Certified	40–49
Silver	50–59
Gold	60–79
Platinum	80–110

Table 13.2 LEED-NC certification categories

Category	Available points
Sustainable Sites (SS)	26
Water Efficiency (WE)	10
Energy and Atmosphere (EA)	35
Materials and Resources (MR)	14
Indoor Environmental Quality (IEQ)	15
Base points	*100*
Innovation in Design (ID)	6
Regional Priority (RP)	4
Bonus points	*10*
Total points	*110*

To deliver a successful LEED project to the building owner, the team must meet LEED certification requirements and provide a building that meets the owner's project requirements (OPR). The LEED AP will assist the owner and design team in determining the best path to take for a particular LEED level by providing assistance in understanding and implementing the LEED scoring system. The design team will work toward delivering designs capable of achieving the LEED points expected from each of their respective areas of expertise. The commissioning authority will assist the owner in developing the owner's project requirements and then verify that the design and construction meet the owner's project requirements. This is a complicated process which requires much coordination, cooperation, and communication.

13.1.2 Commissioning

To ensure receipt of their expected value at project completion, the owner often contracts a third-party commissioning authority to act as technical advisor and to oversee the quality of the design and construction by administering the commissioning process. The American Society of Heating, Refrigerating and Air-Conditioning Engineers (ASHRAE) define the commissioning process as "A quality-focused process for enhancing the delivery of a project. The process focuses upon verifying and documenting that the facility and all of its systems and assemblies are planned, designed, installed, tested, operated, and maintained to meet the Owner's Project Requirements" (ASHRAE 2005).

The commissioning process consists of several steps throughout the design and construction process then extends into building occupancy and beyond (Fig. 13.1). Critical to any project, particularly those seeking LEED certification with high expectations for sustainability and efficiency, is determining what the owner requires and how the design team will respond to those requirements. Two key terms are defined by ASHRAE: (1) owner's project requirements ("A written document that details the functional requirements of a project and the expectations of how it will be used and operated. These include project goals, measurable performance criteria,

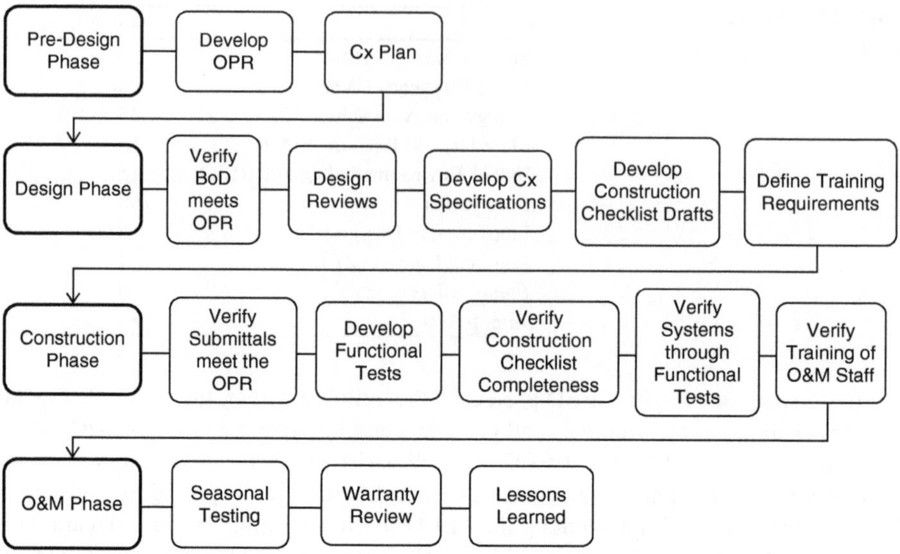

Fig. 13.1 Cx process flow chart

cost considerations, benchmarks, success criteria, and supporting information") and (2) basis of design (BoD) ("A document that records the concepts, calculations, decisions, and product selections used to meet the owner's project requirements and to satisfy applicable regulatory requirements, standards, and guidelines. The document includes both narrative descriptions and lists of individual items that support the design process") (ASHRAE 2005).

The owner's project requirements are considered a living document and are subject to change throughout the design and build. Changes may occur at any time during the project and must be approved by the owner. The commissioning authority must then update the owner's project requirements document accordingly. The basis of design is also a living document and will very likely change as the design progresses and should also be updated as needed. Since both of these documents are likely to change, it is important to have a methodology in place to trace proposed changes back to the owner's project requirements to analyze how the change will impact the other owner's project requirements and the project as a whole. An adapted four-phase quality function deployment (QFD) model and house of quality (HOQ) will provide the means for tracing such changes.

13.1.3 Quality Function Deployment

Quality function deployment has been successfully used in product development to capture the voice of the customer, translating it into engineering characteristics and then carrying the parameters into production and service to ensure the voice of the

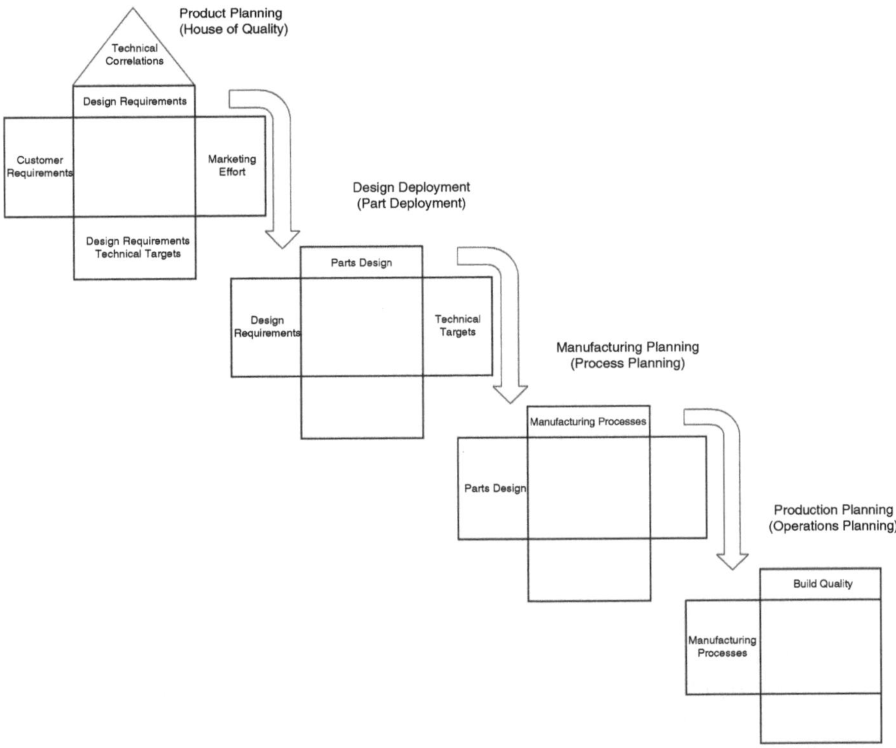

Fig. 13.2 QFD four-phase model

customer is met with the final product. The four-phase quality function deployment model for product development proposed by Cohen (1995) is illustrated in Fig. 13.2.

The house of quality, the first matrix in the four-phase model, is a tool within the quality function deployment process that provides a means of matching the product's design with the voice of the costumer or customer requirements. Figure 13.3 illustrates the basic house of quality. Customer requirements are what the customer desires of a particular product based on marketing studies. The design response is how the designers will meet the needs of the customer. Design correlations are used as a means of understanding if one design response has an impact on another design response. The body of the matrix holds the relationships, or how well each design response addresses the customer requirements. The marketing matrix and design data matrix are used by the marketing and design teams for developing and prioritizing the whats and hows. In short, quality function deployment is designed to gather the customer's needs and desires of a product, weigh those needs and desires against the needs and desires of the company, verify that engineering designs the product according to those requirements, and verify that manufacturing can produce the product as designed. The goal is a product that will appeal to as many customers as possible.

Fig. 13.3 HOQ model

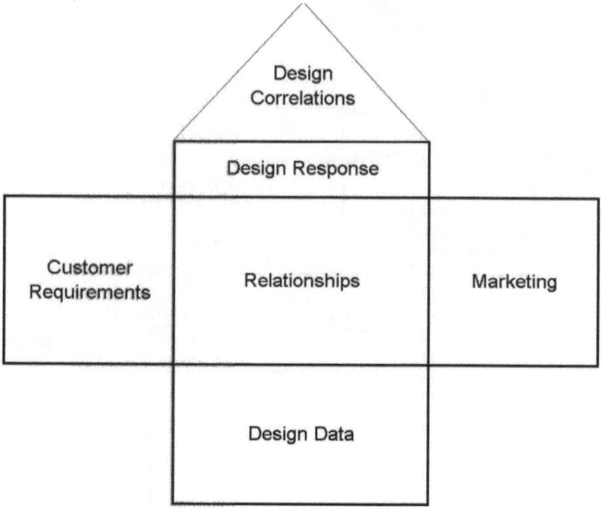

The product development quality function deployment model is well suited for adaption to commissioning of new-building construction, and the house of quality is a perfect tool for understanding the complexities of the LEED for New Construction system.

Whether the owner's motivation for going green is to be a good steward of the environment, to reduce the cost of operating the building, or simply for good public relations, each of these will promote an eco-friendly building with reduced negative impact on the environment. Regardless of the motivation, it must be determined early what the owner wants of the building and what is important to them in the finished product. There is no guarantee that LEED certification will provide added value or an eco-friendly building.

Sustainability, efficiency, and value will largely depend on the owner's project requirements and whether those requirements are met by the finished project. Achieving LEED certification at any of its four levels may not ensure added value in the eyes of the owner if their needs are not met with the final building. There are many combinations of credits that can be utilized to accomplish an owner's LEED certification goal, but which combination best fits the owner's project requirements and the owner's definition of value is a question that must be answered early.

Each credit of the LEED system is designed to improve the eco-friendliness of the building, but some credits are easier and less costly to achieve than others. For example, a credit to provide designated parking spaces near the building expected to promote either carpooling or use of eco-friendly vehicles such as hybrids can be easily achieved by placing signs at those parking spaces. But if they are not utilized because either employees do not carpool or they cannot afford or do not want to purchase an expensive hybrid, the effort is lost and the building is no more eco-friendly than it would have been without the designated parking spaces. Therefore, if the goal of the owner is to have a green or eco-friendly building, it is essential that the LEED credits to be achieved align with the owner's desires and are implemented in the design. LEED certification is part of the owner's project requirements, but the design may drift from the owner's project requirements and simply focus on meeting the LEED certification.

The goal of the proposed methodology is to provide a new integrative methodology that adapts quality function deployment and the house of quality to the commissioning process for new-building construction with emphasis on those projects seeking LEED certification. The LEED house of quality will provide a method for identifying the impact each LEED credit has on the other credits as a means to improve design decisions and ensure the LEED credits pursued meet the owner's project requirements. The complete model will provide an effective method for linking each of the commissioning activities to assist the commissioning authority in managing the quality process and ensuring the owner's project requirements are achieved.

The remainder of the chapter will discuss the background, which provides the motivation for developing the methodology, an explanation of the methodology for adapting the four-phase quality function deployment model to the commissioning and LEED processes, and the concluding remarks.

13.2 Background

In a final report to the US Green Building Council (USGBC), the New Building Institute gathered and analyzed data with regard to the "Energy Performance of LEED for New Construction Buildings" (Turner and Frankel 2008). At the time of the study (2006), 552 buildings had been certified under LEED for New Construction. All were invited to participate with the stipulation that building owners would provide 1-year post-occupancy energy data. One hundred and twenty-one building owners participated and provided the necessary information. Findings indicate that, on average, LEED-certified buildings do reap the benefit of energy savings. Buildings achieving higher LEED certification levels, on average, had higher energy efficiency. A troubling note is that half of the buildings had performance that significantly deviated from the design intent. Twenty-five percent had performance that was lower than intended, but the cause of the poor performance was not investigated.

As a follow-up to the Turner and Frankel (2008) report, Newsham et al. (2009) took a more rigorous approach to analyzing the Turner and Frankel (2008) data. A full statistical analysis comparing the LEED buildings to carefully matched non-LEED buildings produced the same results that energy performance is improved for LEED-certified buildings and, on average, LEED buildings use 18–39 % less energy per floor area than conventional buildings. Newsham et al. (2009) also tested energy performance versus the LEED energy credits received and energy performance versus additional commissioning and measurement and verification credits and found that neither correlated. The study shows that as the certification level of these building increased, so did the credits obtained in the energy areas. However, the additional effort and funds to obtain these credits had no effect on energy performance. The authors suggested that this could be caused by the fact that the energy data provided by the building owners was for the first year. This is typically the period when the deficiencies are worked out of the building systems. They also questioned whether the commissioning process was properly conducted but did not provide an answer.

It should be pointed out that one of the primary purposes of the commissioning process is to deliver a building to the owner that is already free of deficiencies. Commissioning is expected to ensure the building systems are performing to the level of efficiency that was expected from the design and the owner's project requirements.

When the team, including the owner, meets to determine which LEED credits to target, they typically refer to the list provided in Table 13.3 and discuss whether each credit is worth attempting. Much of what is discussed is cost, impact to project budget, sustainability and efficiency goals, and whether particular credits are even achievable. For example, if a new piece of property is purchased in a rural area, it is very unlikely Sustainable Sites, Credit 2, Development Density and Community Connectivity will be achievable. However, this purchase may provide an excellent opportunity to achieve Sustainable Sites, Credit 5.1, Site Development – Protect and Restore Habitat. There are trade-offs to be made. Another example might be questioning the higher cost for achieving Sustainable Sites, Credit 7.2, Heat Island Effect – Roof, which could be accomplished by installing a vegetative or "green" roof.

What is not always clear during these discussions is the total impact of one choice, based on either the ease of obtaining the credit or cost, on the entire project or other credits. Many credits can have a positive or negative impact on other credits and budget. To illustrate, consider the information in the previously mentioned studies. Based on that research, it can be seen that the commissioning process does provide energy savings and that many building owners have not realized the investment made in a LEED-certified building with respect to energy savings. The question of whether to attempt the enhanced commissioning credits must be discussed, and based on findings of the research, many owners would decline spending the money. What is not fully understood is what this means to the other credits and life cycle cost if enhanced commissioning is not pursued.

Commissioning affects numerous LEED credits and a total of 48 % of the possible base points. These credits and points are shown in Table 13.4. A significant portion of the construction budget will be allocated to these areas.

Within the Energy and Atmosphere category is EA Prerequisite 1, Fundamental Commissioning of Building Energy Systems. The commissioning authority is responsible for ensuring the systems within the EA and IEQ credit categories are designed, constructed, and operating as specified, along with developing the commissioning plan, commissioning requirements, and completing the commissioning report. "Commissioning process activities must be completed for the following energy-related systems, at a minimum:

Table 13.3 LEED credits and associated points (USGBC 2009)

Sustainable Sites (SS)		*26 points*
SS Prerequisite 1	Construction Activity Pollution Prevention	Required
SS Credit 1	Site Selection	1
SS Credit 2	Development Density and Community Connectivity	5
SS Credit 3	Brownfield Redevelopment	1
SS Credit 4.1	Alternative Transportation – Public Transportation Access	6
SS Credit 4.2	Alternative Transportation – Bicycle Storage and Changing Rooms	1
SS Credit 4.3	Alternative Transportation – Low-Emitting and Fuel Efficient Vehicles	3
SS Credit 4.4	Alternative Transportation – Parking Capacity	2
SS Credit 5.1	Site Development – Protect and Restore Habitat	1
SS Credit 5.2	Site Development – Maximize Open Space	1
SS Credit 6.1	Stormwater Design – Quantity Control	1
SS Credit 6.2	Stormwater Design – Quality Control	1
SS Credit 7.1	Heat Island Effect – Nonroof	1
SS Credit 7.2	Heat Island Effect – Roof	1
SS Credit 8	Light Pollution Reduction	1
Water Efficiency (WE)		*10 points*
WE Prerequisite 1	Water Use Reduction	Required
WE Credit 1	Water-Efficient Landscaping	2–4
WE Credit 2	Innovative Wastewater Technologies	2
WE Credit 3	Water Use Reduction	2–4
Energy and Atmosphere (EA)		*35 points*
EA Prerequisite 1	Fundamental Commissioning of Building Energy Systems	Required
EA Prerequisite 2	Minimum Energy Performance	Required
EA Prerequisite 3	Fundamental Refrigerant Management	Required
EA Credit 1	Optimize Energy Performance	1–19
EA Credit 2	On-Site Renewable Energy	1–7
EA Credit 3	Enhanced Commissioning	2
EA Credit 4	Enhanced Refrigerant Management	2
EA Credit 5	Measurement and Verification	3
EA Credit 6	Green Power	2
Materials and Resources (MR)		*14 points*
MR Prerequisite 1	Storage and Collection of Recyclables	Required
MR Credit 1.1	Building Reuse – Maintain Existing Walls, Floors, and Roof	1–3
MR Credit 1.2	Building Reuse – Maintain Interior Nonstructural Elements	1
MR Credit 2	Construction Waste Management	1–2
MR Credit 3	Materials Reuse	1–2
MR Credit 4	Recycled Content	1–2
MR Credit 5	Regional Materials	1–2
MR Credit 6	Rapidly Renewable Materials	1
MR Credit 7	Certified Wood	1
Indoor Environmental Quality (IEQ)		*15 points*
IEQ Prerequisite 1	Minimum Indoor Air Quality	Required
IEQ Prerequisite 2	Environmental Tobacco Smoke (ETS) Control	Required

(continued)

Table 13.3 (continued)

IEQ Credit 1	Outdoor Air Delivery Monitoring	1
IEQ Credit 2	Increased Ventilation	1
IEQ Credit 3.1	Construction Indoor Air Quality Management Plan – During Construction	1
IEQ Credit 3.2	Construction Indoor Air Quality – Before Occupancy	1
IEQ Credit 4.1	Low-Emitting Materials – Adhesives and Sealants	1
IEQ Credit 4.2	Low-Emitting Materials – Paints and Coatings	1
IEQ Credit 4.3	Low-Emitting Materials – Flooring Systems	1
IEQ Credit 4.4	Low-Emitting Materials – Composite Wood and Agrifiber Products	1
IEQ Credit 5	Indoor Chemical and Pollutant Source Control	1
IEQ Credit 6.1	Controllability of Systems – Lights	1
IEQ Credit 6.2	Controllability of Systems – Thermal Comfort	1
IEQ Credit 7.1	Thermal Comfort – Design	1
IEQ Credit 8.1	Daylight and Views – Daylight	1
IEQ Credit 8.2	Daylight and Views – Views	1
Innovation in Design (ID)		*6 points*
ID Credit 1	Innovation in Design	1–5
ID Credit 2	LEED Accredited Professional	1
Regional Priority (RP)		*4 points*
RP Credit 1	Regional Priority	1–4

Table 13.4 Credits and points affected by the commissioning process (USGBC 2009)

Credit	Title	Available points
SS Credit 8	Light Pollution Reduction	1
WE Credit 1	Water-Efficient Landscaping	2–4
WE Credit 2	Innovative Wastewater Technologies	2
WE Credit 3	Water Use Reduction	2–4
EA Credit 1	Optimize Energy Performance	1–19
EA Credit 2	On-Site Renewable Energy	1–7
EA Credit 5	Measurement and Verification	3
IEQ Prerequisite 1	Minimum Indoor Air Quality Performance	–
IEQ Credit 1	Outdoor Air Delivery Monitoring	1
IEQ Credit 2	Increased Ventilation	1
IEQ Credit 5	Indoor Chemical and Pollutant Source Control	1
IEQ Credit 6	Controllability of Systems	2
IEQ Credit 7	Thermal Comfort	2
	Total possible points	*48*

- Heating, ventilating, air-conditioning and refrigeration (HVAC&R) systems (mechanical and passive) and associated controls,
- Lighting and day lighting controls,
- Domestic hot water systems, and
- Renewable energy systems (e.g. wind, solar)." (USGBC 2009)

Those systems will be commissioned for every project under the EA Prerequisite 1.

The commissioning authority has the responsibility of ensuring the owner's definition of value is realized. When the systems are designed and installed as specified, as well as operating as intended, commissioning optimizes the energy efficiency of the building systems. "Properly executed commissioning can substantially reduce costs for maintenance, repairs, and resource consumption, and higher indoor environmental quality can enhance occupants' productivity" (USGBC 2009).

EA Credit 3, Enhanced Commissioning, is only worth two additional points; it places additional responsibility and requirements on the commissioning authority but can have a significant impact on the opportunity for the building to be sustainable and energy efficient for its life cycle. Commissioning processes for EA Credit 3 will be expanded to include:

- Documenting the commissioning review process
- Reviewing contractor submittals
- Developing the systems manual
- Verifying the training of operations personnel
- Reviewing building operation after final acceptance (USGBC 2009)

To improve the quality of LEED projects, Tseng (2005) expresses that any LEED credits to be pursued for a project must be integrated into the owner's project requirements to allow the design team an early opportunity to incorporate LEED goals into a design that will meet the owner's project requirements.

Ellis (2009a) discusses the relationship between commissioning and energy conservation stating that one does not necessarily guarantee the other. LEED projects require a specified energy conservation measure above an industry baseline and energy modeling. The owner must express their energy conservation goals in a quantifiable measure for the design team and commissioning authority to utilize. This provides a solid base for design and verification and must be included in the owner's project requirements. There is then a greater opportunity for the commissioning authority to assist the building owner in receiving what the owner truly desires in the finished building.

Barber (2008) explains the importance of the key commissioning documents, the owner's project requirements, and the basis of design. These documents are often seen as expensive and not necessary, and, therefore, these invaluable tools are under used. Documenting the owner's project requirements has many valuable attributes, one of which is to minimize conflicting owner directives. Design teams often receive different priorities and expectations from groups within the owner's organization. In order to fully develop owner's project requirements, the different groups must come to agreement on common functional requirements. The basis of design is the confirmation that the owner's project requirements are understood and provides a description of how the requirements will be met with the design.

Enke (2010) expresses the need for a holistic approach to commissioning. This begins with full implementation of ASHRAE Guideline 0-2005 and places additional emphasis on the owner's project requirements and basis of design development, as well as the review(s) to verify these two align. Enke points out that the basis of design requirement is often overlooked and without this critical information, it is nearly impossible to know the owner's project requirements have been met. High-performance buildings have two key elements: maintainability and measurability. Provisions for maintainability and measurability need to be addressed in the owner's project requirements and basis of design. The design phase of commissioning will carry the long-term measures required through the entire process well into the operations and maintenance phase.

Ellis (2009b) describes a noticed trend of design documents being issued for bid prior to design being complete. It is explained that no design is perfect and the design and construction process has methods in place for dealing with the imperfections, but imperfect and incomplete are completely different issues. Top reasons for this occurrence include the design team's belief that some incomplete designs can be finished later without affecting progress or budget, the owner's inability to express project requirements, and to not delay the construction start date. Commissioning can identify the incomplete design aspects and clearly state to the owner what appears to be missing and the impact or risks involved with not waiting for a complete design prior to bidding. The goal is full transparency for all involved.

Commissioning and LEED certification can benefit any building project and provide additional value to the owner through long-term sustainability and efficiency if both processes are administered properly. LEED certification can improve building efficiency if the proper credits related to energy efficiency are achieved. This will only happen if the owner has determined that (1) a particular level of efficiency is the goal, (2) the budget will allow it, and (3) the design and construction follow through on those requirements. If the LEED credits to be achieved are identified early, ideally in predesign, and are determined to not conflict with each other and jeopardize the certification level, commissioning can improve the opportunity to achieve the desired LEED certification level and expected efficiencies.

To provide the next level of eco-innovation for commissioning and LEED, the four-phase quality function deployment model and the house of quality will be adapted to, and integrated with, the commissioning process.

Researchers over the past several years have demonstrated ways quality function deployment can also be a beneficial tool for the construction industry. Mallon and Mulligan (1993) present quality function deployment as a means for meeting the customers' requirements in construction projects, providing an example of such for a computer workroom. Eldin and Hikle (2003) conducted a pilot study, using the design of a classroom as a case, to determine the effectiveness of using quality function deployment as a means of developing conceptual designs in the preliminary phase of construction projects. Ahmed et al. (2003) utilized quality function deployment for a civil engineering capital project. Yang et al. (2003) integrate fuzzy set theory into the quality function deployment process for use as a decision-making aide in construction design. Alarcon and Mardones (1998), Gargione (1999), and Abdul-Rahman et al. (1999) used it in the design phase to capture the customer needs. Dikmen et al. (2005) examined a completed building to capture customer needs for marketing purposes. Delgado-Hernandez et al. (2007) considered nursery development/design. Each of these are primarily looking at the design and capturing

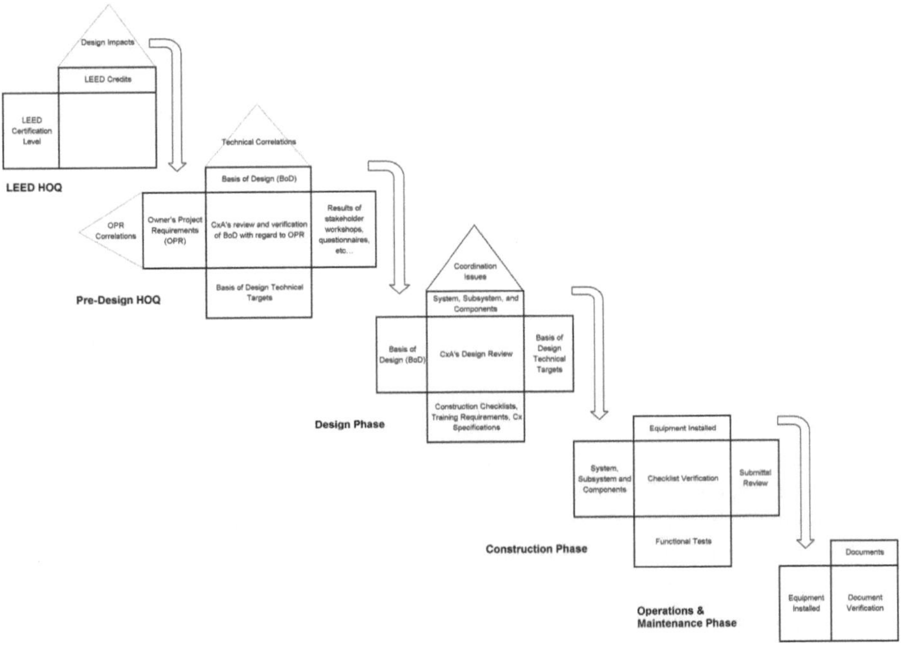

Fig. 13.4 Adapted 4-phase model with LEED HOQ

the needs of the customer. These processes appear to be structured for use by the design team or could be used by the owner where the owner would then specify the design criteria. This may have a tendency to constrain the design creativity. Kamara et al. (2000), Pheng and Yeap (2001), Lee and Arditi (2006), and Lee et al. (2009) used quality function deployment as a tool for concurrent engineering and design/build applications.

The objective of this methodology is not to provide an additional method of how quality function deployment could be used by a design team within the construction industry as a means to improve their final product (design only), rather to introduce a quality function deployment model that has been specifically tailored for use by the commissioning authority during the entire design and construction process. The adapted four-phase quality function deployment model along with the LEED house of quality is presented in Fig. 13.4. This will consist of two houses of quality matrices, one specifically for LEED projects and one serving as the predesign house of quality. Three other matrices, design, construction, and operations and maintenance, will be used to link the design and construction activities back to the owner's project requirements and LEED requirements.

It is assumed that each member of the architect/engineer (A/E) team, as well as the construction contractors, will have their own internal quality processes and will strive to provide the owner with a quality end product, but the commissioning authority represents the owner specifically and applies a blanket quality process to the entire project. This is expected to benefit all involved, including the architect/engineer and contractors, by reducing late changes and rework, therefore saving time and resources. The ultimate goal is to have a satisfied owner at building occupancy and through operations and maintenance with a long-term efficient and sustainable building.

13.3 Methodology

13.3.1 Introduction

The methodology will be illustrated using an abbreviated set of an owner's project requirements. A typical construction project would obviously have many more requirements than can be displayed in this format. The majority of the owner's project requirements used here are LEED related along with a few non-LEED-specific requirements. These owner's project requirements were chosen to assist in illustrating the ability of the adapted four-phase model to link these requirements through each of the phases of the design and construction process.

An emphasis is placed on LEED, as this is a complex system that can be more easily understood with the development of the LEED house of quality. The LEED house of quality provides a compact format for understanding the design impacts one LEED credit may have on another.

13.3.2 Predesign Phase (House of Quality)

Modifying the house of quality for use in both understanding the LEED credits and for the predesign phase of the project will greatly improve the opportunity to achieve the desired LEED certification level and ensure that the sustainability and efficiency goals are accomplished. The first step in the commissioning process is developing the owner's project requirements. There are many methods for assisting the owner in developing the owner's project requirements, some of which might include workshops with key stakeholders, questionnaires, group meetings, or nominal group technique. Based on the focus of this research, these methods will not be discussed here with the exception of those owner's project requirements that include LEED certification.

Information gathered from the owner's project requirements development workshops will be entered into the marketing matrix and whats sections simultaneously. This will provide an opportunity to prioritize the owner's project requirements based on importance to the owner. Once the owner's project requirements have been established, prioritized, and approved by the owner, the information is finalized in the whats area of the house. When LEED certification is an owner's project requirement, the team must now navigate the LEED certification requirements to determine which

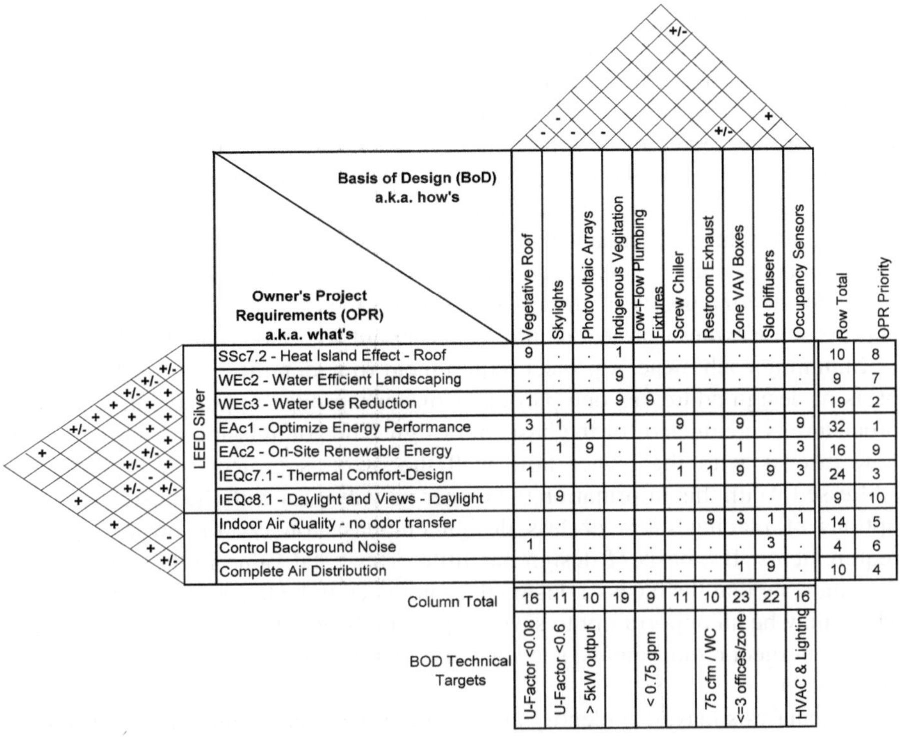

Fig. 13.5 Pre-design HOQ

LEED credits to pursue in order to meet the LEED certification goal and to meet the other approved owner's project requirements. An additional house of quality is developed for this purpose and will be explained later.

To illustrate the predesign house of quality, it has been determined that the project will seek LEED certification with a goal of achieving a level of Silver. Several credits the owner has decided should be attempted are Heat Island Effect – Roof, Daylight and Views – Daylight, Optimize Energy Efficiency, Water-Efficient Landscaping, and Thermal Comfort – Design, feeling these will contribute to the LEED goal and align well with other owner's project requirements. These LEED credits are integrated with other owner's project requirements determined during the development process. The approved owner's project requirements, along with the priorities, will then be provided to the architect/engineer team for use as a reference and compliment to the programming and design effort.

The basis of design, based on the approved owner's project requirements, will be generated by the architect/engineer typically during their schematic design and will be provided to the owner and commissioning authority. The basis of design is entered into the hows area of the house. Figure 13.5 represents the predesign house of quality with the integrated LEED credits.

The commissioning authority will use the body of the matrix during the review of the basis of design to verify it meets or addresses each of the owner's project requirements. Using a numbering scheme of 9 (strong relationship), 3 (medium), and 1 (weak) in each cell to specify how well each basis of design meets a particular owner's project requirement will be used. A "." is used to populate a cell in which the combination has been reviewed and found to have no relationship. This is meant to reduce the duplication of work which may occur if the cell is left blank.

It is important to understand that these entries are the professional opinion of the commissioning authority and are not intended to be a critique of the design per se, rather to identify points of discussion between the owner and design team if it appears some owner's project requirements are not well represented in the design.

Summing each owner's project requirement row identifies how well the entire basis of design addresses one particular owner's project requirement. The same procedure is conducted for the basis of design. The commissioning authority can now discuss with the owner how well the owner's project requirements are addressed. With this information the owner may discuss with the design team the options of placing more or less design emphasis on certain owner's project requirements. If a basis of design has little or no total value in the column, this may indicate that the basis of design does not need to be taken to the final design. There may be an opportunity to save project funds or shift funds toward different owner's project requirements, such as those that promote long-term sustainability and efficiency.

This analysis may also identify an owner's project requirement that needs to be modified or eliminated. The owner making these decisions at this phase can save significant time, money, and effort. Any owner's project requirement or basis of design that is changed must be updated within the matrix. The owner and the design team now know their ideas and efforts are aligned and can agree to proceed to the next phase of design. The design correlations area (roof) of the house of quality is used to understand the impact, negative or positive, each design criteria has on another. The correlation matrix identifies the impact one basis of design has on another basis of design. For example, the impacts are identified using a "–" for a negative impact, a "+" for a positive impact, and a "+/–" if it could be either positive or negative. This roadmap of impacts can also be used to analyze the impact of proposed changes in later stages of the design and construction. The same process as the basis of design correlation matrix is used to populate the owner's project requirement correlation matrix. It can be seen that most of the LEED credits can have both positive and negative impacts on the other credits.

Key benefits of the predesign house of quality: The owner will provide the owner's project requirements to the designers and can be assured the designers will have what is necessary to begin the schematic design. The designers can be assured they are beginning the design with approved owner's project requirements, which have been developed by the owner with the assistance of technical experts from the commissioning authority team. After the basis of design is added and the analysis of the matrix is complete, the designers can move forward knowing that their basis of design has been approved as satisfactorily meeting the owner's project requirements and will likely reduce late changes. The owner's project requirements and basis of design roofs also will provide a quick method for analyzing the effects of proposed design changes in later phases of the project.

13.3.3 LEED House of Quality

The complex system of LEED credits requires the design disciplines to coordinate, cooperate, and communicate if the required LEED certification level is to be achieved. There are many combinations of credits that can be utilized to accomplish an owner's LEED certification goal, but determining the combination that best fits the owner's project requirements can be a difficult task and must be accomplished early in the project. An additional house of quality used to analyze the LEED certification requirements of the owner's project requirements will be constructed. The LEED certification level and the desired credits will become owner's project requirements and be integrated into the owner's project requirements section of the predesign house of quality. A building owner may require a LEED certification level of Silver, for example, and the design team will work toward meeting that goal; however, in many cases the credits attempted are not approved by the certification authority (the Green Building Certification Institute), and a lower level is actually achieved or the project is not capable of meeting the certification level because of other conflicting owner's project requirements.

13.3.3.1 Building the LEED House of Quality

At this point all of the certification levels and the required points are added as owner's project requirements, as the final level that will be achieved is not known this early in the process. The LEED credits will now be added as basis of design responses. Figure 13.6 represents an abbreviated LEED house of quality. The remaining credits would be entered in the same manner, but are not able to be displayed in the format of this chapter due to the matrix size. The associated points for each credit are added to the target value (LEED points) row above the credits categories. LEED categories and credits along with the associated points can now be quickly referenced. The row above the target value is used to symbolically represent whether the target value should be maximized or is the target. Many credits only offer one point level (all or nothing), while others offer a range of points. If the points are "all or nothing," a "x" is used to indicate this is the target. If the points are a range, a "▲" is used to identify that the goal is to maximize points. The next row above is used to represent the difficulty of achieving the target or goal.

In this case, the commissioning authority's experience in the design and construction field is used to rank the difficulty of each credit from 0 to 10, with 10 representing "extremely difficult." These values are not critical to the process, but can be a valuable tool for generating talking points during meetings between the

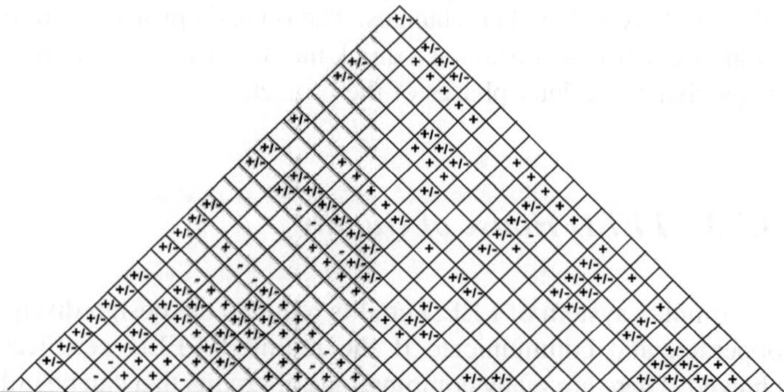

LEED Credits / LEED Certification Level (Point Requirement)	SSc1 - Site Selection	...	SSc5.2 - Maximize Open Space	SSc6.1 - Stormwater Quantity Control	SSc6.2 - Stormwater Quality Control	SSc7.1 - Heat Island Effect - Non-Roof	SSc7.2 - Heat Island Effect - Roof	SSc8 - Light Pollution Reduction	WEc1 - Water Efficient Landscaping	WEc2 - Innovative Wastewater Techniques	WEc3 - Water Use Reduction	EAc1 - Optimize Energy Performance	EAc2 - On-Site Renewable Energy	EAc3 - Enhanced Commissioning	...	EAc6 - Green Power	MRc1.1 - Building Reuse	MRc2 - Construction Waste Management	...	MRc7 - Certified Wood	...	IEQc6.1 - Controllability of Systems - Lighting	IEQc7.1 - Thermal Comfort - Design	IEQc8.1 - Daylight & Views - Daylight	IEQc8.2 - Daylight & Views - Views	IDc1 - Innovation in Design	RPc1 - Regional Priority
Difficulty of Accomplishment (0 = Easy, 10 = Hard)	10		7	7	5	6	3	3	4	6	3	10	8	5		4	8	2		4		2	6	4	5	9	7
Maximize (▲) or Target (×)	×		×	×	×	×	×	×	▲	×	▲	▲	▲	×		×	▲	▲		×		×	×	×	×	▲	▲
Target Value (LEED Points)	1		1	1	1	1	1	1	4	2	4	19	7	2		2	3	2		1		1	1	1	1	5	4
LEED Catergory	SUSTAINABLE SITES								WATER EFFICIENCY			ENERGY AND ATMOSPHERE					MATERIALS AND RESOURCES				INDOOR ENVIRONMENTAL QUALITY						ID	RP
Certified (40-49)	9		1	1	1	1	3	1	1	1	3	9	3	9		3	3	1		1		1	3	1	1	3	1
Silver (50-59)	9		3	3	3	3	3	1	3	3	3	9	3	9		3	3	3		1		1	3	3	3	3	3
Gold (60-79)	9		3	3	3	3	3	3	9	3	9	9	9	9		9	9	9		3		3	9	3	3	9	9
Platinum (80-110)	9		9	9	9	9	9	3	9	9	9	9	9	9		9	9	9		3		9	9	9	9	9	9

Fig. 13.6 LEED HOQ

commissioning authority and the owner. Primarily the difficulty represents the level of cost, time, and resources required to accomplish the associated LEED points. The matrix between the certification level and basis of design is now populated.

This can only be completed with an understanding of the LEED process and the requirements for achieving the LEED credit points. Generally though, without that knowledge, it can easily be seen that in order to achieve LEED Platinum, which requires a minimum of 80 points from the available 110, attempting most of the credits will be necessary since achieving 80 points is not an easy accomplishment.

The same numbering scheme used in the predesign house of quality of 9 (high), 3 (medium), and 1 (low) will be used to populate the relationship matrix to indicate how important it might be for this credit to be attempted in order to meet the level of certification. Analyzing the opportunities or necessities for LEED Platinum, it can be seen that nearly every credit may need to be attempted to accomplish 80 points, particularly if all available points for those credits that have a range of points are not achieved. This matrix provides insight into the difficulty of achieving certification and assists the owner in determining their LEED certification goal and which credits to attempt to meet that goal.

One of the most valuable components of any house of quality is the roof as it represents the affects, either negative or positive, each design criterion has on another. It offers a quick visual for a commissioning authority and owner to understand the impacts one design discipline's decision will have on another design discipline's work. The same strategy used to populate the basis of design and owner's project requirement correlation matrices in the predesign house of quality is used here. This can also provide a quick reference for any designer to understand that communication with another designer or team will be necessary if a +, –, or +/– is found in the cell.

Some design decisions may affect or impact more than one design area or credit. In this situation a chain reaction of impacts occurs. An example might be the following: Sustainable Sites, Credit 7.2, Heat Island Effect – Roof, exposes a number of relationships that should be considered. There are many design strategies that may achieve this credit but let's look at the vegetative or "green" roof option. First, the cost to install a "green" roof would need to be compared to a more conventional roof which would also accomplish the credit (budget impact). Second, the "green" roof will likely have greater mass and may require an increased structural system in the building (budget impact and possibly a reduction of the interior volume, which reduces the available space for the building systems and occupants). Any roof that qualifies for SS Credit 7.2 will reduce the heating and cooling loads on the building and, therefore, reduce the amount of energy required to maintain thermal comfort (life cycle cost). Another advantage with reduced heating and cooling loads is that the equipment required to heat and cool the building can be reduced (first cost reduction).

Since the HVAC system is smaller, so too will be the electrical system necessary to supply the equipment (smaller electrical gear and lower first cost). This is a vegetative roof so it will require irrigation to keep the plants alive (higher water usage and utility bill). This will negatively affect the Water Efficiency (WE) of the building (WE Prerequisite 1). With this there is an opportunity to plant indigenous vegetation and attempt WE Credit 1, Water-Efficient Landscaping, or SS Credit 5.1, Site Development – Protect and Restore Habitat.

Adding a skylight(s) to this roof will reduce the area of the vegetation, the heat island effect will still be reduced, the required irrigation will be reduced, but now the additional sun introduced through the skylight will contribute to achieving Indoor Environmental Quality, Credit 8.1, Daylight and Views – Daylight. However, the insulating factor of the vegetative roof is now reduced and solar heat gain is introduced, which will increase the heating and cooling load, HVAC system requirements, and electrical system requirements.

The house of quality roof as designed for product development cannot take into account all of these LEED credit impacts at one glance, but does still have the ability to expose the possible path of the effects and provides a method to navigate the possibilities. Knowing that one credit will have an effect on another can still lead to the understanding that a third and fourth credit is affected.

Based on other owner's project requirements, some of the LEED credits can or will be eliminated. For example, the owner may have decided that in no circumstance shall the new building be located on a property that is considered a brownfield site. This would eliminate LEED opportunity Sustainable Sites, Credit 3, Brownfield Redevelopment. This is a simple and obvious example, but as some credits are eliminated, it reduces possible impacts that might be imposed on other credits.

Additionally, some credits may be eliminated because they are impossible to achieve given certain circumstances. Reviewing the LEED house of quality roof, it can quickly be seen that nearly all of the Sustainable Sites credits will have positive or negative impacts on each other. The site selection is critical to many other credits and may be outside the control of the design team. Often an owner will have purchased a piece of property long before contacting an architect, design team, LEED Accredited Professional, or commissioning authority. The location and size of the property will immediately either eliminate the possibility of achieving or promote achievement of other credits. For example, if a site is selected in a rural setting, outside city limits perhaps, it is likely that it will eliminate the possibility of SSc2, Development Density and Community Connectivity, and SSc4.1, Public Transportation Access. What begins to happen is that the list of possible credits shortens and the need to maximize points of other credits heightens. Looking at the relationships area of the LEED house of quality, it can be seen that these numbers will have to increase, as fewer possibilities are available. Now many if not all of the relationships will be rated as a 9. It may be determined that it is impossible to achieve LEED Platinum or even Gold.

When this process is complete and the LEED certification level and credits to be attempted have been determined, the credits will be integrated with the owner's project requirements as a sub-requirement to the LEED level.

The complete and accurate owner's project requirements approved by the owner are then provided to the design team. Now the design team's basis of design response to the owner will be directed toward meeting those specific LEED credits as well as the other owner's project requirements. It is recommended that the full LEED house of quality (all of the credits are listed) be kept intact for future review of impacts caused by late design change proposals. The commissioning authority and owner can quickly go back to the roof and determine if a proposed change will negatively impact the original credits, thus providing a greater opportunity to determine if the change has value to the owner.

13.3.4 Design Phase

The typical commissioning process requires a minimum of three design reviews during the design development and construction document phases of the process. A primary goal is to verify the basis of design is being met by the proposed design. The body of the design phase matrix (Fig. 13.7) is used at each review to verify the

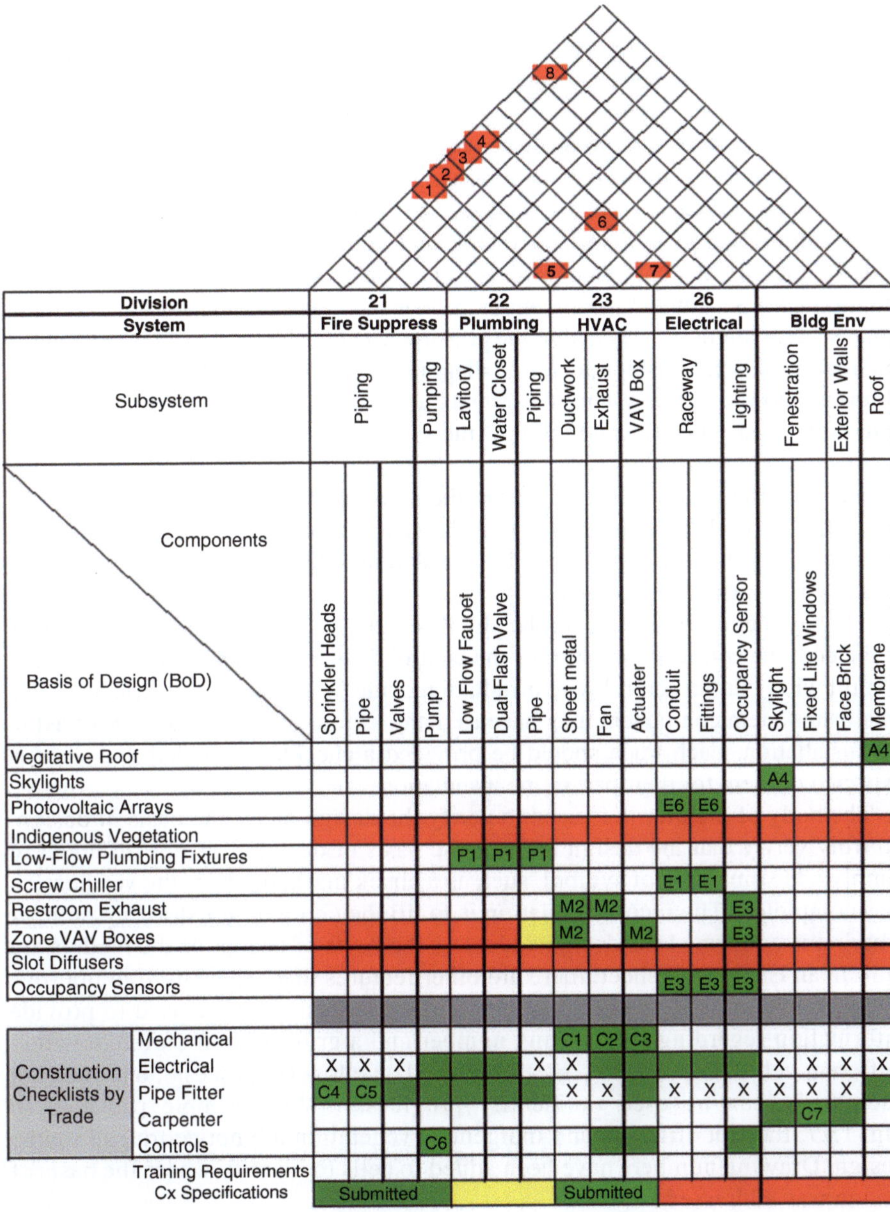

Fig. 13.7 Design phase

design meets the basis of design. The basis of design is now transferred from the predesign matrix to the whats section. This provides the connection between the predesign and design phases.

Beyond the owner's project requirements and basis of design, the commissioning process is typically concerned with the energy consuming systems within the building and generally verifies the design and installation of those systems and components. The commissioned systems would have been identified in the contract with the owner. Future analysis focuses on those systems. Typical systems include mechanical, electrical, plumbing, life safety, and building envelope. These systems are entered as hows in the technical response section. The goal is to enter these by division and break them down to system, subsystem, and component levels. The division is added to identify the requirements for commissioning specifications in the architect/engineer team's technical specifications. Subsystems and components are important because they are used to generate the construction checklists. The hows section is entered in the same fashion as a tree diagram is constructed, first by the division number, followed by the system, subsystems, and finally the components. The purpose is to provide easy identification of how the system is broken down and which trade is constructing the system.

Many of the trades will overlap on the systems, and now a checklist can be developed for each trade that has responsibility for the subsystem components. This ensures that each trade involved in the installation will be assigned a checklist. An example is a variable air volume (VAV) box. The sheet metal crew installs the box, which may include a reheat coil, the plumbers attach the piping and control valves to the coil, the controls crew installs the control devices and may run the necessary control wire, but the control wire installation could be left for the electricians. This small subsystem component may have as many as four different trades involved in the installation. Each trade should be provided a checklist to understand what is expected of them for their part of the installation.

The body of the matrix is used to verify the design meets the basis of design. To truly verify that the design has met the basis of design, it is not sufficient to simply add some type of symbol, such as a check mark, to claim the verification was completed. The recommendation is to fill the cell with the drawing number where it is confirmed the design addresses the basis of design. If the models are built in an Excel worksheet, there are other features that can be used to improve the documentation. For example, the comments feature can be used to provide information regarding the drawing number and a grid location on the drawing. Different cell colors, such as green (verified), yellow (in process), and red (not addressed), can be used to identify progress. In the example provided in Fig. 13.7, the slot diffusers and indigenous vegetation are not addressed by the design. Drawing numbers have been added to cells to identify where the basis of design was addressed.

Construction checklists are developed based on the system, subsystem, or components identified. For the focus of this research, these methods will not be discussed here. Templates are available from several sources, two of which are Building Commissioning Association (BCA) (www.bcxa.org) and The Building Commissioning Handbook (Heinz and Casault 2004). These templates can be modified to suit specific project parameters. The goal is to make certain each need for a checklist is addressed. Each checklist developed should have a unique identifier which will be added to the cell to provide a quick reference back to the checklist. If no checklist is required, an "X" should be placed in the cell so it is clear that nothing was overlooked.

The cells used for the commissioning specifications need to be checked off twice, once when the commissioning specifications are created and again when the commissioning specifications have been verified to have been added to the technical specifications. Again color coding is used, green (complete), yellow (in process), and red (not started), to identify progress. After the commissioning specifications have been submitted to the architect/engineer team, the word "submitted" is also added to the cell. Commissioning specifications require expertise in the development of specifications. The Building Commissioning Association (www.bcxa.org) also offers commissioning specification templates that can be used as a starting point. For the focus of this research, the methods to refine these templates will not be discussed here.

If the operations and maintenance staff is known and available at this phase of the project, the staff should be interviewed and the proposed systems discussed. Any necessary training for the operations and maintenance staff will be identified by comment in the cell below the proper system or subsystem. An example might be "one week on-site training provided by the boiler vendor" entered below the boiler subsystem. No systems requiring training have been added to the example matrix. If the operations and maintenance staff is not known during this phase, a conversation with the owner and the architect/engineer team can help identify possible needs. The training requirements are then added to the technical specification for bidding purposes.

The roof in this phase is again used to identify correlations among the hows, but since this is a review of the complete design, the consideration will be on potential construction coordination issues. Once these are identified the owner can discuss these issues with the design team so they can determine if there truly is an issue that requires design modifications. If no design changes are required, the owner should discuss the locations of possible coordination issues with the contractors once they are hired. These locations should be reviewed during site visits. The roof in Fig. 13.7 identifies a few possible coordination issues. Consecutive numbers are used in the cells along with the fill color of red to make identification easier. A separate numbered list, corresponding to those numbers in the roof, is created to provide a narrative describing in detail the possible issues.

13.3.5 Construction Phase

The construction phase matrix is illustrated in Fig. 13.8. Begin building the construction matrix by transferring the building systems, subsystems, and components from the design matrix to the whats area. Include the divisions as provided in the design phase.

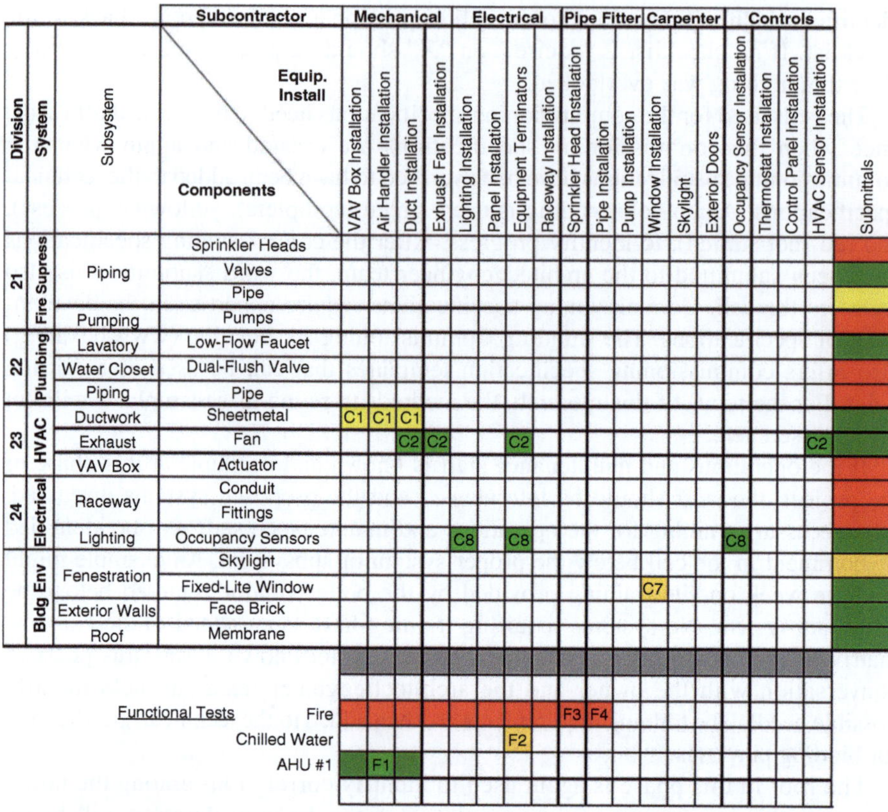

Fig. 13.8 Construction phase

The hows area will hold information regarding the equipment installations and is broken down by construction trade. Enter each trade which will be required to install the systems within the design.

The construction checklists developed in the design matrix can be transferred to the body of the matrix for use in linking the whats with the hows. This process identifies that each trade required for a subsystem installation has a checklist assigned. If any were overlooked in the previous matrix, they should be developed now.

The contractors have their own internal methods and quality control, and the commissioning authority has no authority to request that they change their process; therefore, the relationship matrix focuses on what is needed to verify that the contractors' methods are sufficient to meet the expectations of the owner. This is accomplished by developing construction checklists which will be filled out by each subcontractor during the installation process. The required checklists were identified and developed in the previous phase. The construction matrix is a means for tracking the verification process. Subcontractors will be given their respective checklists prior to beginning the construction process. Their input and approval will be necessary to ensure buy-in and completion. As the checklists are completed, this should be documented, and the completed checklists should be randomly sampled for accuracy by verifying during site visits that each item on the checklist has been completed properly. The suggestion is to mark the matrix to indicate which checklists have not been started, which are in process, and which are complete. Different symbols can be used or a cell color, such as green (complete and without issue), yellow (in process), red (not started), and orange (complete but with unresolved issues).

As the general contractor presents submittals to the design team and commissioning authority, the submittals must be confirmed to meet the owner's project requirements and final design before being checked off as verified. A suggestion is to house the submittals in electronic folders by division and provide a link to the folder and file for each submittal. This will provide for quick reference later. This can also be the basis of an electronic operations and maintenance manual and systems manual. The cells are color coded green (verified and approved), yellow (in process), and red (not started).

The area below the body matrix will be used to identify those equipment installations and systems which require functional tests and later for verification that the tests are complete and without issue. As the submittals come in from the contractors, and the equipment is approved, the functional tests can be developed. A unique identification for each functional test should be developed and entered into the appropriate cell at the bottom of the matrix. The sources mentioned previously can provide templates for the functional tests. For the focus of this research, the methods to refine these templates will not be discussed. Again, links can be provided to electronically store functional tests for quick reference and for use later in the systems manual. Color coding is again used to identify the status of the tests.

13.3.6 Operations and Maintenance Phase

The equipment installation entries from the construction phase are then transferred to the whats in the operations and maintenance matrix (Fig. 13.9). The hows are the documents that need to be collected and provided to the operations and maintenance staff at building occupancy. Documents to be tracked are functional tests, construction checklists, operations and maintenance manual, systems manual, and training documents. Color coding the relationships area is used to identify the status of the documents as they are provided by the contractor. The entire matrix is red at the beginning. Yellow represents those documents that are under review and green represents the approved documents that are in the commissioning authority's possession.

The operations and maintenance phase of the model is simply a checklist for gathering all of the necessary documentation for the operations and maintenance staff at turnover. Having the complete documentation allows the staff to maintain and operate the building at optimal performance. The information contained within

Subcontractor	Documentation Collected & Assembled / Equipment Installation	Function Test	Construction Checklist	Training	O&M Manual	Systems Manual
Mechanical	VAV Box Installation					
	Air Handler Installion					
	Duct Installation					
	Exhaust Fan Installation					
Electrical	Lighting Installation					
	Panel Installation					
	Equipment Termination					
	Raceway Installation					
Pipe Fitt	Sprinkler Head Installation					
	Pipe Installation					
	Pump Installation					
Carpent	Window Installation					
	Skylight Installation					
	Exterior Door Installation					
Controls	Occupancy Sensor Installation					
	Thermostat Installation					
	Control Panel Installation					
	HVAC Sensor Installation					

Fig. 13.9 O&M phase

this documentation provides a method for continued communication among the operations and maintenance staff and occupants. One example of an opportunity to make use of the owner's project requirements and basis of design documentation at building turnover is to develop a training session for the building occupants. Often the occupants are not aware of the efficiency requirements that have been incorporated into the design and construction. Field experience has shown that the efficiency standard that is most often noticed by occupants and the cause of many calls to maintenance is the control of the HVAC system. Most have experienced a cold or warm period in their work space and have attempted to adjust the thermostat to improve the conditions. If it is cold in the office, the thermostat is adjusted to a higher temperature and the expectation is that the space will become warmer. This is not an unreasonable expectation, but one that is often not realized. This may be caused by the HVAC control strategy. For example, ASHRAE Standard 55.1-2010, Thermal Environment for Human Comfort, informs the designer that most occupants will not be uncomfortable if the temperature is approximately 70–76 °F. The HVAC controls the set points for the temperature and are locked to this temperature range. Meaning if the temperature is already 76° and the thermostat is increased to a value above 76, the system will not respond by making the space warmer. This frustrates occupants that are feeling cold and frequently generates a call to maintenance with the complaint that the heat is not working. Other occupants in the area might appreciate the fact that the space is not getting any warmer. This scenario is also true for the opposite state when more cooling is desired.

If the owner's project requirements and basis of design are clear and provide information as to why this control strategy was pursued, the occupants can be informed of this when they move in. With a bit of understanding as to how the system is designed to operate, the complaints may be reduced. The few occupants who prefer a temperature outside the set points will know to dress accordingly to personally adjust for their own comfort before calling maintenance with a complaint. This is a result of simple communication that is typically not accomplished. The commissioning process is about quality, but quality requires communication, particularly through documentation. Detailed and accurate knowledge and information must be transferred through the project so each team has the best opportunity to deliver what is expected of them.

13.4 Conclusion

The traditional four-phase quality function deployment model and house of quality used in product development were modified to integrate with the commissioning and LEED processes. This new eco-innovative quality function deployment model provides a greater opportunity for the commissioning authority to ensure the owner realizes their sustainability and efficiency goals with the final building construction.

Emphasis was placed on improving communication and transferring knowledge among the different project teams and the project phases by linking the key commissioning activities. Early detection of potential issues in the predesign and design phases is critical to reducing costs associated with late issues and for achieving the owners expectations; therefore, three key modifications were made to the model in the early phases. The addition of a roof to the owner's project requirements, used to identify potential conflicts among the different owner's project requirements, is essential for bringing the basis of design into alignment with the owner's project requirements. The traditional house of quality already places a roof on the basis of design to identify potential conflicts among the design criteria, so if an item in the basis of design must change to accomplish owner's project requirements alignment, it can quickly be seen if that basis of design change will impact other design criteria. This ability to investigate potential conflicts must also be available if it is decided that an owner's project requirement must change to accomplish alignment.

With this additional roof a change in owner's project requirements can be investigated for how it will impact other owner's project requirements. Finally, for those projects seeking LEED certification, an additional house of quality was modified to analyze the LEED certification goals and potential credits to pursue. Critical to the LEED house of quality is the use of the roof to understand the design impacts each of the credits has on the others. This understanding allows the owner and commissioning authority to carefully specify which credits to attempt, to integrate with the owner's project requirements, and to provide to the design team.

The owner's project requirements and basis of design are considered living documents and are subject to change. Two activities in the commissioning process, "verify basis of design meets the owner's project requirements" and "verify construction checklist completeness," are typical points in a project when the owner's project requirements and basis of design are frequently required to change. Late in predesign or early in the design phase, the design team provides the basis of design to the owner for review. If the basis of design and owner's project requirements do not align well, it must be determined which document or both must change. To make an informed decision as to whether to allow the change, it must be analyzed for how it will affect the owner's project requirements.

This eco-innovative model provides a means to accurately define the owner's eco-goals and to link the design and construction activities to the owner's project requirements and LEED certification goals. This improved quality assurance methodology can drive improved long-term success of the building.

References

Abdul-Rahman H, Kwan CL, Woods PC (1999) Quality function deployment in construction design: application in low-cost housing design. Int J Qual Reliab Manag 16(6):591–605

Ahmed SM, Pui San L, Torbica ZM (2003) Use of quality function deployment in civil engineering capital project planning. J Constr Eng Manag 129(4):358–368

Alarcon LF, Mardones DA (1998) Improving the design-construction interface. In: Proceedings IGLC '98, Guaruja, Brazil

ASHRAE (2005) ASHRAE guideline 0-2005, the commissioning process. American Society of Heating, Refrigerating and Air-Conditioning Engineers, Inc., Atlanta

Barber K (2008) Commissioning documents: necessary evil. Consult Specif Eng 43(6):51–57

Cohen L (1995) Quality function deployment: how to make QFD work for you. Addison-Wesley, Reading

Delgado-Hernandez DJ, Bampton KE, Aspinwall E (2007) Quality function deployment in construction. Constr Manag Econ 25(2):597–609

Dikmen I, Birgonul MT, Kiziltas S (2005) Strategic use of quality function deployment (QFD) in the construction industry. Build Environ 40(2):245–255

Eldin N, Hikle V (2003) Pilot study of quality function deployment in construction projects. J Constr Eng Manag 129(3):314–329

Ellis R (2009a) Commissioning & energy conservation. Eng Syst 26(10):20

Ellis R (2009b) Commissioning and incomplete design documents. Eng Syst 26(5):16

Enke HJ (2010) Commissioning high performance buildings. ASHRAE J 52(1):12–14, 16, 18

Gargione LA (1999) Using quality function deployment (QFD) in the design phase of an apartment construction project. In: Proceedings ICLC-7'99, Berkley, CA

Heinz JA, Casault RB (2004) The building commissioning handbook. APPA, Alexandria

Kamara JM, Anumba CJ, Evbuomwan NFO (2000) Establishing and processing client requirements – a key aspect of concurrent engineering in construction. Eng Constr Archit Manag 7(1):15–28

Lee D, Arditi D (2006) Total quality performance of design/build firms using quality function deployment. J Constr Eng Manag 132(1):49–57

Lee D, Lim T, Arditi D (2009) Automated stochastic quality function deployment system for measuring the quality performance of design/build contractors. Autom Constr 18(2009):348–356

Mallon JC, Mulligan DE (1993) Quality function deployment – a system for meeting customers' needs. J Constr Eng Manag 119(3):516–531

Newsham GR, Mancini S, Birt BJ (2009) Do LEED-certified buildings save energy? Yes, but.... Energy Build 41(8):897–905

Pheng LS, Yeap L (2001) Quality function deployment in design/build projects. J Archit Eng 7(2):30–39

Tseng PC (2005) Commissioning sustainable buildings. ASHRAE J 47(9):S20–S24

Turner C, Frankel M (2008) Energy performance of LEED for new construction buildings. http://new.usgbc.org/resources/energy-performance-leed-new-construction. Accessed 12 Mar 2013

USGBC (2009) LEED reference guide for green building design and construction. US Green Building Council, Washington, DC

Yang Y, Wang S, Dulaimi M, Low S (2003) A fuzzy quality function deployment system for buildable design decision-makings. Autom Constr 12(4):381–393

Chapter 14
Light Island Ferries in Scandinavia: A Case of Radical Eco-innovation

Mette Mosgaard, Henrik Riisgaard, and Søren Kerndrup

Abstract This chapter shows how a radically new technology can be developed. Through a case study based on interviews and action research, this chapter deals with displacement island ferries in European waters where a radical innovation is made in the shift from steel designs to a lighter carbon-fiber composite alternative. A key characteristic for eco-innovation is that it combines techniques, practices, and knowledge across existing boundaries. Networking and collaboration therefore become important for creating ideas and implementing these in order to get the environmental innovations on the market. The analysis focuses on three main principles that together constitute the radical change in the technology, namely, (1) light construction inspired from yacht racing; (2) to leave ashore what is not needed at sea, also adopted from yacht racing; and (3) to make a modular design that makes the use of the ferries more flexible, which is adopted from the naval sector. The case study shows how new actors on the market create this radical innovation and build a network to support the solution. The new actors, even though they are to some degree competitors, have chosen to collaborate and to access the ferry sector, a sector that they have not previously targeted as their primary sector. The actors have experience with carbon composite technology and are not fixed by a production based on steel. This allows them to introduce this technology as a disruptive innovation that challenges and changes the way ferries are produced.

Keywords Eco-innovation • Radical innovation • Transformative technology • Island ferries • Innovation • Energy efficiency • Networks • Technology development • Carbon-fiber composite

---

M. Mosgaard (✉) • H. Riisgaard • S. Kerndrup
Department of Development and Planning, Aalborg University, Aalborg, Denmark
e-mail: mette@plan.aau.dk; henrik@plan.aau.dk; soeren@plan.aau.dk

S. Azevedo et al. (eds.), *Eco-Innovation and the Development of Business Models*, Greening of Industry Networks Studies 2, DOI 10.1007/978-3-319-05077-5_14,

14.1 Introduction

Transport is one of the major global contributors of greenhouse gasses. In 2010, the sector was estimated to contribute 14 % of the total anthropogenic greenhouse gas emissions in the world, and the emissions are still increasing (Love et al. 2010). To date, the main eco-innovation incentives with the purpose of reducing the greenhouse impact of transport in Europe have been on cars, trucks, and trains (Harvey 2013). Compared to these forms of transportation, maritime transport has a low level of greenhouse gas emissions per weight freight transported (Buhaug et al. 2009). Being a less polluting alternative might explain why the technological development in regard to energy efficiency has been limited. Nonetheless, maritime transport represents 3.3 % of the world's total CO_2 emissions (Buhaug et al. 2009).

While eco-innovations in the maritime sector have been few, those innovations that have taken place have mainly focused on large vessels and their reduction in fuel consumption by improved operation (Lindstad et al. 2011). This chapter does not address these large vessels but instead focuses on one of the less addressed technologies, namely, small island ferries. Opposed to the larger vessels, the small ferries have a high lightweight/deadweight ratio, meaning that the ferry itself stands for half of the total weight in operation. As the energy efficiency of ferries, to a large degree, is dependent on the weight of the water to be displaced, the energy consumption for small ferries is largely dependent on the weight of the empty vessel (Hjortberg 2012). This also means that radical innovations within the small island ferry sector call for including weight reductions.

Lightweight carbon composite constructions have, for decades, provided winning solutions to the navy, as well as to super yachts and extreme sports vessels. Research shows that the same solutions that enable extreme speed can give a 50 % fuel saving potential if applied in small island ferries in Denmark and abroad (Watson and Schmidt 2012).

To make the next generation of ferries greener, life cycle costs and environmental performance have to be reflected in tendering documents and contracts. This could be done, for example, by introducing weight-based financial incentive structures which make lighter solutions more competitive. Perhaps new, green business models are needed, where investments with potential energy savings are financed through the guarantied future savings.

Small island ferries in Denmark, on average, operate with a deficit. Primarily, they are owned and operated by municipalities as a part of securing an appropriate infrastructure to the small Danish islands. There are 70 Danish island ferries currently in operation, with an average age of 26 years. This means that many of them are to be replaced in the coming years (Krag and Trolle 2012). Due to this expected demand for new ferries, it is interesting to look into how ferries can be developed to have a radically lower environmental impact and to investigate how this radically new technology can become the future technology in the sector.

Therefore, the research question that is analyzed in this chapter is:

> How can a transformative technology, which constitutes a radical innovation in the ferry sector, be developed to become a part of the future ferry technology?

The focus is therefore an example of a more general intention to understand how a new technology can be developed into an attractive alternative to the existing technology. In this context, we understand technology as the combination of techniques, knowledge, products, and organization (Müller 2003).

14.2 Background

It is a challenge to develop and implement eco-innovation because eco-innovations often are complex innovation processes combining different forms of knowledge, technologies, and actors. Eco-innovations therefore often have a disruptive effect on the practices of innovation and use. In this chapter we first discuss eco-innovation and its disruptive effects and set up an analytical framework.

14.2.1 Eco-innovation and Radical Innovations

When discussing a new ferry technology as an eco-innovation, it is important to define what actually constitutes an eco-innovation. We apply Rennings' definition of eco-innovations, which states:

> Eco-innovations are all measures of relevant actors (firms, politicians, unions, associations, churches, private households) which;
>
> - develop new ideas, behavior, products and processes, apply or introduce them and
> - which contribute to a reduction of environmental burdens or to ecologically specified sustainability targets. (Rennings 2000)

In this chapter we focus on eco-innovation which has a disruptive effect. Disruptive eco-innovations have been defined by the EU's eco-innovation observatory as:

> Innovations that lead to shifts in a paradigm or in the functioning of an entire system are often referred to as **disruptive eco-innovations**. They can lead to reconfiguring entire markets, consumer behaviour and technological systems. Systemic changes resulting from such innovations can make some existing products or services redundant. (Eco-innovation Observatory 2011)

The 2011 briefing report warns that eco-innovations must be scrutinized to assure that they really have an improved environmental performance and do not just lead to problem shifting or environmentally negative rebound effects. One such way to scrutinize eco-innovations that is often suggested in EU environmental policy is to perform a life cycle assessment (LCA) that follows certain minimum requirements defined by the international standard ISO 14040. It is therefore important to integrate and use tools as life cycle assessment and life cycle costing as active instruments for constructing eco-innovation.

Eco-innovations can constitute both smaller and larger environmental improvements, but it is essential that the innovations actually lead to environmental improvements.

Table 14.1 Examples of eco-innovations in the ferry sector and their fuel saving potential

Technology	Fuel saving potential	Source
Sail → steam	Negative	
*Kappel tip-fin propeller	3–5 %	Hochkirch and Bertram (2010)
*Rudder bulb	2 %	Hochkirch and Bertram (2010)
*Pre-swirl duct (Mewis)	7–8 %	Mewis and Guiard (2011)
*Pre- and post-swirl fins	3–6 %	Hochkirch and Bertram (2010)
Adjusted hull design (small UK trawlers)	40–50 %	Rihan et al. (2010)
*Rudder resistance	2–8 %	Hochkirch and Bertram (2010)
Improved hull coatings	Up to 5 %	Stenzel et al. (2011)
Drag reduction through microbubbles of air on hull	10–15 %	Kumagai et al. (2010)
Sails and kites	10–35 %	Sidhartha and Kumar (2012)
Rotating Flettner cylinders	16 %	Traut et al. (2012)
10 % speed reduction	20–30 %	Corbett et al. (2009)
Steel → light carbon composite structures (small ferries)	50 %	Watson and Schmidt (2012)

In this chapter we consider an innovation that is introduced in the ferry sector, but is it actually a radical innovation?

A literature review shows that there are many types of eco-innovations in ship design, which differ in relation to technological radicalness and environmental effects. Often they are according to their energy saving potential.

Table 14.1 shows that many technologies exist that can reduce the fuel consumption of a ship. The fuel savings are not additive and, especially those innovations marked with *, cannot be mounted simultaneously as they are all addressing the same hydrodynamic elements around propellers and rudders.

As can be seen from the table, light carbon composite structures are among the most effective innovations with a large potential to reduce the environmental impacts related to fuel consumption.

But how did this radical innovation come into play in ferry development? The following section addresses the case of the chapter, namely, the Eco Island Ferry project.

14.2.2 Theoretical Framework for Developing Transformative Eco-innovations

The purpose of this section is to develop a framework for understanding the type of transformative eco-innovations that forms the Eco Island project, eco-innovations that are developed within a niche of actors with limited institutional, economic, and technological resources.

Eco-innovations seem to have an increasing influence in many industries due to the environmental challenges that the industries, the consumers, and the society as a whole are facing. There is a growing recognition of eco-innovations as a means to limit increasing environmental impacts, but each innovation needs to have a characteristic that breaks the dominant ways of developing and improving products, processes, and activities. This is increasingly realized through the development of end of pipe solutions, product technologies, and life cycle thinking to develop more transformative and disruptive solutions (Scrase et al. 2009; Smith et al. 2010). As a supplement to these perspectives, we emphasize a new type of eco-innovations, and the way they are developed, in order to contribute to the understanding of how eco-innovations and their dynamic development can be understood through a combination of innovation theory (Hargadon 2002, 2003), network theory (Håkansson et al. 2009), and institutional theory (Hargadon and Douglas 2001). By doing this we contribute to an understanding of how it is possible for actors and organizations to develop radical new solutions through combining activities, knowledge, and actors, solutions that break the dominant (existing) way of developing technologies. This understanding allows us to contribute with policy suggestions that facilitate eco-innovations.

14.2.2.1 Eco-innovations as Transformative Innovations

The terms that are often applied in the understanding of the potential environmental improvements created by innovations are incremental and radical innovations. An alternative approach is to distinguish between sustaining innovations and disruptive innovations (Scrase et al. 2009).

The distinction between incremental and radical innovations takes a point of departure in the level of change. Incremental innovations reduce the environmental impact, but are based on existing knowledge, technology, behavior, and organizational framework (Conway and Steward 2009). Radical innovations, on the other hand, are based on the application of new knowledge, technology, and organizational framework in a way that interrupts the existing practices in the sector (Conway and Steward 2009). When the interruption occurs, the concept of technological development is central to the development of new products, new processes, and new business models.

Another approach is to distinguish between sustaining and disruptive innovations (Christensen 2013). A disruptive innovation helps create a new market and value network. It is called disruptive as it disrupts an existing market and value network by improving a product or service in ways not expected by the market (Christensen 2013). This means that disruptive innovations change the organizational and institutional framework on the market and are introduced by actors and institutions that are not part of the dominant market actors. In this approach, innovations are not seen as a narrow technical solution but as a change in the framework for innovations. Sustaining innovations, on the other hand, aim to sustain the existing framework for development of products and markets. Sustaining innovations can be both incremental and radical technological changes, but the crucial point is that they sustain the dominant organizations and institutions on the market.

We have chosen to combine these two approaches as we look at transformative innovations as innovations that contribute to changing the existing concepts, technologies, and institutions so that the way we understand and develop ferries is changed. As for the theory of disruptive innovations, we understand transformative innovations as innovations that can be both radical and incremental, as the important criterion is not the radicalness in the potential change in environmental impacts but the way the innovation changes the way ferries are designed, built, and operated (Scrase et al. 2009). This means changes that revise the interplay between organizations, institutions, and regulation.

14.2.2.2 How Are Transformative Innovations Established?

Establishing transformative innovations is an organizational challenge as they disrupt the dominating knowledge, technology, and actors on the market. This disruption is created by combining knowledge, technology, and actors across existing organizations and institutions that can facilitate the development and implementation of the technology.

Hargadon (2002, 2003) has described those processes as consisting of the following two phases. The first phase is the idea generating and conceptual phase where actors, technologies, and ideas are combined across different organizations, sectors, and knowledge domains. It is a decisive point that transformative changes (which Hargadon describes as breakthroughs) often consist of well-known knowledge that is already present in the organizations involved. New and transformative technology then relates to the new constellations of knowledge, techniques, and organizational structure that facilitate the development of new solutions that, in a crucial manner, disrupts the existing ways of combining knowledge, techniques, and organizations. Transformative technologies have a potential for changing our perceptions of how to solve the environmental challenges that we face. There are however some challenges for these potentials to come into play, as the technologies challenge the existing solutions and interests, and might be counteracted by the existing actors. It is therefore important to build a platform or create a network which can support the development and implementation of new technological solutions. This means that it is important to "create meaningful activities and valuable links between previously unconnected people, ideas and things" (Hargadon 2003). Therefore, a second phase of network building is necessary to build a network that can support the transformative technological, economic, and environmental practices and interest. If this network is not built, the transformative technologies will be counteracted by the dominant concepts and their actors and institutions – and will fail in its implementation. The institutions are very important because the actors' behavior is institutionalized in the way they are practicing economy, technology, and environment, especially their concerns about the possibilities and limitations of the new technologies where these are giving value and meaning in relation to practices based on the "old," existing technology. One group type of limitations which are also an important part is regulations as these often become obstacles to the new forms for disruptive solutions.

The development of transformative technologies as a combination of different ideas, technologies, and actors is seen as an important part of understanding how the concept of carbon-fiber ferries is developed here in this case. Three actors combine their activities and competences related to design and construction of ships and couple these to their knowledge from yacht racing and maritime ships.

In the construction of networks that facilitate the possibilities for implementing the carbon concepts in the ferry sector, both institutional capacity building and the establishment of new networks are needed. The actors construct activities, networks, and resources that are crucial for innovation development from a concept or idea through to an actual innovation.

14.3 Case Study: The Eco Island Ferry Project

The "Eco Island Ferry" project includes a number of partners from the project industry, government agencies from both Denmark and Sweden and research institutions. The project is a noncommercial initiative built on open-source principles with all project materials available on a web page. The aim of the project is to develop and demonstrate opportunities for new energy-efficient ship technologies through the construction and building of vessels in lightweight materials. The project period is 2010–2013 (Hjortberg et al. 2012).

14.3.1 The New Technology

The project has had its focal point around designing a new ferry for the Tunø route in Denmark; see Fig. 14.1. The ferry is designed as an "all things being equal" alternative which makes it possible to adapt and to measure the impact compared to the existing institutional set-up. The aim is to demonstrate that the energy efficiency can be improved significantly and, thereby, the new technology constitutes a significantly better alternative both in terms of environmental impact and cost. The capacity is the same as for the existing ferry: It has the same type of gangway and has a capacity of 200 passengers, 3.075 t of cargo, and 6 cars. The calculations are made based on the same timetable as well, the ferry being in operation 4 h a day with a speed of 9.5 kts.

The "all things being equal" alternative allows for a comparison between the new ferry technology and the traditional steel techniques. The comparison is performed by the use of life cycle assessment and life cycle cost analysis.

The key differences between the new ferry and the reference ferry are presented below:

In Table 14.2, the radically new elements of this technology are indicated, and the main point is that the displacement is much lower due to the lighter weight, and this means that the energy consumption is significantly reduced. Another

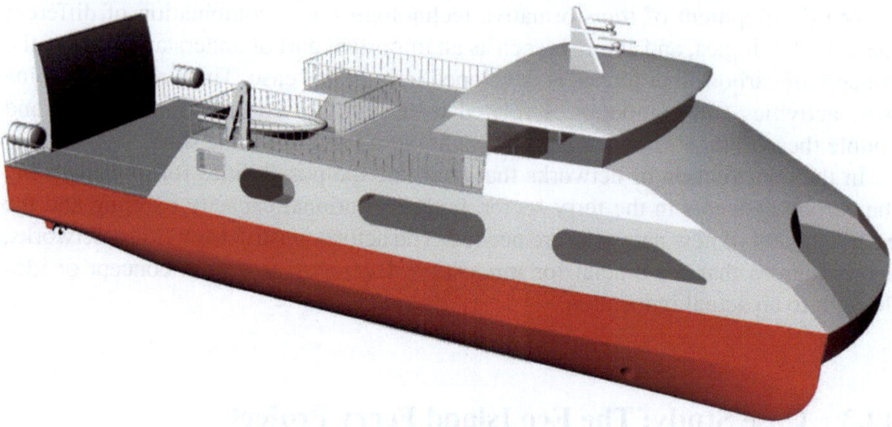

Fig. 14.1 Sketch of the Eco Island Ferry (Hjortberg et al. 2012)

Table 14.2 Key differences between the carbon-fiber ferry and the reference ferry (Hjortberg et al. 2012; Mosgaard et al. 2014)

	Existing steel ferry	New carbon-fiber ferry
Hull design	Single hull	Catamaran
Lightweight	250 t	77 t
Ballast	34 t	0 t
Displacement	340 t	125 t
Engines	2×294 kW	2×110 kW
Fuel consumption	100 l/h	53.1 l/h

environmental improvement is the reduced use of zinc anodes that are traditionally used to protect ships from corrosion. Furthermore, the reduced displacement also reduces the wet surface which again reduces the environmental impacts of anti-fouling paint.

Based on a life cycle assessment that compares the alternative ferry design with the reference ferry, it is concluded that the Eco Island Ferry has a lower impact in all environmental impact categories, and for most impact categories, the potential impact is around 50 % of the impacts of the traditional steel-based ferry (Watson and Schmidt 2012).

A life cycle cost analysis shows that the production and the documentation are more expensive for the carbon-fiber ferry, but it has lower operational and maintenance costs (Lindqvist 2012). Salaries are also an important expenditure, but they are the same for the two ferries. The break-even point is calculated to 8.6 years after operations start – then the additional investment in the carbon-fiber ferry is paid back, but the expected lifetime is more than 30 years. This means that the carbon-fiber ferry has a lower expected life cycle cost than the reference ferry (Lindqvist 2012).

The institutional set-up had very important influence/effect on the possibilities for creating a new technology. One of the obstacles that the project has concentrated their work on is fire and safety measures. Due to previous fires on ferries, such as the Scandinavian Star in 1990 with 158 fatalities, fire and safety is not only an important factor for those buying the ferries, but strict safety regulations are also included in the European ferry directive. However, the regulation is designed for ferries made of steel, and this has shown to be a challenge when designing ferries of carbon fiber. As the mechanical engineer that has been one of the central actors of the design team puts it:

> The biggest challenge has been the risk of fire, because the tradition for getting anything that can burn out of the ferries goes back to the 50'ies. There are still countries that will not consider carbon-fibre ferries at all. Then we can say: Do you build ferries of steel? That can't float, that'll sink! I think that it is an absurd argument for not having carbon-fibre ferries. (Sørensen 2013)

Being challenged a bit on this argument, he continues:

> It is not a matter of whether the ship can burn or not. First of all the fire should never occur. The challenge is what we can do to secure that the fire never occurs, and if it does then how will we fight it? (Sørensen 2013)

He actually thinks that the present regulation is too strict for small island ferries:

> You can fly in a plane 10 km above the ground that is made of 80 % carbon-fiber. How can we accept this safety level and at the same time set so strict demands for a 5 minutes ferry ride to an island? (Sørensen 2013)

The challenge is that the European Ferry Directive stresses in the introduction that it only applies to vessels and that "they shall be constructed of steel or other equivalent material" (EC Directive 2009/45/EC). This means that the carbon-fiber ferries cannot be constructed according to this directive and therefore are instead designed according to national legislation with a national approval. This is inexpensive; however, it impedes easy reexportation of the ferries to other EU countries (Hjortberg et al. 2012).

Another possibility is to build according to the requirements of the SOLAS Convention (International Convention for the Safety of Life at Sea). This is a more expensive solution as the rules are intended for ferries in unrestricted service (Hjortberg et al. 2012). SP Technical Research Institute of Sweden has been responsible for applying a "rule 17 analysis." This is a procedure to analyze and assure safety compliance when deviating from traditional ship design according to the SOLAS convention (Hjortberg et al. 2012). Having an open-source demonstration project where materials and analyses are publicly available, it is hoped that it can influence the general rule making in the EU. One possibility is to make the future directives more in line with the SOLAS convention, opening for a faster dissemination of small light ferries with reduced environmental impact.

14.3.2 The Actors That Are Developing the New Technology

The project was initiated by three small- and medium-sized private actors working with design and construction of ships. Their main experiences lie within yachts, navy vessels, industrial ships, and to some degree fishing boats, but traditionally not ferries. They do, however, have a lot of experience with lightweight materials for ship construction.

Niels Hjørnet Yacht Design is a Danish, one-man business. The owner is an active yachtsman in his spare time. Among other activities, he designs yachts and optimizes sailboats of different varieties, from classic yachts to modern racers. In the Eco Island Ferry project, Niels Hjørnet Yacht Design has calculated the drag, developed the hull, and designed the propulsion.

Danish Yachts is a Danish company with 100 employees. The actor involved in the project is a mechanical engineer that has previous experience from the naval industry and is also an active yachtsman in his spare time. The company is specialized in building modern motor yachts and sailing yachts. The main task of the mechanical engineer in the project has been the layout of the ship. Besides this he has worked with the interpretation of rules and regulations.

Coriolis AB is a Swedish, one-man business. The owner is also an active yachtsman, and he has experience with building vessels in lightweight materials. The business supports both ship owners and shipyards with technical advice on new builds as well as with retrofit projects. They also advise in matters related to rules and regulations. In the Eco Island project, Coriolis has performed calculations and design regarding the stability of the ferry but also other ship design calculations. Besides this, Coriolis has performed a market analysis.

These three actors constitute the core of the network, but other actors have been involved as well. One actor is Aalborg University that has had a role in applying for funds to cofinance the project and as a research partner participant in meetings and conferences. The life cycle assessment and life cycle cost assessment of the ferries are also made by private actors. Both Danish and Swedish maritime authorities have participated in meetings with the aim to comment on and discuss the rules and possible interpretations of these related to carbon-fiber ferries.

Together these actors have had the knowledge and experience to be able to introduce a radically new ferry technology, by applying their knowledge from other industries. Eco-innovations often address incremental changes in product design and to a lesser degree radical innovations. The various experiences from other sectors and also their personal interests in sailing have facilitated the possibility for transferring technologies from other sectors to this ferry project and thereby facilitating a radical innovation.

14.3.3 Method

The empirical basis of this chapter is a case study of a development project including a number of actors, as presented in the introduction. The analysis focuses on how radical innovations can be implemented in the ferry sector through the introduction of a transformative technology. The focus is therefore on how the different actors play a role in introducing this transformative technology and on the interaction with other actors in order to introduce this technology on the market in the future. This demands a change in the way ferries currently are evaluated by customers and by maritime authorities. This focus on a transformative technology only represents a part of the project, but it is chosen with the aim to focus on the research question.

The transformative technology introduced by the project constitutes what Flyvbjerg defines as an extreme case (Flyvbjerg 2001). An extreme case is applied to obtain unusual information which can be special in a more closely defined sense (Flyvbjerg 2001). In this case study, the case is extreme in two ways. First of all, it constitutes a radical change in the technology applied in the ferry sector with a large environmental potential. This might be explained by the rigidity in the ferry sector over time that is now challenged by introducing new actors. Secondly, it represents a strongly regulated sector that counteracts eco-innovations.

As one of the authors has been involved in the Eco Island Ferry project from the beginning, the methodology of the case study is based on an action research approach. Action research has the goal to solving practical problems in a real world through collaboration between research institutions and industry (Denzin and Lincoln 2000).

The action research approach has made it possible to gain thorough knowledge of the complex conditions that characterize the age-long collaboration to develop the ferry concept. Due to the close collaboration between the researcher and the rest of the actors in the project, trust has been built between the actors that allows access to internal documents. A bias might be introduced when the researchers are part of the project, a bias that would not occur if we were objective third-party observers. To counteract this potential bias, two researchers that have not participated in the project as such have conducted the data collection and co-authored the chapter.

We follow a case study research approach and therefore include multiple sources of evidence for the data collection and analysis (Eisenhardt 1989; Yin 2009). The main data collection has happened through participation in the project: in meetings, written correspondence, and conferences. This data is supplemented with semi-structured interviews. The main elements in the interview guide are structured by the theoretical framework and are thus: (a) personal background of the interviewee, (b) characteristics of the new technology, (c) the basic design principles applied, and (d) the future potential of the new technology.

In the network, the plurality of previous experiences of the different actors was used as a way to supplement each other and develop a new technology. One of the challenges in action research is that "the researcher should not subject the research population to embarrassment, harm or other material disadvantage" (Lewis et al. 2009). To counteract this potential problem, the results of the analysis were presented to the central actors involved. A potential risk in action research is that the researchers become trapped in the action and therefore are not able to make objective reflections (Argyris et al. 1985). The way this is counteracted is by choosing a case study design that produces context-dependent knowledge but relies on numerous other data sources that are of a more objective character than just the perception of the researchers. Yin (2009) recommends multiple data sources to establish as much internal and external validity as possible.

One of the main data sources to document the technology developed is document analyses of project material such as minutes from meetings, reports, internal communication, as well as presentations from conferences. We read closely through the material while having the theoretical framework in mind. We thereby identified interesting perspectives related to radical innovations and the transformative technology. To create valid results based on the interviews, the interviewer was explicit about his or her own assumptions before conducting the interviews. This makes it possible for the respondents to either verify or reject the assumptions. To secure as much objectivity as possible in the interviews, they were prepared and conducted by a researcher not involved in the project.

Based on the action research approach and the multiple data sources, the analysis is presented in the following section.

14.4 Case Study Analysis

In this section we present the preconditions for implementing a radically new technology in the ferry sector; how can this disruptive innovation be seen in practice? Afterwards the three main principles for changing the technology are identified and analyzed.

There is radically higher energy efficiency in changing the way in which ferries are constructed by using lightweight construction materials. The necessary technology already exists in other maritime sectors (yachting and the navy), and there is a demand and a need for energy-efficient ferries by the customers.

The radical element in this solution not only addresses the change in technology itself, but the entire concept of how to construct a ferry and the optimization factors, in this regard, are revised. This also means that the ferries are developed first and foremost as island ferries that have limited size and travel short distances in relatively calm waters, as the ferries are less stable at sea than the traditional more heavy constructions. As shown by Clayton Christensen (2001), new technologies can often enter a niche in a market and gain a competitive advantage within this niche. This might be the case with the carbon-fiber ferries as the technology demands other construction facilities than those that are present at the existing traditional shipyards involved in ferry production.

In this process, a number of actors have bonded by collaborating on common activities and adapting their resources (Mosgaard et al. 2014), and common for these actors is that they have working experience from sectors where carbon composites are applied for ships, namely, the naval industry and yacht construction and design. The main actors are two micro companies and one small company, and from each of the companies, one enthusiastic actor has been the main driver of the project. Without these three enthusiasts, the project would neither have

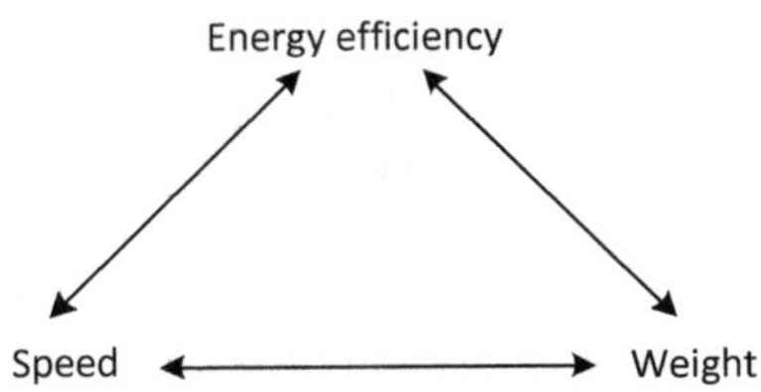

Fig. 14.2 Three interlinked optimization factors for ferry construction. In this case (Eco Island Ferry) energy efficiency and lightweight have been optimized, whereas the speed has been chosen so that it matches the reference ferry

been initiated nor completed. Besides these designers and constructors, a number of research institutions have worked with documenting the economic and environmental potential in lightweight ferries seen in a life cycle perspective, as well as conducting a "rule 17 analysis": a procedure to analyze and assure safety compliance when deviating from traditional ship design according to the SOLAS convention. To promote the idea of the lightweight ferries to both the potential customers and financial institutions, more communicative actors have also been involved (Mosgaard et al. 2014).

In this process, a "them versus us" approach to discussing the technology has occurred, where "them" are other shipyards and actors that design and produce ferries in steel and "us" as those that try to change this by applying lightweight techniques.

The light ferry is constructed based on the fundamental understanding of the link between speed, energy efficiency, and weight; see Fig. 14.2. To increase the energy efficiency, it is possible to either build a lighter ferry or reduce the speed, or if you wish to increase the speed, you can either make the ferry less energy efficient or reduce the weight.

After a long and institutionalized technology focus on steel (Mosgaard et al. 2014), it now seems to be possible to have a radical shift in technology as the customers have increased focus on the life cycle costs of running a ferry as well as the impact on the climate, both of which call for applying lighter technologies.

14.4.1 Disruptive Innovation in Practice

The case is an example of a transformative technology with a disruption of the fundamental way that ferries are constructed, since the focus shifts from stability and few crew requirements to lightweight construction and energy efficiency. As a part of this, there is a disruption with the materials that are used as well, shifting from steel to carbon fiber.

As Christensen has shown for other sectors (Christensen 2001), a disruptive technology is often introduced by new actors on the market, as the existing actors do not have the incentive to introduce it. In the ferry sector, this means that the existing shipyards that focus on steel constructions are rationally driven by their existing markets and the demands from their main customers, namely, to produce ferries of steel. When Eco Island Ferry was established as a project, the intension was to make a disruptive technology with new qualities compared to the old ferries, as this is what constitutes a market potential in the future.

Having this in mind, the new technology is not necessarily a "better solution" if it is evaluated on the premises of the existing ferries, as they might be less stable or brisker at sea. This does not mean that they are less safe, but that the passengers could experience ferries that move more at sea or that the operation of the ferries would demand a more alert staff. These potential drawbacks related to material choice are counteracted by switching to an alternative catamaran design. What these lighter ferries then can offer is a reduced environmental impact and better economy in a life cycle perspective.

Unfortunately, the existing institutional set-up will not automatically favor ferries that have lower life cycle cost. The reason is that the traditional way of organizing the public procurement process of ferries is by publishing a tender document that holds the technical specifications and then deciding on the lowest bid – meaning the lowest initial investment that supplies the requested functionality. If the tendering procedure instead would include a weight-related penalty – or the opposite, an innovation-spurring bonus payment for lighter-than-expected vessels – the bid would reflect the total cost of ownership if the incentive corresponded to the calculated life cycle costs. The Eco Island Ferry project team addressed also these barriers by proposing alternative business models including leasing.

One of the explanations of why this disruption can occur is that the actors have not previously built ferries, but have had a main focus on yachts. This means that they are not dependent on steel as their main construction material as other shipyards are. In an existing shipyard, this change in technology is less likely to occur as it means that most of the production facilities have to be refurbished. Additionally, both designers and employees would require education in working with the new material. Furthermore, the fact that the designers have previous experiences for applying carbon fiber as a material for building ships makes it easier for them to come up with the idea for applying it on ferries as well.

The actors involved in this innovation process seek confrontation with the existing technologies, as it can give them a potential part of the market share in the ferry sector. This means that a radical change is necessary, as small incremental changes in the technologies would not provide them with arguments for implementing new technologies.

Below a sketch of the iterative process of designing a ferry is presented, and it is shown how the designers are influenced by principles from previous experiences in related sectors.

The main difference between the innovation process sketched in Fig. 14.3 and traditional ferry innovation processes is the appliance of three new concepts. The first concept is lightweight construction that is applied with inspiration from yacht racing, and the second is to leave ashore what is not needed at sea, also adopted from yacht racing. The third concept is to make a modular design that makes the use of the ferries more flexible, and this is adapted from the naval sector.

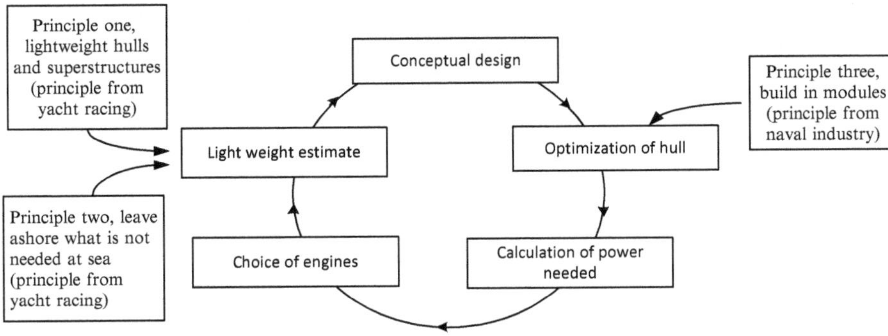

Fig. 14.3 The interactive process of designing a ferry, where inputs from other sectors are illustrated in three external boxes. The development cycle is common for the development of other types of ferries as well

For the actors to put these principles into play in the ferry innovation process, it is essential that they not only know these principles but that they also act upon them whenever choices in the design are made. This is possible as the three main actors not only agree on the principles but also have the previous experience to actually implement them in practice. Their main driver for developing a new lightweight ferry is primarily internal, based on their own previous experiences. This is in contrast to other, more commercial, innovation projects where it is the specifications from the customers that constitute the development framework.

The background and implications of these three principles are analyzed below.

14.4.2 Principle One, Lightweight Hulls and Superstructures

Lightweight structures are the technology that initiated the radical changes in the ferry designs in the first place. Previously safety, stability, and maneuverability have been the main optimization factors, and this means that the weight of the ferries has not been central in the design of the ferries.

Asked why there has been little focus on energy consumption in relation to ferries, the naval architect answers:

> I do not think that they have had their eyes open for the triangle of interlinked factors; weight, speed and energy consumption; like it has been the case in other sectors like the formula one racers, flights and yachts race. Even in the car sector it has been an issue. But it has not been a focus area in the ferry business. Now, demands to reduce the pollution are starting to appear in this sector as well, and how can we manage that? One of the solutions that occur is to make the light weight of the ferries lower than it is the case today. (Hjørnet 2013)

But this technology is actually not new, since it has been applied in the naval industry and race yacht sector for many years. Asked why the combination of lightweight fiber constructions and ferries comes into play now, when it is a well-known technology, the engineer answers:

> Due to economy with rising fuel prices and global warming, it depends on who you ask. (Sørensen 2013)

Asked the same question, the naval architect answers:

> I cannot give you an exact answer but… The price of fuel influences it, it has not been a real problem to consume oil previously… Well, it was just never a real issue before…. (Hjørnet 2013)

One of the points to deduct from this is that the technology does not "sell itself." It is the changed fuel prices and thereby an increased focus on the operation expenses from the customers that gives a potential for implementing the lighter technology. In other words there has to be a market that demands the new technologies.

Another way to reduce the lightweight, besides making a lighter construction, is to leave ashore what is not needed at sea, the second principle.

14.4.3 Principle Two, Leave Ashore What Is Not Needed at Sea

Somehow it seems obvious not to bring things aboard a ferry, and leave them there, if they are not needed on the trip. Nonetheless, this is to some degree practiced today. Asked how this new principle of reducing the weight by removing the items that are not needed at sea has originated, the technician answers:

> It is a really good question, I think that it has its origin in our daily routines at the shipyard; we had a boat built and ready for delivery just before Christmas but then it was 7 tonnes too heavy when it was launched. That is something you learn from. (Sørensen 2013)

Asked if it is possible that his experience from yacht racing has also influenced this, the naval architect answers:

> Yes! We never bring anything that is not necessary. Sometimes we even bring ashore loose seat cushions. (Hjørnet 2013)

Asked to give some examples related to ferries, the naval architect (and engineer) answers:

> It does not make sense to sail with 58 kg of paint, 115 kg of detergents and… I could keep on giving examples like that. It costs money. But the biggest issue for now is ballast water; it does not make sense to sail with water on board. (Hjørnet 2013)

And this issue regarding ballast water has shown to be an important point. Ferry operators tend to sail with quite a lot of ballast water permanently in the tanks, even when this is not necessary for stability or safety purposes. This means that the energy consumption ends up being higher than necessary.

The new focus on lightweight can have even further applications than ballast water, paint, detergents, and other similar items, as the entire ferry infrastructure design can be revised. For example, if the landing stage is only needed ashore, then why is it a part of the ferry design instead of a part of the harbor design?

To secure that the items actually needed at sea can be brought on board, a modular design can be a solution.

14.4.4 Principle Three, Build in Modules

The third principle is to build different modules for the ferries that they can choose to bring on board when they are needed. One incentive for this modular approach is to reduce the weight of the ferry, and another is that some modules can be shared among several ferries, thereby reducing the expenses.

Asked if it is the engineers' naval experiences that have led to the idea of building in modules, the answer is:

> This might be something that has waited in the back of the head [since my naval experiences, eds.], since they could not bring it all they had three stations in the ships where we could put different containers with different functions such as missile launchers. (Sørensen 2013)

If a ferry is optimized in regard to its weight and a specific ferry route, it means that it might not carry enough energy capacity to, e.g., sail to a shipyard for renovation. This can be solved by having a mobile pack that can be placed aboard the ferry once or twice a year if it is needed. This is a cheaper solution for having the needed capacity, as several ferries can share one of these packs, but also makes the lightweight of the ferry even lower.

Another example is a heating system. During the winter season in Scandinavia, it is sometimes necessary to deice the deck as well as heat up, e.g., toilets and the bridge. An oil burner can be chosen as a module that is brought aboard when it is needed, but left ashore during the warmer season.

Some modules can be specific for the individual ferry and others can be shared. The modular approach can be a way of making the ferries more flexible so that they can operate on different service routes if necessary. Still, this approach constitutes a disruption in the way ferries are constructed, as the tradition leans to the solution to have everything on board that you might need at some point.

14.4.5 Why Radical Innovation Occurs Through Transformative Processes

As defined in the introduction and the theoretical framework, a radical change is something that makes drastic changes compared to the small daily steps that constitute incremental changes. The radical changes in this innovation process lie especially within the energy efficiency potential that relates to the lightweight ferries. A transformative process relates to a disruption of the dominating knowledge, technology, and actors in a given sector, and for this case study, this relates to the three principles presented above. But why is it necessary for these radical changes to occur through transformative processes?

First of all, the existing shipyards that produce ferries do not have an incentive to change their technology; they have a well-known concept that the customers ask for. This means that there is no need to make radical changes in the technology. The transformation then occurs because new actors enter the market bringing new ideas and concepts for what constitutes an appropriate technology. These actors bring their knowledge, organization, and techniques from other sectors and are able to combine them in a way that makes it possible to develop an idea for radical innovation. This equals the first step in Hargadon's innovation model (Hargadon 2002, 2003).

Secondly, the market is changing as the small island ferry operators (namely, the municipalities) have had an increasing focus on energy consumption and climate impacts (Elkjær 2011; Østergaard et al. 2010), and this opens a possibility to create a niche in the market, where the customers actually ask for energy-efficient ferries (Mosgaard et al. 2014).

Finally the Eco Island project has introduced a new way of organizing the innovation process as several actors are brought together in a network with the purpose of making a transformative change in the ferry technology. The approach of a noncommercial project with open access to the data has facilitated the radical changes, as several actors have not only shown interest but also participated with knowledge and ideas for the ferries. This related to the second phase of Hargadon's innovation model (Hargadon 2003), namely, to establish a network where new, transformative concepts can actually be developed into transformative innovations.

14.5 Conclusion and Policy Recommendations

The existing technology of light ferries can be seen as a niche production that mainly addresses small island ferries that operate over short distances. If the technology is developed further in the future, this might change and the technology might enter the market for larger vessels as well.

The analysis shows how a transformative technology can be innovated as an alternative to traditional steel ferries. The eco-innovation is only possible through a disruptive innovation that challenges and changes the organizational, economic, and institutional basis for developing ferries.

The organization is challenged by getting new actors into the innovation process, actors that bring experiences from other sectors and are not limited by the technologies used at the existing shipyards. The economic basis is challenged both by the open-source, noncommercial approach taken in the project and by the financing of the ferries where the investment costs are significantly higher than for a traditional ferry but operation costs are lower. The institutional challenges are addressed by involving research institutions and authorities as active partners when addressing the implications this new technology has for the regulation and general perception of safety of ferries.

One of the interesting perspectives for the innovation process is that the actors to a high degree work not only against but also with other actors in order to change the institutional settings for ferries. This means that in practice they need to collaborate, for example, with the authorities to gain the safety approval for the ferries. This collaboration among new actors and the established institutional setting is what facilitates the possibility for gaining a niche where the lightweight technology can enter the market. Three central actors have driven the project, and without their initiative and enthusiasm, the carbon-fiber technology would not have the potential of becoming a transformative technology in the ferry sector.

It is still too soon to say whether this innovation will actually be disruptive for the ferry sector as this is an extreme case and there are still barriers to overcome before the technology is implemented, but without "outsiders" that apply new technologies, potential disruptions like this would not occur.

Due to the strict top-down regulation regarding safety in this sector, technological developments have stagnated and a technological backlog has formed that creates a huge environmental potential in the new technology. This is one of the main opportunities for the technology to be implemented, as it has both environmental and economic potential.

The direct implications for policymakers are to influence the development of the EU ferry directive so as not to keep specific demands for steel-like construction of ferries but to open towards other materials that can comply with the same functional demands.

Other policy barriers are identified in the tendering legislation as well as tendering practice.

The first policy recommendation to facilitate the introduction of light ferries is to apply a financing approach where potential future energy savings, and thereby a reduced cost for operation of the ferries, are used to finance the investment in the new technology. This might be done by providing loans for more expensive ferries if they have lower energy consumption. In general this means that the life cycle cost is taken into consideration and not just the initial investment in the ferry itself.

Another policy recommendation relates to the European ferry directive which in its present form impedes the development of energy-efficient light ferries. A suggested adjustment is to provide a possibility to apply alternative materials as it is the case for the SOLAS Convention (Hjortberg et al. 2012).

The International Maritime Organization (IMO) has initiated requirements for an Energy Efficiency Index for new builds and for a Ship Energy Efficiency Management Plan for all ships. These are important steps towards more energy-efficient ships, but it is also important to address the crew that operates and maintains the ship. To support "energy management" through regulation can, e.g., change the practice of sailing with ballast water when it is not necessary.

The Eco Island Ferry project was founded as an open innovation project, and we recommend continuing the support of demonstration projects that facilitate collaboration between relevant actors. This can be one way to involve existing knowledge among various actors and thereby facilitate future innovations, but it is important that these projects are open innovation projects so that the knowledge becomes accessible to all interested parties.

This type of project has wider applications by providing examples for others on how private companies can drive an innovation process and facilitate a development towards a radical innovation of a sector.

Acknowledgement The Eco Island Ferry project was partly supported by the Danish Maritime Fund, Region Västra Götaland, MARKIS (EU Interreg IV A), SP Technical Research Institute of Sweden, as well as the three companies involved in the design team: Danish Yachts, Niels Hjørnet Yacht Design, and Coriolis AB.

References

Argyris C, Putnam R, Smith DM (1985) Action science. Jossey-Bass, San Francisco

Buhaug Ø, Corbett J, Endresen Ø, Eyring V, Faber J, Hanayama S, Lee D, Lee D, Lindstad H, Markowska A (2009) Second IMO GHG study 2009, vol 24. International Maritime Organization (IMO), London

Christensen CM (2001) The past and future of competitive advantage. Sloan Manag Rev 42(2): 105–109

Christensen CM (2013) Disruptive innovation. In: Soegaard M, Dam RF, Friis R (eds) The encyclopedia of human-computer interaction, 2nd edn. The Interaction Design Foundation, Aarhus

Conway SH, Steward F (2009) Managing and shaping innovation. Oxford University Press, Oxford

Corbett JJ, Wang H, Winebrake JJ (2009) The effectiveness and costs of speed reductions on emissions from international shipping. Transp Res Part D: Transp Environ 14(8):593–598

Denzin N, Lincoln Y (eds) (2000) The handbook of qualitative research, 3rd edn. Sage, Thousand Oaks

Eco-innovation Observatory (2011) Introducing eco-innovation: from incremental changes to systemic transformations. Eco-innovation observatory. http://www.eco-innovation.eu/index.php?option=com_content&view=article&id=275&Itemid=208

Eisenhardt KM (1989) Building theories from case study research. Acad Manag Rev 14(4): 532–550

Elkjær J (2011) Udvikling af tilgange til kommunal Strategiske energiplaner. Roskilde University, Roskilde

Flyvbjerg B (2001) Making social science matter: why social inquiry fails and how it can succeed again. Cambridge University Press, New York

Håkansson H, Ford D, Gadde LE, Snehota I, Waluszewski A (2009) Business in networks. Wiley, Chichester

Hargadon AB (2002) Brokering knowledge: linking learning and innovation. Res Organ Behav 24:41–85

Hargadon A (2003) How breakthroughs happen: the surprising truth about how companies innovate. Harvard Business Press, Boston

Hargadon AB, Douglas Y (2001) When innovations meet institutions: Edison and the design of the electric light. Adm Sci Q 46(3):476–501

Harvey LDD (2013) Global climate-oriented transportation scenarios. Energy Policy 54:87–103

Hjørnet N (2013) Interview with engineer and naval architect Niels Hjørnet the 3rd of April 2013 [Online]

Hjortberg M (2012) Lightweight structures at sea in practice, Power point presentation edn, Coriolis at LIWEM 2012, Gothenburg

Hjortberg M, Hjørnet NK, Sørensen JO (2012) 20-12-2012-last update, http://eco-island.dk/. Available: http://eco-island.dk/. 18 Apr 2013
Hochkirch K, Bertram V (2010) Engineering options for more fuel efficient ships. In: Proceedings of first international symposium on fishing vessel energy efficiency. Vigo, Spain, p 1
Krag A, Trolle JG (2012) Danmarks Småfærger – En fælles standard. Svendborg International Maritime Academy, SIMAC, Svendborg
Kumagai I, Nakamura N, Murai Y, Tasaka Y, Takeda Y, Takahashi Y (2010) A new power-saving device for air bubble generation: hydrofoil air pump for ship drag reduction. In: Proceedings of the international conference on ship drag reduction SMOOTH-SHIPS. Vigo, Spain
Lewis P, Saunders MN, Thornhill A (2009) Research methods for business students. Pearson, Harlow
Lindqvist Å (2012) Life cycle cost analysis – eco-island ferry. SP Technical Research Institute of Sweden, Borås
Lindstad H, Asbjørnslett BE, Strømman AH (2011) Reductions in greenhouse gas emissions and cost by shipping at lower speeds. Energy Policy 39(6):3456–3464
Love G, Soares A, Püempel H (2010) Climate change, climate variability and transportation. Procedia Environ Sci 1:130–145
Mewis F, Guiard T (2011) Mewis Duct®–new developments, solutions and conclusions. In: Second international symposium on Marine Propulsors (SMP-II), Hamburg, Germany
Mosgaard M, Riisgaard H, Kerndrup S (2014) Making carbon-fibre composite ferries a competitive alternative – the institutional challenges. Journal Article edn, In review at Inderscience
Müller J (2003) A conceptual framework for technology analysis. In: Culture and technological transformation in the south: transfer of local innovation. Forlaget Samfundslitteratur, Copenhagen
Østergaard PA, Mathiesen BV, Möller B, Lund H (2010) A renewable energy scenario for Aalborg Municipality based on low-temperature geothermal heat, wind power and biomass. Energy 35(12):4892–4901
Rennings K (2000) Redefining innovation – eco-innovation research and the contribution from ecological economics. Ecol Econ 32(2):319–332
Rihan D, O'Regan N, Deakin B (2010) The development of a "green trawler"
Scrase I, Stirling A, Geels F, Smith A, Van Zwanenberg P (2009) Transformative innovation: a report to the Department for Environment, Food and Rural Affairs, Science and technology policy research (SPRU). University of Sussex, Brighton
Sidhartha J, Kumar MSP (2012) Kite technology (pull shipping to greener future). Int J Innov Res Dev 1(10):16–39
Smith A, Voß J, Grin J (2010) Innovation studies and sustainability transitions: the allure of the multi-level perspective and its challenges. Res Policy 39(4):435–448
Sørensen JO (2013) Interview with mechanical engineer Jens Otto Sørensen 12th of April 2013
Stenzel V, Wilke Y, Hage W (2011) Drag-reducing paints for the reduction of fuel consumption in aviation and shipping. Prog Org Coat 70(4):224–229
Traut M, Bows A, Gilbert P, Mander S, Stansby P, Walsh C, Wood R (2012) Low C for the high Seas Flettner rotor power contribution on a route Brazil to UK. In: International conference on technologies, operations, logistics and modelling for low carbon shipping. The University of Newcastle, Newcastle
Watson J, Schmidt JH (2012) Eco Island Ferry – comparative LCA of island ferry with carbon fibre composite based and steel based structures. 2-0 LCA consultants, Aalborg, Denmark
Yin RK (2009) Case study research: design and methods. Sage, Thousand Oaks

Chapter 15
BioTRIZ: A Win-Win Methodology for Eco-innovation

Nikolay Bogatyrev and Olga Bogatyreva

Abstract Current attempts to resolve ecological problems by only social or technological means clearly have their insufficiencies; otherwise the ecological crisis would not be so obvious. Eco-sound innovations and nature conservation projects are not profitable, because there is no reliable methodology that allows bio- and eco-approaches to contribute to industry and the economy to the full extent. Technology and ecology are often mutually contradictive and mismatching domains: bio- and techno-worlds speak different languages. A methodology for eco-innovation called BioTRIZ is presented in this chapter. BioTRIZ is the only methodology which is fully capable of dealing with contradictions between biology and technology, because its main mechanism is based on revealing conflicting requirements and a win-win resolution. The authors' 30-year experience in ecology, biomimetic design, and theory of invention led to the formulation of basic axioms and rules for eco-innovation that are presented in this chapter. As an example of the application of these axioms and rules, a case study on eco-innovation is presented: the design of an eco-park that takes into account technological requirements, generates profit, and recovers damaged ecosystems.

Keywords Biomimetics • Eco-design • Theory of Inventive Problem Solving (TRIZ)

N. Bogatyrev • O. Bogatyreva (✉)
BioTRIZ Ltd., University of Bath, Bath, UK
e-mail: nikolay@biotriz.com; olga@biotriz.com

S. Azevedo et al. (eds.), *Eco-Innovation and the Development of Business Models*, Greening of Industry Networks Studies 2, DOI 10.1007/978-3-319-05077-5_15,

15.1 Introduction: Why Does Knowledge of Living Nature Help in Eco-innovation?

Currently our economy globally faces social and environmental issues: public health problems; social unrest; migration; and water, energy, and financial crises. Because of these wrenching changes faced by contemporary society – from ecological crises and climate change to health-threatening pollution of the environment – it has appeared that conventional engineering and organization management is not necessarily working as effectively and efficiently as we would hope. These issues have started to attract the attention of governments, media, architects, designers, engineers, and the general public.

Human activity at all levels – from individual behavioral patterns to the whole economic system – does not take into account the "interests" of the environment that we live in. Our shortsighted linear models of "cause-*immediate* effect" reactive behavior[1] do not fit the extremely complex natural world, and they therefore exclude us from ecosystem cycles. We have huge problems with the utilization of waste, and yet there is no such thing as waste in nature. We declare an energy crisis, although energy exists everywhere in nature and "Life" does not use as much energy as we do – it is at least four times more efficient than current technology (Vincent et al. 2006). We devote a lot of resources to producing new and diverse materials – and yet living nature has created numerous different materials with a wide range of properties (from extremely hard to soft and liquid) using only two basic polymers: proteins and polysaccharides. Humans use more than 350 different polymers, and yet "we" have still not been able to match the variety of materials that exists in living nature. And last but not least, living systems typically avoid problems, resolving them in advance before they even occur.

All the factors mentioned above make living systems the main source for ideas for self-sustainable cycles of materials, products, and services. If Life is "so perfect," why can we not take ideas from living systems and implement them straight away into macro- and microeconomic reality? Using knowledge from biology, and especially ecology, is very tempting. But apart from great interest, there is nothing for business to rely on – no procedures that guarantee success, making such projects risky and expensive. At the moment, there is no solid methodology or procedure for making biological mechanisms applicable to implement within nonbiological reality such as engineering, business, and economy. Inspiration from biology at present is an irresponsible strategy, because there is no single proof that biological mechanisms being literally applied to the economy will not cause new economic disasters. Moreover, there are plenty of strategies in biological systems that are not acceptable for application in contemporary human society due to cultural, moral, economical, and other reasons.

[1] The ecological response to our activity is often sufficiently delayed: this has its own advantages, but an accumulative and postponed reaction can be unexpected and catastrophic.

Before we start borrowing natural strategies from ecosystems for sustainability or manipulating them for our benefit, it would be worth answering some basic questions to make an exact definition of what we actually need:

- Does the strategy minimize waste?
- Does the strategy increase speed of economic growth?
- Does the strategy provide social stability?
- Does the strategy keep living standards high?
- Does the strategy reduce risk of crises?

People have already realized that if a manufacturing process is linked with consumption and recycling in a closed loop, it is desirable and favorable strategy, but if we look at natural ecosystems, we will see a great variety of cycles. So we can ask ourselves – what sort of cycles do we need:

- Long or short cycles? Fast or slow?
- Reversible or irreversible?
- Single cycle or multiple interacting cycles?
- Interrupted or continuous "rotation"?
- Simple or complex?
- Cycles of what – material, energy, information, or anything else?

We need to clearly know the result we want to achieve as we already have cycles in economy, which we do not appreciate at all – the well-known cycles of crises! Thus we can say that there are some cycles that are undesirable. How do we recognize them?

Currently, we cannot guarantee that, without appropriate adaptation, the use of knowledge from biology will bring the required result because:

- The economy is not strictly speaking a living system as it includes technology as well: laws that are applicable for ecosystems may not work there.
- The economy is a complex system that comprises social and technological components. The social component may be considered a direct analog to natural societies, but the technological component's behavior and evolution is very much different to biology (there is only a 12 % similarity in approaching challenges in biology and engineering (Vincent et al. 2006)), and even some trends for the directions of the evolution process in technology and biology are opposite (Bogatyrev and Bogatyreva 2009).
- Successful experience of eco-innovation in one industry may not be directly applicable to industries from completely different domains – products and consumers are very different, and it is not obvious that a successful circularization model taken from one industry will work for a different industry as well.

There is certainly a demand for a reliable method that can interpret and transfer data from biology into technological procedures in a manner which would not cause harm to business and which would allow a transformation from the current linear model to an eco-friendly circular one. We are going to fill this gap with a solid methodology, called BioTRIZ (Bogatyrev and Bogatyreva 2012), which is capable of working with the incompatible requirements that economy and ecology represent. Moreover, this methodology is universal; in other words, it is acceptable in all domains of human activity – from medicine and pharmacology to education, culture, engineering, science, politics, and sociology.

So, in this chapter you will find:

- The answer to the questions of how, where, why, and in what circumstances do we need eco-innovation
- A brief introduction to BioTRIZ methodology
- Axioms and rules for eco-innovation
- A case study on eco-innovation to show how BioTRIZ rules and axioms work: the layout architecture for an eco-park that takes into account technological needs, generates profit, and provides the revitalization of damaged ecosystems

15.2 Ecology and the Economy: Defining the Goals and Issues of Eco-innovation

Economy and ecology have a common linguistic root – "eco," which means "home." Our home is our planet, so all of human society and its economy should be associated with the natural ecosystem. Economy and ecology are the most complex and sophisticated spheres. Sciences that study them suffer from the lack of formal methods and procedures that are expected to lead us to predictable outcomes. If this were not true, we would never experience any economic crises, ecological disasters, or technological catastrophes. When extremely complex systems such as technology, ecology, and society come to interact and interconnect, some interacting components, which may be useful and harmless within their own domain, suddenly generate long-lasting disasters when they interact with other components. For example, transport development has changed our lives – people move across continents for holiday and business. Such mobility of people opens the opportunity for international cooperation in industry and cross-cultural enrichment, but it unlocks the way for local microorganisms, plants, insects, and viruses to spread globally. Our transport system thus works like a mixer in the world ecosystem, and we deal with its consequences every day.

A natural ecosystem is a complex system of interactions between living organisms, their groups, and populations plus the abiotic environment. These interactions – relying on closed and open cycles, feedback loops, redundancies, adaptability, and variability – define the ecosystems' performances and yield biomass production, energy efficiency, and resilience. On the other hand, in industrial production and consumption systems, we tend to rely on standardization and specialization, linear processes, and maximization of the performance – which leads to greater vulnerability, less resilience (capacity to recover), and poor optimization of energy and material use and reuse (Bogatyreva and Bogatyrev 1998).

To achieve a performance similar to natural ecosystems in economy is the way to sustainability. This means that there is need for a new interdisciplinary approach with solid methodology, which will provide for the economic prosperity of humankind, aiming beyond simple survival in the destabilized ecological conditions on our planet. This innovative approach should involve most of the scientific and social disciplines as well, as actors from private, public, and political sectors. We need to go beyond isolated and singular initiatives and experiment in different areas – industry, agriculture, art, architecture, etc. – and develop *systematic tools* for eco-innovation that integrate all the most efficient methodologies and promote these methods in industry, education, and the whole culture. A multidisciplinary approach makes this grand task possible but also raises challenges. Developing a common ground methodology for such a large number of different domains is not a straightforward task (Bogatyrev 2000). Economy and ecology, although similar in terms of complexity and involvement of animate and inanimate components, possess starkly different properties for these components.

15.2.1 Biology and Technology: Two Contradictive Design Strategies

Biomimetics is a relatively young branch of engineering, which take ideas for design from living nature. Those who wish to trace its roots may find many historical attempts to copy living nature (Vincent et al. 2006). For example, Leonardo da Vinci observed animals and plants and foresaw the possibility of converting biological principles into technological ones. Later, more emphasis was placed on the need to increase the functional capability of engineering devices. Today, biomimetic ideas are also driven by the concepts of sustainability and nature-friendly engineering and can be considered an eco-innovation approach. These ideas are getting popular but do not have much support from business due high project costs and lack of design methodology that can guarantee successful outcomes and reduce risk of investments.

To fill this gap, we decided to have a look at the major differences between animate nature and technology and consider how useful it is to follow the principles of living nature (Vincent et al. 2006; Bogatyrev and Bogatyreva 2009). To compare the strategies that technology and biology follow in their development, we have analyzed more than 2,500 biological phenomena, covering over 266 functions at different levels of the biological hierarchy – from molecule and cell level to ecosystem level (Fig. 15.1). To enable us to present huge amount of information from biology in a way that it becomes useful to engineers, we have established a logical framework based on the "mantra": "things do things somewhere." This establishes six fields of operation in which all actions with any object can be executed: "things" (substance, structure), "do things" (requiring energy and information), and "somewhere" (space, time).

The data presented in the upper diagram on Fig. 15.1 show that at size levels of up to 1 m, where most traditional technology (or maybe better to say – tools) is

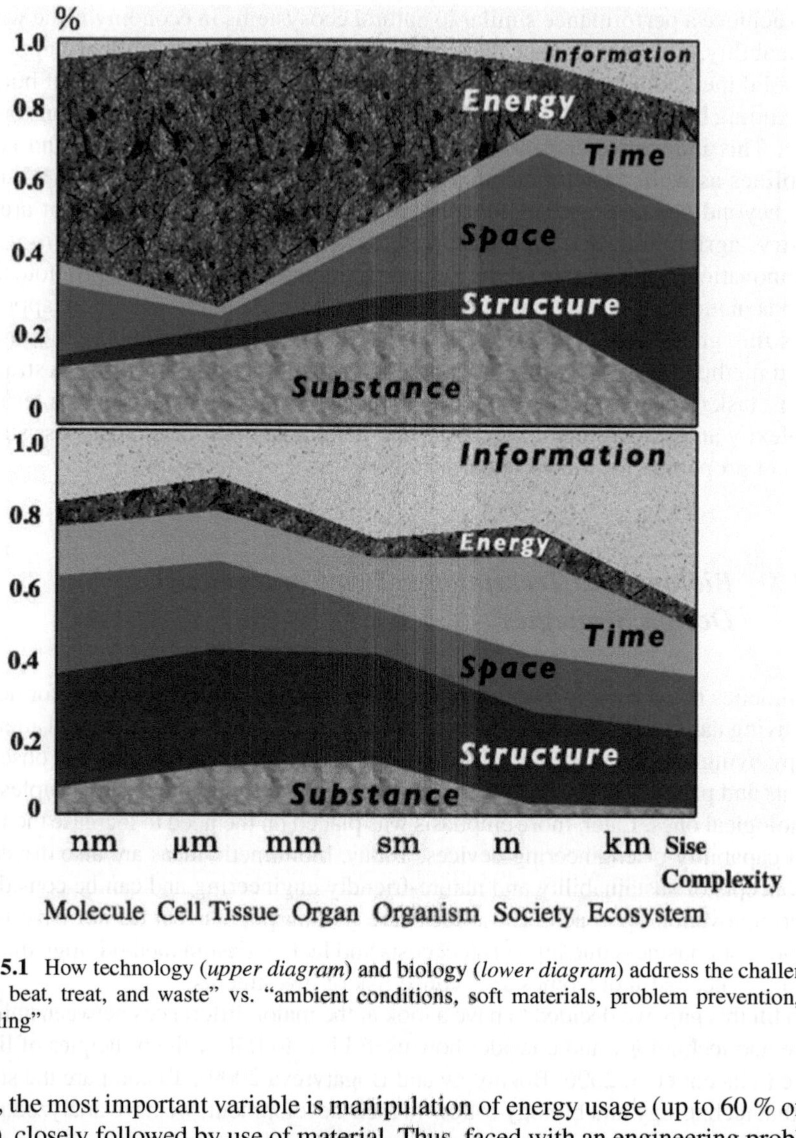

Fig. 15.1 How technology (*upper diagram*) and biology (*lower diagram*) address the challenges: "heat, beat, treat, and waste" vs. "ambient conditions, soft materials, problem prevention, and recycling"

sited, the most important variable is manipulation of energy usage (up to 60 % of the time), closely followed by use of material. Thus, faced with an engineering problem, our tendency is to achieve a solution by changing the amount or type of the material or changing (usually increasing) the energy requirement. In contrast, in biology the most important variables for the solution of problems at these scales are information and space (Fig. 15.1 lower diagram). This can be illustrated by comparing the functionality of biological and man-made polymers, proteins, and polysaccharides. It appears that biological systems have developed relatively few synthetic processes at low size, at which the contribution of energy is significant. For biological systems, the main variety of function is achieved by manipulations of shape or combinations of materials at larger sizes achieved by high levels of the hierarchy where energy is not an issue. This is a very subtle biomimetic lesson: instead of developing new materials each time, we want new functionality, and we should be adapting and combining the materials we already have. Although this is already done to some extent, it is unclear whether this is adequately recognized as a significant route. We can also realize the potential of nanotechnology, which is to escape from the nano-approach as soon as possible, and progress to the creation of larger structures, which can self-assemble in a programmed manner. This is one of the most important trends that contemporary eco-innovation should follow.

If we explore the graphs (Fig. 15.1) deeper, we may notice that there are major differences between technology and biology at the minimum and maximum complexity scale. This means that these two extremes in complexity are largely contradictive in the way they are functioning in technology and biology, and therefore they are the most sensitive to any misbalance caused by human activity. This means that eco-innovation touches the most sensitive point on the largest scale[2] of this graph, where differences in functioning can cause global problems and can potentially be a source of real disasters at all levels of our activity – from environmental pollution and people's health problems to political conflicts and social unrest.

15.2.2 The Ideal Eco-innovation Strategy: A Win-Win Balance of Biological and Technological Interests

Let us now focus on the similarities between ecology and economy. Many professionals hope to resolve the numerous problems of humankind by following the principles that can be found in biology. One of the most popular ideas that have emerged recently – a circular economy concept – is based on the phenomenon of cycles of matter and energy, a well-known ecological mechanism (MacArthur 2012). In fact, this approach was manifested in the past in different countries and cultures. But now, the scale of both capabilities and problems of contemporary technology has increased enormously. The complexity and unpredictability of existing and emerging engineering systems has grown dramatically. Meanwhile extremely complex systems do exist in living nature – natural ecosystems. Their adaptability, sustainability, self-perpetuating properties, and the long life span of some of them can serve as a model for arranging human mechanisms (social, economic, political, and cultural) in a win-win fashion and allowing us to get rid of undesirable, dangerous, and harmful practices.

Contemporary technology does not have many of the necessary features that we find in living systems (Tables 15.1 and 15.2), but then again living nature is not totally useful for us – complexity, high unpredictability, and vulnerability are not features we would like to transfer to our economy and society. Now we can clearly see the challenge for eco-innovation: to merge the advantages from both domains

[2] One of the popular trends in contemporary engineering is nanotechnology, which touches another sensitive point on the lowest scale of the graph (Fig. 15.1).

Table 15.1 Living systems: desirable and harmful features for engineering

Features of living systems which are *useful* for engineering	Features of living systems which are *not useful* for engineering
Reliable	Complex
Adaptable	Slowly evolving
Self-repairing	High uncertainty in behavior
Self-regulating	Adapted to narrow-specific conditions
Self-reproducing	Mortal
Recyclable	

Table 15.2 Technology: advantages and shortcomings

Technology: useful features	Technology: undesirable features
Limited functionality	Limited functionality
Fast changes and development	Complexity
Deterministic and automatic	Slow or limited adaptation
Functionality beyond the limits of living creatures	Require maintenance
	Short range of efficiency ("here, now, and at any price")

and get rid of the disadvantages. For Tables 15.1 and 15.2, we would like to capture the leftmost features while casting off the rightmost ones.

In eco-innovation products, we would like to jointly capture all the advantages of technology and living systems, even if they are apparently contradictive. Some possible contradictive requirements to an eco-innovation product (machine, process) are:

- Limited functionality, but reliable and self-repairing
- Adapted and adaptable
- Fast changes and development, but precisely adapted to functioning conditions
- Functionality beyond the limits of living systems, but should be recyclable – operating within the limits of an ecosystem
- Complex and easy manageable
- Recyclable (mortal) and everlasting

We suggest that our future ideal eco-innovation product or process should possess all the advantages of animate nature together with our current traditional technological achievements (Bogatyrev 2006). To merge the contradicting requirements, we use the BioTRIZ methodology described in the next section.

15.3 Methodology: Brief Description of Tools and Capabilities

Hundreds and thousands of years of technological development were driven by trial-and-error methodology, insight, serendipity, and random discoveries. Eventually this approach proved that it couldn't provide predictable, reliable, and safe results – especially when technological systems become so complex that they start demonstrating dangerous and unexpected emergent effects. Only radically novel systematic methodology can provide the pathways for an inventive process with a guaranteed result. Such systematic methodology started to emerge in the twentieth century, and one of the most effective appeared to be TRIZ (known by the Russian acronym for Theory of Inventive Problem Solving) (Altshuller 1973). The following features make TRIZ a powerful analytical technique suitable as the basis for eco-innovation:

- *Novelty*: TRIZ derives its strength from a set of inventive principles that have been defined through the analysis of more than 30,000 successful patents (Altshuller 1973; Online TRIZ Journal 2013). Its solutions are never average but radically innovative, by aiming higher than other creative approaches because it searches and deals with contradicting requirements on a win-win basis.
- *Reliability*: TRIZ leads us toward a solution systematically; therefore the result is guaranteed.
- *Capability*: TRIZ has been successfully applied to solve not only complex problems but also problems perceived as otherwise "unsolvable," where conflicting requirements are extreme (e.g., an object should be both long and short, be present and absent at the same time, etc.).
- *Breadth*: Its problem-solving capabilities are superior, since its embedded innovative principles mirror a range of approaches to creative thinking. Moreover, its underlying process takes the user all the way from problem analysis to solution.
- *Depth*: As a creative methodology, TRIZ helps "imagine the unimaginable" by going beyond perceived limitations and widening the horizon by probing deeper into a topic. Thus, TRIZ helps prepare for as yet unexpected successful solutions and also helps identify and mitigate unconventional threats.
- *Flexibility*: TRIZ has been shown to be adaptable to many different domains beyond its original remit in mechanical engineering. Its basic truths are transferable and applicable to any problem-solving domain. The content of the online TRIZ Journal clearly supports this statement.

The vast sphere of innovation that deals with the consequences of negative effects from technology on the environment uses pre-systematic (or "pre-TRIZ" if you like) methodology – trial-and-error, inspiration, and random search for the solution. These cure symptoms – not causes. The direct application of TRIZ to bio-systems, according to issues described above in this chapter, is not always relevant (Bogatyrev and Bogatyreva 2009), in spite of the seemingly "obvious" similarity of the domains involved. We have resolved the contradictive requirements from biology and technology (economy) described in Sect. 15.2.1 and created a method to operate with systems that comprise living and nonliving interfaces and natural and artificial parts (e.g., agriculture, forestry, medicine, etc.). We develop and use the BioTRIZ method in various projects with industry and teach this method at our workshops. For example, one construction company in the UK successfully applied the BioTRIZ method to solve a problem in building construction. There was a need for free cooling without losing the insulation properties of houses in extremely continental climates where days are hot, but nights are cold. The BioTRIZ method led to the development of two new environmental building technologies: an infrared transparent insulation and a concrete formwork, made from biodegradable starch foam to achieve a much better end-of-life profile compared to other materials (Craig et al. 2008).

15.3.1 BioTRIZ Axioms for Eco-innovation: A Win-Win Resolution of Conflicts Between the Economy and Ecology

Just after biomimetic approach declared its new attitude to engineering[3], at the beginning of the seventies, a holistic and constructive approach to ecosystem engineering was undertaken with the appearance of *permaculture* (*perm*anent + agri*culture*). It was the first method for conscious design of artificial ecosystems that possess the productivity and benefit of conventional agricultural systems combined with sustainability and self-serving features of natural ecosystems (Mollison 1988). Achievements of permaculture are impressive, and it appeared that this approach can be applied not only in agriculture (the initial realm of application) but also in forestry, fishery, architecture and construction, energy supply, transport, and even to economy, education, and culture. Both permaculture (eco-engineering) and biomimetic (biological engineering) represent two different methodological platforms for eco-innovation. As any methodology starts with axioms and theorems, we can consider BioTRIZ axioms for eco-innovation, framing them into two methodological platforms mentioned above.

The means of introducing biology into engineering is indicated in the left column of Table 15.3, showing the different approaches to innovation in biomimetic engineering.

The introduction of technological features into biological systems (forestry, parks, agriculture, etc.) requires a different approach than biomimetic engineering and is shown in the right column in Table 15.3. The BioTRIZ axioms make the two directions of the knowledge-transfer process mutually compatible and less contradictive. Biomimetics and eco-engineering can be considered as two parts of the eco-innovation strategy as they both target to get maximum benefit for economy knowledge about living nature (Table 15.3).

15.3.2 BioTRIZ Rules for Eco-innovation from Ecology

To develop BioTRIZ rules for eco-innovation, we used the very basic concept from ecology: r- and K-development strategies of survival of organisms/species or ecological systems (Pianka 1983). These strategies relate to cycles, energy, and matter resource management and even indicate system readiness for change (Bogatyreva and Bogatyrev 1998).

[3] Biomimetic engineering is discussed in Sect. 15.2.1 of this chapter.

Table 15.3 BioTRIZ axioms for eco-innovation

Biology – to engineering: biomimetics	Engineering – to biology: eco-engineering
Axiom of simplification: reduce the functionality of a biological prototype for an engineering design	Axiom of maximization of useful function: add as many beneficial functions as possible to the ecosystem
Axiom of interpretation: instead of copying, interpret the essence of a biological mechanism/structure/function or strategy into engineering language	Axiom of interpretation: engineering functions and strategies that we would like to see in the ecosystem need to be interpreted into biological language
Axiom of ideal result: if you still want to copy, do not copy the means of providing a function, but copy the result of the function	Axiom of ideal result: maximum benefit/profit for humans should be achieved only with super-optimal functioning of the ecosystem
Axiom of contradictions: translation of "What?" (engineering target) into "How?" (answers from biology) should be done via aggravated statement of conflicting requirements	Axiom of contradictions: eco-requirements and human requirements are usually contradictive. Any engineering manipulation may cause damage if interests of the ecosystem are not taken into account

The r-strategy is characterized by fast-reproducing, short-lived, and "cheap" (in terms of resources spent on raising an offspring) individuals. The r-strategy is also characterized by fast adaptation and is therefore more common in unstable environments with abundant resources. This strategy is common for the initial stages of ecological succession – the beginning of the cycle of an ecosystem life. The K-strategy is characterized by successfully competing for scarce resources, a constant size of population/system, and being constantly close to the carrying capacity of the environment. Slower reproduction and long life expectancy are also traits of the K-strategy. Both the K- and r-strategy strike a balance between opposite trends in productivity and production: the beginning of the cycle ecosystem consists of highly productive species with a cheap brood but with low biomass production (e.g., annual plants, grass meadow, microscopic algae). Closer to the climax of an ecosystem cycle, it consists of K-strategy species with high biomass production, but lower productivity (trees, large predators – such as bears, tigers, whales, etc.). Knowledge of these phenomena can be used in designing new systems or revitalizing existing ones and should be the main guide for eco-innovation.

Following the maximum profit ideology of our economy, we need both types of strategies: maximum production and maximum productivity, durability and easy recycling, and cheap production and high demand – which is produced with minimum resources, but is a valuable resource itself. All these contradicting requirements are resolved by the following BioTRIZ rules for eco-innovation:

- Everything is a resource, if and only if it is there at the right time, in the right place, in the right dose, in the right mode, and for the right consumer. If any of these compulsory conditions are not fulfilled, a valuable resource turns into pollution, waste, direct damage, or potential harm. So the rule for waste management is to find the right time, place, dose, mode, or user so that any waste becomes a valuable resource.

- Manufacture a product in minimum number of steps, but use it in maximum number of steps.
- Any product should be either inert/everlasting (K-strategy) or easily degradable (r-strategy). Analyze the properties of a product, classify them into K- and r-categories, and modify the product or process properties to follow the rule above.
- Products of degradation should not be harmful for the ecosystem (see resource definition in rule 1) and should be included as natural ingredients in natural cycles.
- To avoid a misbalance in the ecosystem, use only local resources that are already present or available within the given ecosystem.

What we should avoid:

- Overexploitation of a resource, system, or process.
- Under-exploitation of the resource, system, or process.
- Blind mechanistic imitation of successful models in living nature.
- Maintaining the same state of a system in a changeable environment.
- A new system should not aggravate the existing ecological status quo.

15.4 The Eco-innovation Project BOMBORETUM as an Illustrative Example of BioTRIZ Axioms in Action

Following BioTRIZ axioms we have developed BOMBORETUM, which is the layout architecture for an eco-park that takes into account technological needs, generates profit, and provides the revitalization of damaged ecosystems. The origin of the name BOMBORETUM is in two words: "Bombus" is the Latin name for a bumblebee and "arboretum" is a nursery for trees, bushes, and shrubs. This case study is an example of eco-innovation in the context of introducing technological, engineering features to living nature.

Eco-parks can play an important role in aiding the sustainability of ecosystems that bear the load of an urban environment. Any terrestrial ecosystem is based on plants, which provide the primary organic resource for the whole ecosystem as the result of photosynthesis.[4] To make maximum yield possible and provide for the well-being of flowering plants, it is necessary to guarantee the pollination process. One of the most essential elements of terrestrial ecosystems is a wild pollinator. These beneficial insects can be called "flowers' legs" or "love messengers" that allow plants – which are far away from each other and cannot move to "mate" with each other – to transfer pollen from one flower to another. Thanks to pollination, we get seeds and fruits.

[4] By the way, at this stage CO_2 is excluded from the atmosphere, and it is essential in the contemporary global climate situation.

When agricultural fields are surrounded by a healthy natural ecosystem that consists of all the compulsory elements (including its "reproductive system" – wild pollinators), we take for granted that these crop fields bloom and give yields. Wild pollinators do their invisible job and also help agriculture by pollinating fruit trees, bushes, vegetables, and herbs. When the population of pollinators decreases, crop yields inevitably drop. This happens due to the uncontrollable pressure of human activities (chemical application in agriculture, constant tillage, grass mowing and/or kettle grazing, various types of pollution, etc.). In most cases, "reversing" the scenario is difficult: natural ecosystems are never restored from agricultural landscapes to the initial wild form. But it is possible to compensate for the distorted systems with relatively simple measures by maintaining and facilitating the crucially important parts of ecosystems – its "reproductive organs," or wild pollinators (Bogatyrev 2001).

The simplest way to protect an environment is to not touch it at all. However, this is not a good strategy for damaged ecosystems that have already appeared below the threshold of possibility of being able to recover without our help. Moreover, there are no methods of keeping natural ecosystems above this threshold. BOMBORETUM is the first working model for the active recovery of a damaged ecosystem with added value for the economy (Bogatyrev 1992).

The best pollinators in the world are insects. Bees are among the most adapted for this function (there are 20,000 different species of bees!). Common honeybee apiaries are valuable, but are not able to provide pollination for all plants due to natural limits of these insects (short proboscis, nectar and pollen preferences, climate and weather restrictions – bees do not like rain and extreme temperatures – etc.). That is why we need specially reared wild pollinators for getting seeds and fruit from crops (e.g., the solitary bee *Megachile rotundata* for alfalfa crops and bumblebees for vegetables and berries in greenhouses – cucumbers, peppers, tomatoes, eggplants, raspberries, strawberries, etc.). Industrial companies that grow pollinators yield multimillion profits in the Netherlands, Belgium, Canada, Israel, Japan, etc. The less expensive strategy of obtaining pollinators is to provide them with artificial domiciles in the required numbers and nectar and pollen plants as food sources. This approach was successfully practiced in Canada, Australia, New Zealand, Russia, Poland, France, etc.

To rear wild pollinators in an artificial environment and to then introduce them into the wild is not a good idea – it is too expensive and does not guarantee successful naturalization. To resolve these issues, we decided to create a sustainable and self-reproducing, self-perpetuating eco-mechanism that will work by itself and give the maximum effect (Bogatyrev 2004). Different parts of this park were implemented, tested, and proved in the Central Siberian Botanical Garden of the Russian Academy of Sciences (Novosibirsk, Russia).

Let us follow the BioTRIZ axioms for eco-innovation to design this eco-park.

Axiom of Maximization of Useful Function This means we need to add as many beneficial functions as possible to the ecosystem, providing a super-optimal environment not only for pollinators and plants but also for people – the users of the park, who get an immediate profit (from the fruit and seeds of the crops) plus a nice and educational recreation area. We are going to design a meadow with an additional

Table 15.4 Biological interpretation of engineering functions

Requirements from machinery and management	Engineering function interpreted into ecosystem design
Machinery is designed to process large uniform areas	Minimal outer borderlines to minimize the outer impacts on the inner environment
Less plant diversity is preferable; all the agricultural machinery is adapted to monocultures	Continuously blooming plants during the summer to provide colonies with pollen and nectar
Single harvesting process, after which there are no blooming plants on the field	Plants with different flower depths simultaneously blooming to prevent competition among different bumblebees' species, because different species of bumblebees have different length of proboscis
Park should be a busy place to make a profit	Diverse environment due to segmentation of the foraging area as bumblebees keep to their own territories
Parks are never quiet due to visitors and noisy play areas	Nesting zones should be separated from foraging zones by 20–25 m, because bumblebees do not forage near their nests

unnatural functionality – everlasting blooming, unlimited access for pollinators to food and nesting zones, and high yield of beneficial crops (berries, fruit, herbs, seeds, etc., with efficient technology of harvesting and service). In addition, BOMBORETUM is a recreation area and should possess functionalities to support this aim.

Axiom of Interpretation Engineering functions that we would like to see in the ecosystem need to be interpreted into biological language. Interpreted functions are presented in the Table 15.4.

Axiom of Ideal Result Maximum profit for humans should be achieved only with super-optimal functioning of the ecosystem part of an eco-machine. Let's define what is an ideal eco-engineering system:

- Sustainable: providing the conditions for a self-regenerating ecosystem – plants and their pollinators
- Cost-efficient maintenance: allowing a "mass-production" strategy with the use of existing machinery
- Simple and regular layout that allows safe crowd management and health and safety, as well as convenient operating with the existing agricultural machinery
- Attractive place for human recreation and for nesting and foraging of wild pollinators

Desirable features can be achieved if we replace the natural reserve of these beneficial insects with laboratory rearing, but it is still very expensive: even under industrial breeding, one bumblebee colony's price is $70–150 (USD). This solution does not match the ideal result axiom formulated for humans/engineers. So this means that bumblebees should stay "self-serving," i.e., sustainable mode of existence. This can be achieved by the following measures:

- Decreasing the contrast of the environmental conditions inside and outside of the reserve. This is a typical compromise and is not perfect. It is not a win-win solution at all.

Table 15.5 Ideality in ecological and engineering contexts

Ecosystem requirements (from the position of bumblebees and plants)	Requirements from machinery and management
Individual foraging patches should not be larger than 100 m^2	Machinery is designed to process large uniform areas
To be suitable for different species of pollinators, the variety of plants should provide different lengths of nectar tubes	Less plant diversity is preferable; all the agricultural machinery is adapted to monocultures
Bees require blooming plants during all seasons	Single harvesting process, after which there are no blooming plants on the field
Nesting zones for wild pollinators should be isolated	The park should be a busy place in order to make a profit
Bees should not be disturbed	Parks are never quiet due to visitors and noisy play areas

- Creating a network of small nature reserves instead of one.
- Creating a "nomadic reserve" – alternating reserves of small territories for 1–2 years each.

All these options do not suggest a radical win-win decision, as they do not match both the technical and natural ideal result's requirements. We therefore need to apply the next axiom – the resolution of contradictive requirements – to strike a win-win balance.

Axiom of Contradictions Ecosystems' and human requirements are usually contradictive. Any engineering manipulation may cause damage if the interests of the ecosystem are not taken into account. For example, to protect the natural populations of bumblebees, small nature reserves (from 1 to 5 ha) were established throughout the agricultural landscapes. However, unexpectedly, the number of bumblebees decreased rather than increased. This happened because the accumulation of parasites, natural enemies, and diseases was activated at full strength, and there was a high rate of mortality due to competition and other population growth regulative mechanisms. Thus, the micro-reserve worked as a huge "meat grinder" which was draining the surrounding territories of bumblebees and destroying them constantly (Bogatyrev 2001). In other words, the small-sized nature reserves brought harm and damage instead of the expected benefit. Following the axiom of ideal result, we can create a crucial requirement: the natural reserve must exist and at the same time must not exist, plus some more contradictions presented in Table 15.5. This is a contradiction that the BioTRIZ method is looking for! Now it should be resolved by win-win strategies – 40 inventive principles (Altshuller 1973). Let's first consider the different possibilities.

Now we can start to design "the ideal reserve" layout. To minimize the borderline of the reserve, it should be circular. To minimize the borderline of the nesting zone, it should be circular. To provide the necessary distance from foraging spots, a nesting zone should be in the center of the circle. Thus, the foraging zone with blooming plants should be the wide ring (Fig. 15.2). The foraging zone can be fragmented in the two ways (Fig. 15.2A, B) to provide more plant diversity: bumblebees, bees, and other wild pollinators prefer flowers with different length of nectar tubes

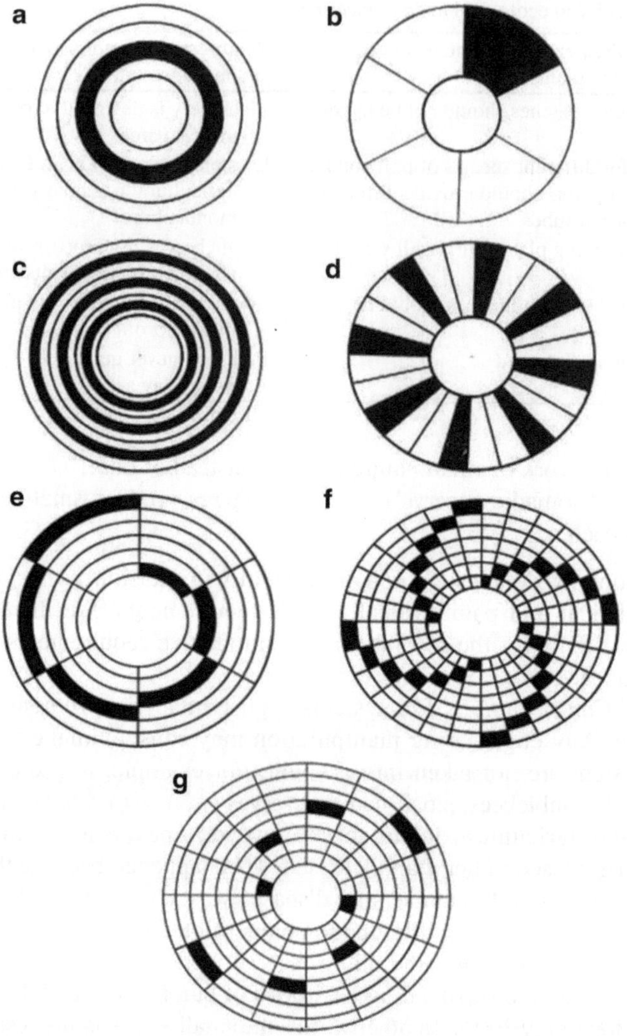

Fig. 15.2 Step-by-step process of the design of a park/recreation area that restores surrounding damaged ecosystems. *Black zones* – areas with blooming plants, *central circle* – nesting zone is isolated from public. Wide ring foraging zone: *A* – concentric, *B* – radial. Large fields are subdivided into *C* (more radial beams) or *D* (concentric rings), *E* (single-spiral layout as a result of merging radial and concentric fragmentation), *F* (plain four-spiral layout), *G* (two-spiral layout with the interruption of spirals along "parallel and meridian")

But bumblebees like small foraging patches! Radial patches or numerous rings are the possible answers to these requirements (Fig. 15.2C, D). A radial highly segmented layout is not good, because bumblebees avoid foraging along lines – according to our observations, spots for individual or group feeding zones have equilateral shape. A concentric highly segmented layout is even worse, because bees will concentrate within the nearest rings and will not use the whole territory evenly, which is detrimental to pollination and crop productivity.

The solution to these contradictions can be achieved if we superpose radial and concentric designs, divide the foraging zone into spots suitable for individual use for bumblebees, and make standard the optimal spot size for bees all over the blooming foraging area to reduce competition for nectar.

Following these ideas, a fragmented spiral appears "automatically." If a park area and therefore foraging zone for bees is large enough, the spots of the fragmented spiral can exceed the optimal size, which is similar for different species of bumblebees (Bogatyrev 2001). In a single-spiral layout, the rear blooming spot happens to be too large and bees may ignore it (Fig. 15.2E). To avoid this, it is possible to increase the number of spirals (Fig. 15.2F). The larger the park is, the more spirals with different blooming plants we can introduce as well as the variants of their fragmentation – along the meridians and/or parallels (Fig. 15.2G). The regular network of radial and concentric lines represents paved paths/roads, which is convenient for agricultural machinery to move along, as well as for visitors of the park.

Thus, BOMBORETUM has become technological system – "invisible rearing machine for natural ecosystem" – while remaining a park and recreation area. Is it simple to operate and control? Is it sustainable (self-repairing, adaptive, self-regulated)? Does it need minimum resources and does it produce minimum waste? The answer to all these questions is "yes" – this is an ideal eco-innovation product that comprises features and follows the "interests" of nature and technology. This makes BOMBORETUM manageable and Profitable.

15.5 Conclusions

Technology is evolving very fast, giving people more and more power to change many aspects of living: lifestyle, health, culture, society, industry, economy and, finally, the ecosystem itself. We still are only a part of the whole biosphere, and taking everyday decisions, we should also consider its "interests" – this is the aim of eco-innovation initiatives. Eco-innovation manifests itself in two contexts: introducing biological principles into technology makes it more life-like and therefore eco-friendly and adding technological features to ecosystems opening the opportunity for using nature to our benefit without causing it damage, misbalance, or ecological catastrophe.

A tremendous difference in the way that human economy and living nature function causes a lot of problems in our days and is the main obstacle for spreading ideas of eco-innovation, making them profitable and attractive for business. We need to honestly face the conflicts between nature and technology and not just proclaim them and at the same time postpone action, leaving aggravated problems for our children to resolve. This happens because there were no methodologies to deal with contradictive requirements that are capable of merging incompatible things. The BioTRIZ methodology of facing and resolving contradictions in management and engineering contributes to eco-innovation initiatives by replacing a passive approach with elements of inspiration and irresponsible belief that nature is always perfect (perfect for whom?) by solid methodology. The BioTRIZ method described in this chapter not only is capable of facing and resolving contradictions to strike a win-win balance but also has tools for searching and revealing hidden conflicts and obstacles for eco-innovation.

The BioTRIZ rules and axioms for eco-innovation were presented, and the practical implementation of BioTRIZ axioms was explained in the case study of an eco-park. The design of the eco-park restores damaged natural ecosystems and produces extra value for agriculture, education, and recreation – essential parts of any economy. This is a valid example of a win-win strategy, when local positive effects proliferate to global constructive consequences. This is what we call sustainability – local action with global positive outcome – and this is exactly what the eco-innovation initiative aims for.

So, if this chapter makes you think first, then act in the following way, our aim has been achieved: do not avoid or dodge conflicting requirements and sidestep seemingly unsolvable problems, face (and even search for!) contradictions and resolve them on a win-win basis.

References

Altshuller G (1973) Algorithm of invention. Moscow Worker, Moscow

Bogatyrev N (1992) BOMBORETUM – a biodynamic model of a nursery reserve for bumblebees. In: International workshop on non-Apis bees and their role as crop pollinators. Logan, UT, p 6

Bogatyrev N (2000) Ecological engineering of survival. SB RAS, Novosibirsk

Bogatyrev N (2001) Applied ecology of bumblebees. City Centre of Education, Novosibirsk

Bogatyrev N (2004) A living machine. J Bionic Eng 1(2):79–87

Bogatyrev N (2006) Biological and engineering parts of biomimetics: “software” and “hardware” of adaptability. In: Proceedings of AISB’06 conference. University of Bristol, 3–6 Apr 2006, vol 2, pp 106–109

Bogatyrev NR, Bogatyreva OA (2009) TRIZ evolution trends in biological and technological design strategies. In: CIRP design conference, Cranfield University, UK, 30–31 March, pp 293–299

Bogatyrev N, Bogatyreva O (2012) TRIZ-based algorithm for biomimetic design. In: TRIZ future 2012. Lisbon, Portugal, 24–26 Oct 2012, pp 251–262

Bogatyreva O, Bogatyrev N (1998) General regularities in social and ecological transformations. Hum Ecol Interact Cult Educ Novosib 2:31–51

Craig S, Harrison D, Cripps A (2008) Two new environmental building technologies, one new design method: infrared transparent insulation, biodegradable concrete formwork and BioTRIZ. In: Proceedings of the conference for the engineering doctorate in environmental technology. University of Surrey, UK

MacArthur E (2012) Towards the circular economy. http://www.ellenmacarthurfoundation.org/. Accessed 5 Aug 2013

Mollison B (1988) Permaculture: a designer’s manual. Tagari NSW 2484, Tyalgum

Online TRIZ Journal (2013) Part of real innovation network. http://www.triz-journal.com. Accessed 5 Aug 2013

Pianka E (1983) Evolutionary ecology. Harper & Row, New York

Vincent J, Bogatyreva O, Bogatyrev N, Bowyer A, Pahl A-K (2006) Biomimetics – its practice and theory. Interface J R Soc 3(9):471–482

ERRATUM

Portraying the Eco-innovative Landscape in Brazil: Determinants, Processes, and Results

Flavia Pereira de Carvalho

S. Azevedo et al. (eds.), *Eco-Innovation and the Development of Business Models*,
Greening of Industry Networks Studies 2, DOI 10.1007/978-3-319-05077-5_7,

---

DOI 10.1007/978-3-319-05077-5_16

The names and affiliations of two co-authors were omitted in the original publication. Those are as follows:

Paulo Savaget
Visiting professor at FDC (Innovation and Sustainability),
partner at Innovare Research, Belo Horizonte, Brazil
e-mail: paulosavaget@gmail.com

Carlos Arruda
Professor of Innovation and Competitiveness,
Associate Dean for Business Partnerships,
and Director of the Fundação Dom Cabral - FDC Innovation
and Entrepreneurship Center, Belo Horizonte, Brazil
e-mail: arruda@fdc.org.br

The online version of the original chapter can be found at
http://dx.doi.org/10.1007/978-3-319-05077-5_7

MIX
Papier aus verantwortungsvollen Quellen
Paper from responsible sources
FSC® C105338

If you have any concerns about our products,
you can contact us on
ProductSafety@springernature.com

In case Publisher is established outside the EU,
the EU authorized representative is:
Springer Nature Customer Service Center GmbH
Europaplatz 3, 69115 Heidelberg, Germany

Printed by Libri Plureos GmbH
in Hamburg, Germany